TEACHERS TALKING ABOUT THEIR CLASSROOMS

Different communities, speaking different languages, employ different naming systems to describe the events, actions, and interactions of the mathematics classroom. The International Classroom Lexicon Project documented the professional vocabulary available to middle-school mathematics teachers in Australia, Chile, China, the Czech Republic, Finland, France, Germany, Japan, Korea, and the United States. National teams of researchers and experienced teachers used a common set of classroom videos to stimulate recognition of familiar terms describing aspects of the mathematics classroom. This book details the existing professional vocabulary in each international community by which mathematics teachers conceptualise their practice, and explores the characteristics, structures, and distinctive features of each national lexicon. This book has the potential to enrich the professional vocabulary of mathematics teachers around the world by providing access to sophisticated classroom practices named by teachers in different countries.

This one volume offers separate, individual lexicons developed from empirical research, the capacity to juxtapose such lexicons, and an unmatched opportunity to highlight the cultural, historical, and linguistic bases of teachers' professional language.

Carmel Mesiti is Research Fellow of the International Centre for Classroom Research (ICCR) and Lecturer in Mathematics Education at the University of Melbourne, Australia. Her research has centred on exploring, through international video-based research, the nature of teaching and learning in mathematics classrooms, as well as the differences in pedagogical lexicons of education communities worldwide.

Michèle Artigue is Emeritus Professor at the Université de Paris and Associate Researcher at the Laboratoire de Didactique André Revuz, France. In 2013, Michèle was awarded the Felix Klein ICMI Medal for sustained, consistent, and outstanding lifetime achievement in mathematics education research and development. Michèle's recent research interest has centred on the building of connections between approaches and theories in mathematics education.

Hilary Hollingsworth is Principal Research Fellow at the Australian Council for Educational Research (ACER), Australia. Hilary has over 30 years' experience working in a wide range of national and international educational contexts including schools, universities, research organisations, government education departments, and private education service organisations. Her research interests include teacher professional learning, video classroom observations, teaching quality, assessing student learning, and communicating student learning progress.

Yiming Cao is Professor at Beijing Normal University and Chair of the Chinese Association of Mathematics Education, China. Yiming has worked throughout his career to improve mathematics education in China. His research interests have included curriculum, assessment, teacher knowledge, and cooperative learning and he currently leads the revision of the Chinese Mathematics Curriculum Standard for compulsory education.

David Clarke was Professor at the University of Melbourne and Director of the International Centre for Classroom Research (ICCR), Australia. David established and led a substantial, internationally-extensive, innovative research programme in video-based classroom research. Over the last 20 years, David's research activity centred on capturing the complexity of classroom practice in more than 20 countries.

TEACHERS TALKING ABOUT THEIR CLASSROOMS

Learning from the Professional Lexicons of Mathematics Teachers around the World

Edited by
Carmel Mesiti, Michèle Artigue, Hilary Hollingsworth,
Yiming Cao, and David Clarke

Routledge
Taylor & Francis Group

LONDON AND NEW YORK

First published 2022
by Routledge
2 Park Square, Milton Park, Abingdon, Oxon OX14 4RN

and by Routledge
52 Vanderbilt Avenue, New York, NY 10017

Routledge is an imprint of the Taylor & Francis Group, an informa business

British Library Cataloguing-in-Publication Data
A catalogue record for this book is available from the British Library

Library of Congress Cataloging-in-Publication Data
A catalog record has been requested for this book

ISBN: 978-0-367-37697-0 (hbk)
ISBN: 978-0-367-37693-2 (pbk)
ISBN: 978-0-429-35562-2 (ebk)

Typeset in Times New Roman
by SPi Global, India

With great respect and affection, this book is dedicated to our colleague, mentor, and friend

DAVID JOHN CLARKE
24/2/1952 – 25/1/2020

Professor David Clarke has been well known internationally for his establishment and leadership of a substantial, internationally-extensive, innovative research programme in video-based classroom research. Over the last 20 years, David's research activity centred on capturing the complexity of classroom practice in more than 20 countries. The combination of theoretical, methodological, and technological advances moved the field towards greater critical self-reflection, and increasingly sophisticated research designs and research tools for understanding complex teaching and learning practices in different parts of the world.

The research programme led by David has also created an active and vibrant international research community; an achievement he was most proud of. The list of active collaborators associated with David's research programme is very large, numbering more than 55 research academics and students in Australia, Chile, China, the Czech Republic, Denmark, Finland, France, Germany, Hong Kong, Israel, Italy, Japan, Korea, New Zealand, Norway, the Philippines, Portugal, Singapore, South Africa, Sweden, the UK, and the USA.

We are richer for having worked alongside David and we remain honoured to have known a wonderful, generous, and inspirational educator. We share the task of continuing his legacy and miss him deeply.

CONTENTS

FIGURES

TABLES

CONTRIBUTORS

Carolina Araya
Adjunct Professor
Pontificia Universidad Católica de Chile

Michèle Artigue
Emeritus Professor
Laboratoire de Didactique André Revuz
Université de Paris, France

Jiří Bureš
Principal
School da Vinci, Czech Republic

Elisa Calcagni
Senior Researcher
Chair for Teaching and Learning
Friedrich-Schiller University of Jena,
Germany

Yiming Cao
Professor
Beijing Normal University, China

Thierry Chevalarias
Research Fellow
IREMS de Poitiers, France

Hyungmi Cho
Lecturer
Gyeongin National University of
Education, Korea

David Clarke
Professor
Director, International Centre for
Classroom Research (ICCR)
University of Melbourne, Australia

Mónica Cortez
Analyst
MIDE UC Research Centre
Pontificia Universidad Católica de Chile

Jenny Christine Cramer
Research Fellow, University of Bremen
Teacher
Bremen, Germany

Florence Debertonne-Dassule
Research Fellow
IREMS de Poitiers, France

Gladys Díaz
Mathematics Teacher, Enseñanza Media
Colegio Southland School, Chile

Tracy E. Dobie
Assistant Professor (Mathematics
Education, Learning Sciences)
University of Utah, United States

Lianchun Dong
Associate Professor
Minzu University of China

Yuka Funahashi
Associate Professor
Nara University of Education, Japan

Daniela Gómez
Associate Professor
Universidad Pedro de Valdivia, Chile

Valeska Grau
Associate Professor
Pontificia Universidad Católica de Chile

Brigitte Grugeon-Allys
Emeritus Professor
Laboratoire de Didactique André Revuz
Université Paris Est, France

Hayato Hanazono
Associate Professor
Miyagi University of Education, Japan

Markku S. Hannula
Professor
University of Helsinki, Finland

Hilary Hollingsworth
Principal Research Fellow
Australian Council for Educational
Research (ACER), Australia

Julie Horoks
Senior Lecturer
Laboratoire de Didactique André Revuz
Université Paris Est, France

Alena Hošpesová
Associate Professor
University of South Bohemia in České
Budějovice, Czech Republic

Rita Järvinen
Mathematics and ICT Teacher, Textbook
Author
Olari Secondary School, Espoo, Finland

Hee-jeong Kim
Assistant Professor
Hongik University, Korea

Jani Kiviharju
Mathematics Teacher
Helsinki Normal Lyceum
University of Helsinki, Finland

Christine Knipping
Professor (Mathematics Education)
University of Bremen, Germany

Katie M. Mayle
Mathematics Teacher
Northwestern University, United States

Carmel Mesiti
Research Fellow
International Centre for Classroom
Research (ICCR)
University of Melbourne, Australia

Hana Moraová
Deputy Headmaster, Duncan Center
Conservatory
Lecturer
Charles University, Czech Republic

Shogo Murata
Research Fellow
University of Tsukuba, Japan

Jarmila Novotná
Professor
Charles University, Czech Republic

Julia Pilet
Senior Lecturer
Laboratoire de Didactique André Revuz
Université Paris Est, France

David D. Preiss
Professor
Pontificia Universidad Católica de Chile

Amanda Sfindilis Reed
Head of Mathematics
Department of Education (DETVic),
Australia

David A. Reid
Professor (Mathematics Education)
University of Agder, Norway / University
of Bremen, Germany

Katherine Roan
Learning Specialist (Numeracy)
Department of Education (DETVic),
Australia

Maarit Rossi
Emeritus Mathematics Teacher and
Principal
Kirkkonummi, Finland

Fritjof Sahlström
Professor
Dekanus, Fakulteten för pedagogik och
välfärdsstudier
Åbo Akademi, Finland

Anna Sfard
Professor Emerita
University of Haifa, Israel

Miriam Gamoran Sherin
Alice Gabrielle Twight Professor of
Learning Sciences
Associate Provost for Undergraduate
Education
Northwestern University, United States

Yoshinori Shimizu
Professor
University of Tsukuba, Japan

Birte J. Specht
Research Assistant (Mathematics
Education)
University of Oldenburg, Germany

Sarah L. White
Doctoral Student
Northwestern University, United States

Guowen Yu
Research Fellow
Beijing Academy of Educational Sciences,
China

Iva Žlábková
Senior Lecturer
University of South Bohemia in České
Budějovice, Czech Republic

FOREWORD

Mathematics, often called "a universal language," has long been considered as perhaps the only domain of human intellectual activity that rises above cultural and linguistic differences. With this view in mind, the issue of the language in which mathematics is done, learned, or taught appears of little importance. After all, what difference could it make whether one deals with numbers, geometric figures, sets or functions in English, Chinese, or Hebrew? In fact, how much dissimilarity may there be between, say, English and Chinese vocabularies used in teaching, in the first place? Aren't these vocabularies directly translatable into each other? But suppose one experiences some difficulties in her attempts to translate the Chinese vocabularies of teaching into English. Even then, is there a reason to suspect that the lexical disparities signal some deeper differences in the teaching itself? And if the impression of language-independence prevailed in the now almost-gone world of impermeable political borders and of national languages that kept cultures from mixing with one another, this impression is obviously growing in strength in our times of globalisation and of the accelerated cultural entropy. Perhaps the best evidence for our sustained belief in the irrelevance of language for how one gets access to the universe of abstract mathematical objects comes in the form of international comparative examinations, such as TIMSS or PISA, in which a single set of questions is used worldwide to test students' mathematical competencies. The fact that the questions are administered to the examinees in different languages does not seem to bother those who carry comparative analyses of results. But is language really as inconsequential as to be negligible in this kind of study?

The editors and authors of this volume are clearly skeptical about this. Indeed, were their answer to this last question "yes," they would have had no reason to undertake the international research project called Lexicon, presented on these pages. Not only does this impressive team refuse to treat the language of mathematics teachers as transparent, but it obviously considers its potential impact as significant enough to justify launching this world-embracing study of ten mathematics teaching vocabularies. The argument with which the conceivers of Lexicon have been defending their belief in the importance of language proved convincing enough to secure funding for the project (true, David Clarke, the leader of the undertaking and its animating sprit, is widely known for his creative thinking and contagious enthusiasm; yet, this fact alone cannot account for the popularity of the project with both researchers and funding agencies).

What are the expected rewards for the Lexicon team's efforts? How can the reader benefit from the resulting introduction to a number of national MEPLs (mathematics education practice lexicons) of different countries? The most significant gains are likely to come in the form of useful understandings of teaching practices in different countries. The possible insights, I wish to claim, are diverse and many, and they range from hints on the kind of *opportunities for learning* the teachers who use a given MEPL tend to create and on *the kind of mathematics* the students are invited to learn, to insights into *power relations* likely to arise in the mathematics classroom and into *identities* the learner may develop in the course of learning. Of course, the teacher's vocabulary – the words and the way they are employed – is not the only key to all these issues, but it is certainly among the most important ones. This space is too tight to illustrate all these categories, but let me give just a few examples. Thus, the proportion between the frequency of words that pertain to processes (verbs) on the one hand, and of those that signify objects (nouns) on the other indicates the *objectification* of the mathematical discourse to which the learner is invited. Although the more objectified the discourse, the more conducive it is to future developments, it also constitutes a greater challenge for the learner. The degree of objectification is thus something to be considered by those teachers who wish to fine-tune their teaching to the needs of their learners. Further, when it comes to teachers' pedagogical vocabulary, a high frequency of words related to remembering and "exercising," especially if coupled with a relative scarcity of terms that cue the learner toward inquiry, imply the teacher's tendency to encourage ritualised mathematising on the expense of genuine explorations. With regard to the issue of power relations and students identities, high occurrence of terms that pathologise difficulty (e.g. "learning disability," "slowness," "low potential") and of those that reify (and idolise!) easiness and success ("gift," "high ability") may turn classroom conversation into an incubator of unhelpful learning identities, leading to a paralysing fear of mathematics.

The rich data offered by the authors of this book in the form of MEPLs may become source of further insights if one examines the relations between different elements of each MEPL. Such study may reveal the basic assumptions underlying the pedagogy of teachers who use these vocabularies. In this kind of inquiry, one must keep in mind that these are *the relations* between words, rather than just their external connections, that play a principal role in making the words meaningful. From here it follows that a translation of a term such as *learning* from, say, English to Chinese, may evoke an understanding of learning – of its nature and possible course – quite different than the one encouraged by *learning*. And even more can be gleaned from the Lexicon project if one examines relations between each MEPL and the national language within which it sits. Locating this inquiry within the context of the relevant culture and its history may bring particularly interesting insights. One question to address in this way is that of metaphors implicitly present in a given MEPL. Different-language names for the same mathematical idea, although considered as mathematically equivalent – after all, their use is determined by mathematical definitions – may carry different metaphorical entailments. Thus, for instance, the English mathematical term *limit* comes from everyday language, where it is used in expressions such as "speed limit" or "there is a limit to my patience". In Korean, on the other hand, the mathematical word for limit comes from Chinese and it is in mathematics classroom that the learner meets it for the first time. This difference has been shown to have important implications for how, and with what kind of difficulties, the English and Korean students learn to use the mathematical word for limit. An even more important question for a thorough exploration is whether a given MEPL constitutes an organic part of its language (and thus of its culture) or is rather a possibly ill-adjusted transplant from somewhere else. The distinction may be made, among others, by finding out how well the original lexico-grammatical features of a given language support those required by its MEPL. Whereas in some cases, the MEPL may appear to its users as a natural part of their language, in some other circumstances the learners may have good reasons to experience it as contrived. In this latter situation, they will encounter special kind of hurdles. This was shown to be the case, for instance, with Tongan speakers who, while not any less competent than English speakers in most areas of mathematics, viewed fractions as an almost insurmountable challenge. A closer inspection of the traditional Tongan language disclosed a telling "semantic void": this old language had no counterparts for fractions and even for just parts of things. Tongan traditional cultural practices were predicated on the exclusive use of whole, unbroken objects and thus produced no need for talking about parts of things, let alone for translating fractioned object into quantities.

Research possibilities created by the authors of this book seem truly inexhaustible. The reader will surely enjoy raising questions about MEPLs and imagine research trajectories she may subsequently travel under their guidance. My last remarks are directed to those future travelers who are concerned with the question of whose lexicons are those offered in this book. According to the editors and authors of this volume, these are professional vocabularies of communities of teachers united by the language they speak and the country they live in. This said, one may wonder about the boundaries of these communities: Are they the same as those of *all* the speakers of a given language from a given country? Probably not. Do they include all the *teachers* in a given country who speak a given language? Implausible either. And if so, what are these boundaries? Another set of worries may be raised with regard to how the MEPLs, as offered in this book, have been influenced by institutions, such as universities or ministries, or by the researchers themselves. I wish to claim, though, that these concerns, although important, do not, in any way, undermine the usefulness of the data. Suffices to remember that the present MEPLs do underlie authentic, living discourses used on an everyday basis within a certain community of teachers. As such, they may be employed as useful samples with which to learn about existing practices and to assess them critically with an eye to necessary future reforms. The editors of this important volume should be congratulated on having opened such broad fertile field in which to grow our understanding of the role of mathematics teachers' language in shaping their classroom practices and, eventually, in supporting their students' learning.

Anna Sfard

1

THE INTERNATIONAL CLASSROOM LEXICON PROJECT

Carmel Mesiti, Michèle Artigue, Hilary Hollingsworth, Yiming Cao and David Clarke

Background

Comparisons of student achievement on international tests have prompted a wide variety of international comparative research projects examining good practice in the mathematics classroom. Since the first TIMSS video study was reported by Stigler and Hiebert (1999), international comparative research in mathematics education can largely be identified with three dominant approaches: large-scale studies of student achievement (TIMSS and PISA); video survey studies of typical classroom practice (Hiebert et al., 2003); and, cross-cultural video case studies of well-taught classrooms, as in the Learner's Perspective Study (LPS; Clarke, Keitel & Shimizu, 2006).

The search for legitimate units of cross-cultural comparative analysis for the LPS led to the identification of *lesson events* with sufficiently universal form to support cross-cultural comparison. One such example is the event *kikan-shido (Jap.) 'instruction between desks'*. This event refers to moments when the teacher walks around the room observing students as they work; a familiar teaching occurrence. An international comparative analysis identified differences in function of this event as illustrated by teachers in different classrooms around the world (O'Keefe, Xu & Clarke, 2006).

Such comparative analyses of *lesson events* have provided insights into local norms and differential learning outcomes (Clarke, Mesiti, O'Keefe, Xu, Jablonka, Mok, & Shimizu, 2008); however, the reasonable translation or adaptation of these events into forms likely to be workable in classrooms around the world remains a challenge. Our LPS colleagues (whose first language is not English) engaged in collaborative work that revealed differences in the way that classroom phenomena are described and interpreted. They found that there are some educational terms, particularly relating to classroom practice, that

a.   have no reasonable English translation, or;
b.   have never made the transition and thus are absent from the educational literature in English.

The following section expands on each of these points.

Terms with no reasonable English translation

Teacher actions recorded by outside observers run the risk of disconnecting instructional acts from the culture, language, and pedagogical traditions that give them meaning and effectiveness. Stengers (2011) argues: "No comparison is legitimate if the parties compared cannot each represent his own version of what the comparison is about; and each must be able to resist the imposition of irrelevant criteria" (p. 56). Consider the difficulty of translating the Russian word *obuchenie* into English. Does one choose the term "instruction" (p. 350) as favoured by Hedegaard (1990) in his translation of Vygotsky, or the term "learning" (p. 90) as found in an earlier translation (Vygotsky, 1930–34/Vygotsky, 1978)? Similar difficulties present with the Dutch term *leren* and the Japanese term *gakushushido* (discussed in Clarke, 2001). These three terms recognise the interdependence of instruction and learning and capture both activities within a single term. In the English language, however, we appear compelled to "dichotomize classroom practice into teaching or learning" (Clarke, 2012a).

The examples above illustrate how the English translations of educational terms originating in non-English languages can misrepresent the meaning of the original terms, and potentially distort the educational practices that the terms are intended to represent. It also illustrates the handicap of doing all our writing and theorising in English (Clarke, 2001, 2006). In particular, the use of English as the international *lingua franca* of the research community in education denies use of many sophisticated terms developed in languages other than English. This means that however productive a collaborative analysis might be, the international search for effective classroom practice is hampered by the universal use of English as the classificatory, analytical, and communicative medium of international research.

Terms absent from the educational literature

Further to terms that have no reasonable English translation, a second consideration is even more significant: There are educational terms, particularly relating to classroom practice, that have no obvious English equivalent, and which have never made the transition into the educational literature in English. Consider the following examples:

i. *pu dian* 铺垫 and *jiao shi jiang ping* 教师讲评
    In 2007, Professor Cao from Beijing Normal University spent some time working with Professor Clarke coding Chinese classroom data. Through lengthy conversation, it became clear that Professor Cao was using Chinese terms for which there were no precise English equivalents: *pu dian* 铺垫, an event occurring ahead of the introduction of new content which can take a variety of forms (Cao, Clarke & Xu, 2010); and, *jiao shi jiang ping* 教师讲评, an activity involving teacher's public evaluation of student work (see also *Chapter 7 Chinese Lexicon*).

ii. *narration de recherche*
    The French pedagogical term, *narration de recherche*, identifies a programme of work that may last several lessons where students are invited in groups to solve a rich mathematical problem. Students are expected to produce a report detailing their avenue of explorations including those that were unsuccessful; they are advised that the teacher is more concerned with the description in the report than the final result. The final

lesson is devoted to a whole class discussion (Sauter, 1998; see also *Chapter* 13 *French Lexicon*).

iii.   *matome*

Matome refers to a teacher-orchestrated discussion, named in Japanese, that draws together the major conceptual threads of a lesson or extended activity, most commonly a summative activity towards the end of a lesson (Shimizu, 2006; see also *Chapter* 17 *Japanese Lexicon*).

Each of the terms discussed above names an activity that has been refined and elaborated over time. These activities are essential components of accomplished practice, and essential elements of teachers' professional vocabulary in their respective national contexts. The terms, and their meanings, cannot be captured accurately with a term (or short phrase); each of these activities has no precise English equivalent.

There are many terms, employed in non-English speaking countries, which describe aspects of classroom practice, but which do not have English equivalents and therefore, are not available to support research, theorising, or teacher reflection in English. To put this simply, the absence of such terms in English makes those aspects of classroom practice less visible to English speakers. And, possibly most significantly, educational theorising regarding classroom practice and learning may be undertaken in English in ignorance of the potential insights and accumulated wisdom embedded in terms which are not English in origin.

The international dominance of English has denied researchers, theoreticians and practitioners access to many sophisticated, technical classroom-related terms used in languages other than English, which might otherwise have contributed significantly to our understanding of classroom instruction and learning. Since our practice and our theory may be constrained by available terminology, a significant challenge that continues to confront researchers is the reasonable translation or adaptation of documented practice into forms likely to be viable in classrooms internationally.

The International Classroom Lexicon Project

The International Classroom Lexicon Project was established to identify the key pedagogical terms from various educational communities around the globe. It employed a novel approach to address the challenges in identifying and naming practice, and, it sought to engage with mathematics research and teaching communities around the world.

The ten research teams whose research is reported in this volume are situated in universities in Australia, Chile, China, Czech Republic, Finland, France, Germany, Japan, Korea, and the United States. Captured in this combination of countries are different educational traditions and pedagogical histories. This book is the product of a concerted effort by each of these research teams to identify a pedagogical lexicon, "the vocabulary of a person, language or branch of knowledge" (Stevenson, 2015), from their community. A significant distinguishing characteristic of this study is the documentation of classroom pedagogical practices in their original language, supported by English descriptive detail and illustrative, classroom examples. Unlike any previous international study, this project has the capacity to compare and contrast identified teacher practices with those from other communities. Such analyses have the potential to contribute significantly to the study and promotion of reflective practice for teachers of mathematics, as constructs found in other communities and which are otherwise absent from one's own may operate as reflective tools.

Although the objective of this project is to document a professional lexicon of middle-school mathematics teachers, the researchers recognise that a professional lexicon is a dynamic, changing and evolving entity, varying in its substance and its use according to the geographic and socio-demographic characteristics of the individual teaching communities in question. As such, a definitive lexicon that would hold for all across a single country is not a plausible or realistic goal. Instead, the aim of this project is to document a lexicon that is a reasonable representation of the language in use by mathematics teachers, validated as reasonable by members local to the national research teams as well as others, nationally, whose opinions the researchers were able to access.

Theoretical framework

The International Classroom Lexicon Project began with the recognition that classroom practice named in one community is not necessarily named in another. Some communities have had their named activities translated into English in ways that misrepresent their true meaning, whilst other named activities have been omitted from the *lingua franca* of research (see also Background). The chapters in this book confirm that lexicons of mathematics teachers from different cultural communities do differ.

The notion that 'language shapes thought,' sometimes referred to as The Whorfian hypothesis (Whorf, 1956) or the Sapir-Whorf hypothesis (Sapir, 1949), has been the subject of much debate. Whorf and Sapir argued that a person's thoughts and actions are determined, and thus restrained, by the language the individual speaks. This strong interpretation of the hypothesis, *linguistic determinism*, has been much less favoured than its weaker version, commonly identified as *linguistic relativity*, specifically, that differences amongst languages may influence thinking and behaviour and hence our lived experience (Boroditsky, 2001; Casasanto, 2008; Levinson, 2003). The theoretical position adopted for this project is that of *linguistic relativity* that differences in language are important; they indicate and shape the diversity of teachers' worldview about the classroom.

The researchers involved in this project have recognised that the languages in which our lexicons are expressed differ in a variety of ways. These include differences in vocabulary, grammatical structure and organisational categories.

Differences in vocabulary

The terms of the ten lexicons differ in content and number (see also Book Structure and Character). The benefit of a 'named' activity is that it can be readily observed and recognised within a classroom setting. In this way, it becomes easier to question how well the activity is executed and how it might be improved. Schoenfeld (2011) makes a related point, "what you see and don't see shapes what you do and don't do" (p. 228). It might follow that a classroom activity that cannot be named is less likely to be observed; the activity is less 'visible' for the purpose of reflective analysis. Given the complexity of classroom instruction teachers will make decisions about where to focus their attention (Sherin et al., 2011); an unnamed practice might deny teachers the recognition of an activity that at least one teaching community has identified as sufficiently important to have been assigned a specific name.

Differences in grammatical structure

The nine languages involved in this project differ in grammatical structure. They differ, for example, in their approach to grammatical gender (a noun class system). The languages *English* and *Finnish* have no grammatical genders as do the languages *Chinese, Japanese,* and *Korean*; however, the latter East Asian languages have noun classifiers (words or affixes that classify nouns). The *French* and *Spanish* languages have masculine and feminine genders whilst the *German* language includes a third gender, neuter. The *Czech* language has more than three grammatical genders. Boroditsky, Schmidt, and Phillips (2003) found that structural differences in gender have an effect on people's descriptions and assessment of similarity between objects thus forcing the speaker to attend to certain aspects of a language. Other ways in which the languages involved in this project differ are discussed in later chapters.

Differences in organisational categories

Language assists in organisation of the world into identifiable labels and categories. Winawer and his colleagues (2007) conducted a study that explored the influence of different colour terms for dividing the colour spectrum. For example, the Russian language compels a distinction between lighter (*goluboy*) and darker (*siniy*) blues. When this boundary distinction was subjected to testing, Russian speakers displayed faster perceptual ability in discriminating between light and dark blues. There is a growing body of research that supports the notion that differences in organisational language appear to influence how the world might be conceived, structured, and divided by speakers of a language (Gelman & Roberts, 2017).

By virtue of having been born into a culture with an accompanying language, thinking and behaviour and lived experience, appear to differ. Different languages vary with respect to vocabulary, structure, and organisational language, and appear to influence thinking about certain aspects of the world.

The research methodology

The research question,

> *What are the terms that teachers use to describe the phenomena of the middle-school mathematics classroom?*

united the collaborators of the International Classroom Lexicon Project in pursuit of the documentation of national lexicons for each participating teaching community. These national lexicons consist of the familiar terms (and short phrases) used by teachers of middle-school mathematics. The focus on middle years was chosen to complement international research studies such as TIMSS (Hiebert et al., 2003; Stigler & Hiebert, 1999) and LPS (Clarke, Keitel & Shimizu, 2006).

The terms featured in each lexicon were included in their original language, supplemented with literal and closest English translations where appropriate, as well as descriptive detail and illustrative examples from classrooms. These lexicons are intended to capture the language that the educator communities of practice (CoP) (Lave & Wenger, 1991) use to talk *about* the phenomena of the middle-school classroom; that is, the language used by teachers in conversation with each other to identify classroom events, actions, and practices. The project shares traits with the study of cultural anthropology (Mercier, 2019) as the lexicons may be regarded as cultural and social artefacts, representative of the vocabulary of a professional discourse amongst teachers of mathematics, and their method of documentation is ethnographic in nature. The researchers allowed the terms (and agreed meanings) to emerge from the ethnographic encounter between the researchers and the teachers who represented our 'insiders' (Hammersley & Atkinson, 1995; Hoey, 2014). The insiders' input was sought to refine and ratify the lexicon in a variety of face-to-face formal and informal meetings.

More broadly, the documentation of the lexicons involved putting into practice a 'negotiative' methodology (Clarke, 2012b). Central to this methodological approach was the incremental, iterative negotiation of the national lexicons both in the progressive aggregation of terms and in the progressive expansion of the community with whom the content was negotiated from local to national in each participating country. This protocol is explained in more detail in the following section.

The research design

In preparation for the identification and documentation of key lexical items, national teams were advised to include 'teachers as researchers' and invited to contribute video material from one middle-school mathematics classroom.

Composition of national research teams

Each national research team comprised both senior and junior researchers and teachers of middle-school mathematics. The only constraint on team membership was the inclusion of a minimum of two experienced teachers with strong preference given to mathematics teachers of years seven to nine who were currently teaching. It was felt that this membership, by varying in expertise, knowledge, and experience, would produce a knowledgeable team capable of meeting the task of documenting the lexicon.

Stimulus package

A project-wide package of nine[1] lessons was assembled to include a middle-school lesson of mathematics from each participating community. Each national team contributed video material and time-stamped transcripts that were then re-configured into a single viewing window (Figure 1.1). This package of lessons, presenting a variety of instructional approaches and classroom settings, was made available to each national team and functioned to stimulate thinking about the candidate terms of the draft lexicon.

The collaborative documentation of key lexical terms, by national research teams, involved three distinct main phases: *Identification*, *Validation,* and *Clarity Check*. Figure 1.2 summarises the significant elements within each phase and illustrates a pathway from the initiation of the documentation process (*Identification*) to its conclusion (*Final Lexicon*). This research design is elaborated through a discussion of each of these phases in the following subsections.

FIGURE 1.1   Video material re-packaged into a single viewing window (illustrative only).

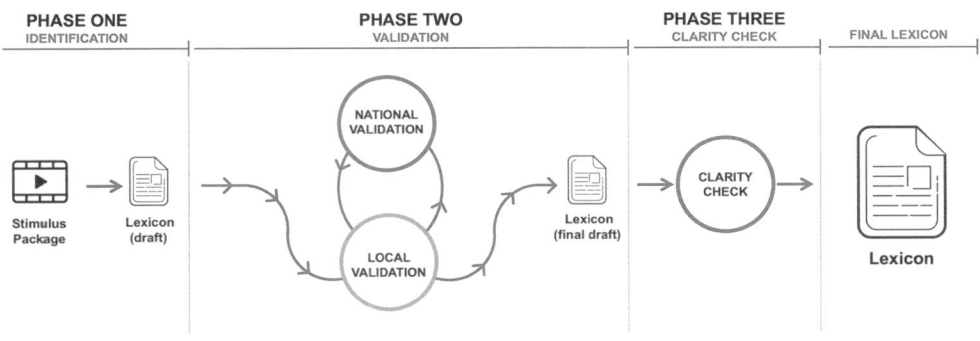

FIGURE 1.2   Main phases for the documentation of national lexicons.

Phase one – Identification

Throughout the *Identification* phase, each research team worked to identify the candidate terms empirically by viewing the videos of mathematics lessons from around the world. These videos had been distributed via the stimulus package and members of each research team proposed a name for the observed phenomena within the classroom. These proposed names were recorded as candidate lexical terms and they additionally served to bring to mind related terms that might also be considered for inclusion in a draft lexicon illustrative of a teacher's professional vocabulary. The initial prompt used for stimulating thought about the video was,

> *What do you see that you can name?*

This general prompt and approach were developed to facilitate what could be named. That is, a researcher might identify a classroom characteristic whilst another might identify an activity engaging a group of students. An important decision was the exclusion of language that:

a. identified material objects in the classroom, such as rulers, textbooks, calculators, and protractors;
b. referred to mathematical domains, such as Geometry and Algebra; and
c. referred to mathematical objects such as equations or functions.

The videos were intended to promote one's thinking about candidate terms, whether present in the video or not, as identifying one classroom phenomenon may call others to mind. Hence, whether or not observable in the video recordings, both terms would be recorded for inclusion in the draft lexicon.

Operational definitions

Once each national research team developed an initial draft assemblage of terms, composite operational definitions were developed for each term. The set of essential elements of the operational definitions included: the term (in the original language); descriptive detail, and, examples and non-examples from the classroom (in the original language and in English). Where possible and informative, some lexicons also included: a literal translation of the term into English; a closest/best-fit English translation; and, a commentary on the origin of the term.

Phase two – validation

Throughout the validation phase, from local to national to local, the opinions of the broadest relevant community were sought regarding the extent to which the draft lexicon was seen as reasonable. The question was,

> *Do different communities within a country, who might employ such a lexicon, see its content and structure as reasonable?*

The onus was then on each national team to demonstrate that they had accessed the voice of communities likely to be familiar with or make use of the phenomena referred to in the lexicon and the terms by which those phenomena were identified within the lexicon.

Local validation

The terms of the proposed lexicons, including the composite operational definitions, were subjected to a local validation process by each national team. The intention of this process was to investigate the extent to which the local community of mathematics education researchers (and, in some cases, their educational colleagues) would endorse the listed terms and any emergent structure, as well as the descriptions, examples, and non-examples. In this case, endorsement was understood as confirmation of the terms as credible and plausible. Each team has outlined their process of local validation, including its membership, in the first of their two chapters within this book.

National validation

Surveys were developed by each research team to collect data about the middle-school mathematics teachers' familiarity with each of the lexicon terms and the extent to which the operational definitions were endorsed. Respondents were invited to comment on the operational definitions with respect to suitability and clarity. The goal of the national survey was to establish that the candidate terms of the draft lexicon were not only familiar to the mathematics teaching community, but also that the meaning of the terms was accepted as being represented appropriately by the mathematics teaching community. The survey included such questions as:

> *How familiar are you with this term?*
> *Do you use this term in conversation with your colleagues?*
> *How might we improve the description of this term?*

The general principle agreed to by all research teams was that a term would be included in a national lexicon if it were familiar to two-thirds or more of national respondents. It was recognised, however, that particular terms might be of sufficient significance for a special case to be made for inclusion outside this rule. Where this occurred, the inclusion of such terms was noted by the relevant national team and justification for the term's inclusion was provided. Following the analysis of responses collected from the national survey the lexicon was re-drafted and subjected to an additional local review.

Phase three – clarity check

The clarity check was the final phase involved in documenting the lexicon. Each national team had been assigned another country's lexicon for the purpose of reviewing the clarity of the description and examples. Any terms whose descriptions were considered problematic, either for reasons of conceptual clarity or clarity of expression were identified to the originating national team for review.

These national lexicons were documented over a number of years and underwent various stages and phases of development, ratification, and refinement as represented in the illustration (Figure 1.2). This protocol involving the incremental and iterative collaborative documentation and negotiation of key terms, by national research teams with the mathematics teaching community, has resulted in the ten national lexicons that are presented in this book.

Book structure and character

This book is structured in 20 main chapters, in addition to this introductory chapter, with each of the ten participating country teams contributing two chapters. The first chapter from each team describes: the process of development and refinement of their lexicon; some analyses of their lexical terms; characteristic features of their lexicon; and, reflections about potential uses for their lexicon. The second team chapter presents each of the terms included in their lexicon; in both original language (including characters where appropriate) and closest English translations, together with accompanying descriptions and classroom examples. Some country teams presented their lexicons together with a structural framework in the second of their two assigned chapters.

As reported earlier, there was an overarching methodological framework used to guide each country team's development of their lexicon. However, there was some flexibility around details of implementation, as well as local analyses conducted, and each international team has a unique story to tell. These unique stories relate to the processes each team used to develop their lexicon, the final 'product' or outcome that is their lexicon, and results of initial analyses of their lexicon.

The generation of the lexicon included four main steps: team members viewing videos from each country and identifying initial lexical terms, descriptions, and classroom examples (the Korean team joined the project at a later stage and only viewed Korean lessons); local reviews and refinements; national review and refinement; and, a final cross-country clarity check. In terms of the processes used by each country team to develop and refine their lexicons, most processes were characterised by strong initial input of a small group of mathematics teachers working together with mathematics education researchers, followed by consultation with the wider teaching community. There were some variations in team composition. For example, the Czech team also included a language teacher and a researcher in pedagogy, and the Korean team included a professor of Korean language education.

With respect to the 'final' lexicons prepared and presented by the country teams, these varied in the number and nature of the terms included. The number of terms included in the lexicons range from 57 to 123 (Czech 57; Australia 61; Germany 65; Japan 70; Chile 74; Finland 99; USA 99; Korea 103; France 116; China 123). Details of the initial numbers of terms identified by each country team and the particular ways that they refined their lexicons to arrive at these final numbers are explicated in each team's chapters.

In regard to the nature of the terms included in the lexicons, country teams described different contextual factors that shaped and influenced the terms they included in their lexicon. These included cultural, historical, educational, and language-based factors. Teams also reported undertaking different analyses of their lexical terms, such as considering how familiar the terms were and how often they were used, and investigating connections and relationships between terms. Country teams also described characteristic features of their lexicons. Some examples of the characteristic features of the different lexicons are outlined below:

> *The Australian Lexicon* includes terms that are generic in nature, with few pertaining specifically to the teaching and learning of middle-school mathematics. The categories proposed for organising the terms reflect this generic quality (*Administration, Assessment, Classroom Management, Learning Strategies, Teaching Strategies*).
>
> *The Chilean Lexicon* has lexical terms with connotations of agency, most referring to actions that are exclusive to teachers. This reflects the prevalence of teacher-led

activities focusing on transmission of knowledge and skills practice with few opportunities for students' contributions.

The Chinese Lexicon has an inherent structure that prioritises connections and relationships between terms, and identifies five main "critical pedagogical behaviours": *Teacher-student-interaction*; *Teacher questioning*; *Teacher displaying*; *Student listening;* and, *Student doing exercise.*

The Czech Lexicon entries focus mainly on pupils' and teachers' observable activities, the mathematical nature of the activities is less significant. This lexicon reflects the discourse used in practice in classrooms and the discourse used by researchers.

The Finnish Lexicon includes terms more focused on teacher-student interaction and lesson organisation than specific aspects of mathematics teaching; this reflects a relative focus on "the act of teaching," with less attention given to student learning, and even less attention to mathematical content.

The French Lexicon has a distinct mathematical orientation and a 'didactic' nature that together create the specificities of mathematics classrooms; this is in line with the mathematical orientation of the professional discourse of French mathematics teachers.

The German Lexicon includes some terms particular to the German language which have no English word translations that capture their precise meaning; these terms are accompanied by synonymous terms in the lexicon.

The Japanese Lexicon is characterised by its evolving nature through its explicit association with the professional practice of 'lesson study'; some terms may be regarded as value-laden, as teachers' aims and intentions are embedded within the description of the terms.

The Korean Lexicon includes terms that focus on mathematics instruction-learning theory and teaching practice (including lesson structure and preparation) that teachers learn in pre-service programs; terms influenced by national education policy; and, terms that originated in foreign countries and have been translated into Korean.

The United States Lexicon includes a significant number of terms focused on discussion, collaboration, participation and assessment; this reflects relevant research literature findings and policy/standards initiatives.

As can be seen in this brief introduction, country teams differed in the ways they identified, organised, and examined the terms they included in their lexicons, and each lexicon has its own characteristic features. It is anticipated that details presented in the 20 chapters prepared by the country teams will stimulate much discussion and learning within and across the ten participating countries, as well as the global education community more broadly.

Reflection

In this section, the authors reflect on this first phase of the International Classroom Lexicon Project, in particular, the conceptual and methodological challenges posed by the production of these ten lexicons.

Prior to The International Classroom Lexicon Project, the national research teams observed, to varying degrees, an absence in their language of a well-documented, pedagogical lexicon for the mathematics teaching profession. This contrasted in many languages and cultures with a great wealth of expressions to describe mathematics classrooms and teaching expertise; a wealth of terminology that the English language, the *lingua franca* of international communication, scarcely expresses. These observations motivated the international partners to commit to this endeavour with much enthusiasm; and they serve to explain the research community's heightened interest. Moreover, these observations confirmed the challenging nature of this undertaking. In order to determine the finer details of our research protocol the research members were required to clarify, as a larger team and within our national teams, our responses to questions such as:

> *What terminology are we seeking to list?*
> *What are our justifications for the inclusion or exclusion of a given word or expression?*
> *How do we decide on its description?*

In addition to the challenge of ten international communities sharing lexicons from nine different languages, the project partners also grappled with communicating the wealth of terminology from these communities with one another. This involved much discussion of their literal translations into English, generating detailed explanations and illustrations from the classroom to ensure accuracy of the terms.

The main chapters in this volume illustrate the difficulties associated with these challenges as well as the process by which these challenges were resolved. Despite similarities, whether culturally or due to common pedagogical and historical influences, each of the ten contexts posed specific problems that required locally-determined decisions. It is strongly recommended that each lexicon chapter and its associated chapter are read as a pair, as individual, cultural contexts and knowledge of the decision-making processes within each, are essential to one's understanding of each lexicon.

The reader may be interested to know that this particular work, producing stabilised and validated lexicons for the nine original countries, took four years to complete. From this point of view, it is encouraging to note that the Korean team, a late admission to the international project team, were able to complete the production of its lexicon much more quickly than the other teams with the support of the accumulated experience of the original nine teams. This suggests that the project may be extended to other countries, and other contexts, all the more easily, since the transmission of its conceptual and methodological foundations will be greatly facilitated by the existence of this book.

The lexicons produced were subjected to a complex process of validation and revision. The project-wide methodology is detailed in The Research Methodology section of this chapter and additional methodological detail is provided in each country chapter. Each of the lexicons are reasonably considered to be shared by mathematics teachers within the given contexts; however, it is also recognised that these are not the only lexicons which could have emerged. These lexicons were naturally influenced by the teams of teachers and researchers who were responsible for their initial documentation. In recognition of this, the validation protocols were both strict and involved, and engaged both members of the local and extended mathematics communities in an iterative process. Readers are encouraged to be mindful that the lexicons themselves are dynamic, subject to change, and by the very nature of language, incomplete.

Future directions

The lexicons now exist in a temporarily stabilised state. They can be used to contribute to the professionalisation of teaching, to the profession's discourse, and to the status of teaching as a profession. Terms that make up the vocabulary of teachers may be used in a productive and purposeful manner, as opportunities emerge for focussed interaction amongst classroom mathematics teachers as well as with teacher educators and policy makers. Alexander (2008) recognised the power of verbal interaction for learning and by extension the 'things' we have words to talk about: "Of all the tools for cultural and pedagogical intervention in human development and learning, talk is the most pervasive in its use and powerful in its possibilities" (p. 92). Recently, the lexicons have been used with teachers and educators to explore different classification and organisational systems and their effectiveness in the coding of lesson episodes. This initial access to the ten lexicons has stimulated thoughts about new and fascinating perspectives for use.

The lexicons can form the basis of comparative research. Currently underway is research focussed on the exploration of new conceptual and methodological constructions. Research interests of the international research team members include:

- Comparing terms:
  - related to assessment, classroom environment, emotion, beliefs, participation, and discursive actions;
  - considered essential for teacher practice;
  - unique to a lexicon;
  - specifically related to mathematics;
  - similarly named but defined differently;
  - that use metaphor;
  - that are value-laden; and
  - that are present (or absent) or emerging in official documents.
- Using terms:
  - in lesson narratives;
  - for the coding of classroom video; and
  - to inform practice.
- Applying different organisational categories and structures to the entire set of lexicons (categories could include 'teacher questioning' or 'whole class discussion')
- Comparing the teaching lexicons with the language of the mathematics researcher community
- Extending the lexicons to include terms related to teacher preparation, planning, and activity beyond the classroom

This project used innovative video technology to investigate the insights encrypted in different pedagogical naming systems and novel methods of collaborative negotiated analysis to explore how these insights might inform instruction and research in classroom settings around the world. With global trends towards educational uniformity mediated by international measures of student achievement, the international community should now look to investigate the pedagogical terms privileged in some communities but not others, and utilise the insights represented by such terms to interrogate and inform our classroom practice, our classroom research, and our theorising about classroom settings. Differences and similarities across lexicons might have significant implications for the translation of research findings for practitioner use.

Concluding remarks

The International Classroom Lexicon Project sought to document professional languages of teachers in order to advance discussion about classroom practice. An empirical identification of lexicons has been the entry point into developing a professional lexicon that is considered familiar and in current use amongst teachers. The lexicons offer the educational community the vocabulary that teachers, from ten communities worldwide, use to talk *about* the middle-school mathematics classroom. The lexicons presented in this book represent robust, coherent, and validated lexicons, both defined and illustrated, and which offer a common point of reference for teachers and teacher educators alike. This project presents an opportunity to evaluate the adequacy of these lexicons to encompass and distinguish the variety of practices and pedagogical and didactical phenomena prioritised by contemporary mathematics education.

If the general aim of an education research community is to support the development of pre-service and in-service teachers, a significant starting point is engaging both groups in a study of the 'terms' that feature in teachers' professional speech when conceptualising the practice of the classroom. Equipped with such lexicons, teachers will be better able to reflect on and improve their practice. The primary intention of this research was to provide insight into the naming system employed by middle-school mathematics teachers in relation to their classroom practice, by documenting and interpreting the constructs that are well-known and more frequently utilised in discussions with others. From this foundation, this research endeavour hopes to inform national and international efforts to provide contemporary mathematics teachers with a sophisticated lexicon to shape their professional practice.

Acknowledgements

We acknowledge the teachers, teacher educators, and students who welcomed us into their classrooms and participated in discussions and surveys to refine our lexicons; without your generosity this work would not have been possible.

This project has been funded by a Discovery Grant from the Research Council of the Australian Government (ARC-DP140101361).

Note

1 The package included nine lessons instead of ten as Korea joined the project at a later date and developed their own stimulus package from mathematics lessons of Korean teachers only.

References

Boroditsky, L. (2001). Does language shape thought?: Mandarin and English speakers' conception of time. *Cognitive Psychology*, 43, 1–22.

Boroditsky, L., Schmidt, L. A., & Phillips, W. (2003). Sex, syntax and semantics. In D. Gentner & S. Goldin-Meadow (Eds.), *Language in mind: Advances in the study of language and thought* (pp. 61–79). Cambridge, MA: MIT Press.

Cao, Y., Clarke, D. J., & Xu, L. (2010). Qifa Shi teaching: Confucian heuristics. In M. M. F. Pinto & T. F. Kawasaki (Eds.), *Proceedings of the 34th Conference of the International Group for the Psychology in Mathematics Education*, (Vol. 1, pp. 232–234). Belo Horizonte, Brazil: PME.

Casasanto, D. (2008). Who's afraid of the big bad Whorf? Crosslinguistic differences in temporal language and thought. *Language Learning*, 58(1), 63–79.

Clarke, D. J. (2001). Teaching/Learning. In D. J. Clarke (Ed.), *Perspectives on practice and meaning in mathematics and science classrooms* (pp. 291–320). Dordrecht, Netherlands: Kluwer Academic Press.

Clarke, D. J. (2006). Using international comparative research to contest prevalent oppositional dichotomies. *ZDM*, 38(5), 376–387.

Clarke, D. J. (2010). The cultural specificity of accomplished practice: Contingent conceptions of excellence. In Y. Shimizu, Y. Sekiguchi, & K. Hino (Eds.), *In search of excellence in mathematics education - Proceedings of the 5th East Asia Regional Conference on Mathematics Education (EARCOME5)* (pp. 14–38). Tokyo, Japan: Japan Society of Mathematical Education.

Clarke, D. J. (2012a). International comparative research into educational interaction: Constructing and concealing difference. In K. Tirri & E. Kuusisto (Eds.), *Interaction in educational domains* (pp. 5–22). Rotterdam, Netherlands: Sense Publishers.

Clarke, D. J. (2012b, November 22–23). *Constructing and concealing difference in international comparative educational research* [Keynote]. *2012 Finnish Educational Research Association (FERA) Conference on Education*, Helsinki, Finland.

Clarke, D. J., Emanuelsson, J., Jablonka, E., & Mok, I. A. C. (Eds.). (2006). *Making connections: Comparing mathematics classrooms around the world*. Rotterdam, Netherlands: Sense Publishers.

Clarke, D. J., Keitel, C., & Shimizu, Y. (Eds.). (2006). *Mathematics classrooms in twelve countries: The insider's perspective*. Rotterdam, Netherlands: Sense Publishers.

Clarke, D. J., Mesiti, C., O'Keefe, C., Xu, L.H., Jablonka, E., Mok, I. A. C., & Shimizu, Y. (2008). Addressing the challenge of legitimate international comparisons of classroom practice. *International Journal of Educational Research*, 46(5), 280–293.

Gelman, S. A., & Roberts, S. O. (2017). How language shapes the cultural inheritance of categories. *PNAS*, 114(30), 7900–7907.

Hammersley, M., & Atkinson, P. (1995). *Ethnography: Principles in practice* (2nd edition). London, UK: Routledge.

Hedegaard, M. (1990). The zone of proximal development as basis for instruction. In L. C. Moll (Ed.), *Vygotsky and education* (pp. 171–195). Cambridge, MA: CUP.

Hiebert, J., Gallimore, R., Garnier, H., Givvin, K., Hollingsworth, H., & Jacobs, J. (2003). Understanding and improving mathematics teaching: Highlights from the TIMSS 1999 Video Study. *Phi Delta Kappan*, 84(10), 768–775.

Hoey, B.A. (2014). A simple introduction to the practice of ethnography and guide to ethnographic fieldnotes. *Marshall University Digital Scholar*, 2014, 1–10.

Lave, J., & Wenger, E. (1991). *Situated learning: Legitimate peripheral participation*. Cambridge, MA: Cambridge University Press.

Levinson, S. C. (2003). *Space in language and cognition: Explorations in cognitive diversity*. Cambridge, MA: Cambridge University Press.

Mercier, J. (2019, January 23). Cultural anthropology. *Encyclopædia Britannica*. https://www.britannica.com/science/cultural-anthropology

O'Keefe, C., Xu, Li Hua., & Clarke, D. J. (2006). Kikan-Shido: Between desks instruction. In D. J. Clarke, J. Emanuelsson, E. Jablonka, & I. A. H. Mok (Eds.), *Making connections: Comparing mathematics classrooms around the world* (pp. 73–106). Rotterdam, Netherlands: Sense Publishers.

Sapir, E. (1949). *Selected writings on language, culture and personality*. Berkeley, CA: University of California Press.

Sauter, M. (1998). Narrations de recherche: Une nouvelle pratique pédagogique [Research narrations: A new pedagogical practice]. *Repères-IREM*, 30, 9–21.

Schoenfeld, A. H. (2011). Noticing matters. A lot. Not what? In M. G. Sherin, V. R. Jacobs, & R. A. Philipp (Eds.), *Mathematics teacher noticing: Seeing through teachers' eyes* (pp. 223–238). New York, NY: Routledge.

Sherin, M. G., Jacobs, V. R. & Philipp, R. A. (Eds.). (2011). *Mathematics teacher noticing: Seeing through teachers' eyes*. New York, NY: Routledge.

Shimizu, Y. (2006). How do you conclude today's lesson? The form and functions of 'matome' in mathematics lessons. In D. Clarke, J. Emanuelsson, E. Jablonka & I. A. C. Mok (Eds.), *Making connections: Comparing mathematics classrooms around the world* (pp. 127–145). Rotterdam, Netherlands: Sense Publishers.

Stengers, I. (2011). Comparison as a matter of concern. *Common Knowledge*, 17(1), 48–63.

Stevenson, A. (Ed.). (2015). *Lexicon. Oxford dictionary of English* (3rd edition). Oxford, UK: Oxford University Press. Retrieved from https://en.oxforddictionaries.com/definition/lexicon

Stigler, J. W., & Hiebert, J. (1999). *The teaching gap: Best ideas from the world's teachers for improving education in the classroom*. New York, NY: The Free Press.

Vygotsky, L. S. (1978). *Mind in society: The development of higher psychological processes* (M. Cole, V. John-Steiner, S. Scribner & E. Souberman, Eds.) (A. R. Luria, M. Lopez-Morillas & M. Cole [with J. V. Wertsch], Trans.) Cambridge, MA: Harvard University Press. (Original manuscripts [ca. 1930–1934]).

Whorf, B. L. (1956). *Language, thought, and reality: Selected writings of Benjamin Lee Whorf* (J. B. Carroll, Ed.). Cambridge, MA: MIT Press.

Winawer, J., Witthoft, N., Frank, M. C., Wu, L., Wade, A. R., & Boroditsky, L. (2007). Russian blues reveal effects of language on color discrimination. *Proceedings of the National Academy of Sciences*, 104(19), 7780–7785.

2

NAMING ASPECTS OF TEACHING PRACTICE

Describing and analysing a lexicon of mathematics teachers in Australia

Carmel Mesiti, Hilary Hollingsworth and David Clarke

Introduction

Researchers have argued that fundamental differences between languages, both linguistic and semantic, influence (Boroditsky, 2001; Levinson, 2003) or indeed determine (Sapir, 1949; Whorf 1956) our lived experience. Marton, Runesson, and Tsui (2004) have similarly argued that "categories … [and] distinctions made by language, not only express the social structure but also create the need for people to conform to the behavior associated with these categories" (p. 28). Our interactions with classroom settings, whether as learners, teachers, or researchers, are mediated by our capacity to name what we see and experience. However, as noted by Marton et al. (2004) the function of language goes beyond mediation. Our capacity and our inclination to act in classroom settings are significantly shaped by what our discourse establishes as conceivable, and therefore as possible. In short, teachers' classroom activity is influenced by those practices they are able to name.

This chapter reports on the work of a team of Australian researchers who set out to document the professional language employed by Australian mathematics teachers to name aspects of their practice in lower secondary mathematics classrooms. Our primary interest is lexical and concerns the actual terms by which teachers name events and activities related to their mathematics teaching. The lexicons documented as part of The International Classroom Lexicon Project constitute empirical frameworks through which phenomena of the mathematics classroom can be described.

The importance of naming professional practice

The preparation of future practitioners of a profession such as medicine, law, nursing or indeed teaching not only concerns itself with the learning of large amounts of theory and content knowledge but also the learning of that which constitutes professional practice. Simply put, each profession characterises its practice differently: lawyers acquire a vocabulary that allows them to engage in particular forms of argumentation; doctors have a terminology that facilitates their clinical reasoning; and teachers develop a professional vocabulary, related to activities that enhance student learning and the instructional orchestration of such activities, and which supports professional discussion of the efficacy of those activities. The forms of teaching and learning in which different professions engage in terms of their professional preparation have become known as "signature pedagogies" (Shulman, 2005). These signature pedagogies refer to the pedagogies by which each profession trains its novices; in education this refers to the training of teacher candidates so they may teach.

For the preparation of professions, a "grammar of practice" (Grossman, Compton, Igra, Ronfeldt, Shahan & Williamson, 2009) with an accompanying vocabulary, framework and category system may provide a number of learning opportunities for novices, but it creates its own set of challenges as well. The ability to decompose practice depends on the existence of a language and structure for describing practice (Grossman et al., 2009).

While some professions are able to boast of a professional language that is well-documented and comprehensively understood by members of its profession, Grossman and her colleagues (2009), in their cross-professional study of clergy, teaching and clinical psychology noted: "among our trio of professions, this language of practice seems particularly well-developed in clinical psychology but less developed in teaching" (p. 2075). This finding is not dissimilar to Lortie's much earlier observation, in his social portrait of the "Schoolteacher," of a lack of "technical language" in teaching (Lortie, 1975). Lampert (2000) concurred, reporting that "no professional language for describing and analysing [teaching] practice has developed in the United States" (p. 90).

In the United States of America, Lampert (2000) has argued that the lack of opportunities to work collaboratively with peers on the problems of practice has resulted in "a language of practice [that] remains flat or nonexistent" (p. 90). Connell (2009) has also noted the absence in Australia of a lively occupational culture in teaching which includes "the informal processes by which practical know-how is passed to new teachers in on-the-job learning" (p. 223). The vocabulary used by Australian teachers to describe their practice does not appear to have been clearly developed, refined, or articulated for shared use among the profession. This situation appears to contrast with a well-articulated structure for teaching vocabulary in China, and strong traditions in Japan of educators and teachers discussing research lessons (Lampert, 2000). This has implications for each education community's conception of accomplished practice and for the preparation and ongoing professional development of their teachers. Our articulation of teaching standards and good teaching models is limited to those practices we can adequately name and describe, and professional learning opportunities are likely to be missed by teachers with only limited language to reflect on teaching practice and the phenomena of the classroom.

The Australian education context

This section describes four characteristics of the Australian education context that might influence or shape the development and use of language by mathematics teachers, and the subsequent development of an Australian Lexicon.

Australian schools, students, and teachers

Schools in Australia vary considerably in terms of setting, size, governance, resourcing, and cultural diversity. Australia is a very large island continent with a relatively small population (approximately 25.4 million). Most Australian people live near the coast; only one-third of the population live in rural and remote areas. Schools are therefore situated in urban, suburban, rural, and remote settings, and the communities that they serve across these contexts differ considerably.

Australians are culturally and linguistically diverse. The most recent census data indicated that 49% of the Australian population were either first-generation Australians (born overseas) or second-generation Australians (one or both parents born overseas). In 2016, 21% of Australians spoke a language other than English in the home; the next most common languages other than English were Mandarin, Arabic, Cantonese, and Vietnamese. The number of Australians identifying as indigenous (being of Aboriginal or Torres Strait Islander origin) was 2.8% of the Australian population (approximately 650,000 people) (ABS, 2016). In 2019, student numbers, primary and secondary, increased to almost four million. In secondary schools, male teachers comprise 39% of the total number of teachers, while in primary schools only 18% of teachers are male (ABS, 2020).

A priority, and indeed a challenge, for schools and teachers is to support the specific learning needs of students from such diverse cultural and linguistic backgrounds. The many and varied contexts within which teachers work are anticipated to influence the language that they use with their students and with other members of the education community.

National education initiatives

In recent years, the Federal Government in collaboration with state and territory governments initiated a number of major national education initiatives, including the development of a national curriculum, national standards for teachers and school leaders, and national literacy and numeracy assessment for students in some year levels (Wernert &Thomson, 2016). The Australian Curriculum was introduced in 2012 and revised in 2015. Prior to 2012, states and territories authored their own curricula. While a new mathematics curriculum for primary (Foundation to Year 6) and lower-mid secondary grades (Year 7 to Year 10) describes what is to be taught and learned, and how content is explored or developed, states and territories who are responsible for their own education administration, are able to interpret this curriculum at their local level. This means that there is variation across state and territory curriculum programs implemented, including the materials and resources that teachers in different locations use, and this may have an effect on the professional language that mathematics teachers are familiar with and use.

Shortage of teachers of mathematics

There is currently a shortage of qualified mathematics teachers available to work in Australian secondary schools (Weldon, 2015). This means that it is not uncommon for teachers who have had no specialised training in mathematics education to be teaching mathematics. This "out-of-field" teaching is reported to occur most often in lower secondary classes (including Year 8, the focus of the Lexicon project), with the most competent and experienced teachers being placed in upper secondary classes to work with senior students undertaking high-stakes testing associated with tertiary admissions and other career pathways (Weldon, 2015). It is possible that out-of-field teaching of mathematics might have an effect on teachers' use of language in classrooms, especially at the lower year levels where out-of-field teaching is more likely to occur. As Weldon (2016) notes, the many tasks of teaching, including those that explicitly involve using language such as giving helpful explanations and asking productive questions, "depend on the teacher's understanding of what it is that students are to learn" (p. 1). Other research has shown that a thorough understanding of the subject to be taught is a key attribute of highly effective teachers (Darling-Hammond, Bransford, LePage, Hammerness & Duffy, 2005; Masters, 2016; Teacher Education Ministerial Advisory Group (TEMAG), 2014).

Teaching in Australia

While all teachers in Australia need to be qualified to work in schools and there are Australian Professional Standards for Teachers (AITSL, 2018) that describe expectations across career stages and enumerate requirements related to provisional and full teacher registration, registration and certification of teachers is undertaken by states and territories, and differences in course and program offerings may contribute to the variety of language used by mathematics teachers. In 2018, states and territories committed to nationally consistent, strengthened accreditation of initial teacher education courses; however, implementation details for each course are decided at the local level (Wernert & Thomson, 2016). Similarly, there is an Australian Charter for the Professional Learning of Teachers and School Leaders, that describes the importance and characteristics of high-quality professional development in improving teacher practice, and an Australian Performance and Development Framework which outlines the critical factors for creating a performance and development culture in schools, with the implementation of these occurring locally rather than nationally. Mathematics teachers' participation in professional development and their opportunities for professional collaboration may be influenced by initiatives of their state or territory education departments, school-based programs, their affiliation with subject associations such as the Australian Association of Mathematics Teachers (AAMT), and programs provided by consulting organisations and publishers. In addition, Australian mathematics teachers are able to engage in professional activities with teaching and research colleagues from across the globe using online platforms. These many and varied opportunities can provide mathematics teachers with opportunities to learn, use, and potentially create language relevant to their mathematics teaching.

Documenting the Australian Lexicon

The project-wide methodology adopted by the ten research teams of the Lexicon Project (Australia, Chile, China, Czech Republic, Finland, France, Germany, Japan, Korea, and the USA) is detailed in *Chapter 1, The International Classroom Lexicon Project*. In summary, the main phases for documenting the lexicons involved:

- national teams contributing a video of a single lesson of mathematics from their community;
- the combining of team video contributions into a stimulus package;
- careful viewing of the stimulus package by team members in order to generate candidate terms for the draft lexicon;
- a process of local validation;
- a process of national validation by survey; and
- a cross-team clarity-check.

The Australian research team worked through each of these phases to document the Australian Lexicon. This section describes the Australian research team and the processes used to generate candidate terms and validate the Australian Lexicon.

The Australian research team

The Australian research team included four university researchers and three practising teachers. The academic researchers varied in status and experience. Two of the teachers had more than 15 years of mathematics teaching experience and the third teacher was a recent graduate. The varied positions and experience of the members of the research team, in combination, produced a vibrant team who brought different levels of knowledge and expertise to the task of documenting the lexicon.

Generating candidate terms

Viewing lessons in the stimulus package

All members of the Australian team viewed the Australian lesson. The remaining eight lessons from the other countries involved in the project were assigned to team members using a matrix structure (see Table 2.1), ensuring that at least one experienced teacher viewed each lesson, and each lesson was viewed by a minimum of four team members. Each research team member generated candidate terms from watching the video of the Australian lesson and their allocated lessons and responding to the prompt, "What do you see that you can name?"

Including a term in the lexicon

The Australian research team met regularly to share and discuss terms, phrases, and short descriptions of familiar activities that were felt to be possible candidates for inclusion in the Australian Lexicon. In order for a term or short phrase to be included in the lexicon, team consensus was required and, if agreement was not reached, authority was accorded primarily to classroom experience and whether it was agreed that the term was in current use by teachers. In other words, the teachers on the team were given final say about whether a term was indeed likely to be familiar to teachers and should be included in the lexicon as a candidate term.

An important matter for the Australian research team was distinguishing the language of discipline (mathematics) from the language of practice (mathematics teaching/learning). On occasions, specific mathematical terms would be suggested for inclusion; however, after discussion these would be rejected because the aim of the lexicon project was to identify terms, short phrases, and activities, with implied reference to instructional function, rather than terms that describe mathematical activity only. An example of this is the familiar activity of *graphing a linear equation*. We can identify and name this activity in the videos as it is clearly recognisable. However, the activity is not regarded as an instructional practice, and therefore it would not be included in the lexicon. The essential point of the lexicon project was to record single words or short phrases with pedagogical features that are consistently and widely used within the mathematics teaching community.

TABLE 2.1 The allocation of lessons to members of the Australian research team.

| Research Team Members | Lessons included in the Stimulus Package* | | | | | | | | |
	Australia	Chile	China	Czech Republic	Germany	Finland	France	Japan	USA
Academic 1	✓	✓	✓	✓	✓	✓	✓	✓	✓
Academic 2	✓			✓	✓			✓	✓
Academic 3	✓	✓		✓			✓	✓	
Academic 4	✓	✓	✓			✓	✓		
Teacher 1	✓	✓	✓	✓				✓	
Teacher 2	✓				✓	✓	✓		✓
Teacher 3	✓		✓		✓	✓			✓

* At this point in time, the Korean team had not yet joined the project.

Additional classroom events

The Australian research team found it useful to include two additional types of classroom events that did not seem to meet the criteria for inclusion:

- <u>Phrases</u> that are recognisable and readily understood, describing familiar classroom phenomena for which there did not appear to be a single, institutionalised name (e.g. *setting a time limit*).
- <u>Familiar Activities</u>, those pedagogical activities which are seldom described or referred to but have a familiar quality to them (e.g. *arranging the seating*).

It was considered useful to record items falling into these two categories, in part, to anticipate the possibility that other communities might name these practices. However, these items were not included in the validation process.

Preparing operational definitions for the terms

The 69 terms in the lexicon were supplemented with definitions which included a description of the classroom practice and illustrated examples from the classroom. A sample of these terms and operational definitions is displayed in Table 2.2.

TABLE 2.2 A sample of terms and operational definitions from the Australian Lexicon.

Term	Description	Examples and Non-examples
guiding	The teacher offers advice or suggestions in the form of questions or comments to assist students in completing and solving tasks.	Example: A teacher asks: "What else might you do?" *Non-examples:* *A teacher says: "Try multiplying". The teacher gives step by step instructions.*
justifying	An activity undertaken by the teacher or students that involves expressing why particular mathematical processes, solutions or theories work, and providing evidence.	Example: The teacher encourages students to explain why a mathematical generalisation holds true. *Non-example:* *The teacher makes a statement about how a particular mathematical idea is true in all cases but gives no explanation related to how or why this is so.*
practising	The activity of repeating a procedure for the purpose of improving efficiency or accuracy in its use.	Example: A student solves ten consecutive tasks all involving the addition of fractions. *Non-example:* *A student attempts to make use of the property of similar triangles in a real-world context for the first time.*

Validating the Australian Lexicon

Local validation

In order to determine whether the local teaching and education communities endorse the lexicon (its terms, descriptions and examples), two groups of people were invited to participate in a local validation of the lexicon: i) Mathematics Education Researchers (Specialists) and ii) Education Researchers.

The intention for recruiting the first group was to investigate the extent to which the local community of mathematics education researchers would endorse the purpose, the structure, and the constituent terms of the Australian Lexicon. This group (8 participants) was, strictly speaking, the group that was "validating" the terms, descriptions, examples, and non-examples from the perspective of the discipline of mathematics education, but not from the perspective of mathematics teachers, which was undertaken separately with a national validation. The second group was recruited to provide a check on the possible cross-disciplinary nature of the mathematics lexicon. This group (11 participants) provided an understanding of how widely used and understood the terms are outside of mathematics education. The lexical terms offered for local validation included the 69 terms from version two of the lexicon (see also Table 2.4 for a description of the different developmental phases of the lexicon).

Interview protocol for the local validation

The researchers involved in the local validation participated in an audio-recorded, one-on-one discussion, in which each interviewee was presented with ten terms for examination and discussion. The combination of terms was organised such that no two people received the same set of terms and the entire set of terms was seen by at least one representative from each group – Mathematics Education Researchers (Specialists) and Education Researchers. The allocation of the set of ten terms to an interviewee was random.

Two sets of cards were prepared with the term presented on one side and the description with examples and non-examples on the reverse. The protocol for the interview involved the participant receiving ten cards and completing these tasks:

- sort the ten terms (by name only) into two piles: Familiar and Unfamiliar
- match the description/example cards with their term
- suggest improvements to the operational definitions (terms, descriptions, examples and non-examples).

The participants suggested improvements to the wording, of descriptions and examples, to improve clarity of these. They were also asked to reflect on whether there might be other term that teachers might use. To conclude, participants were invited to take the entire stack of cards and sort them according to whether they were Familiar or Unfamiliar.

In response to this additional data collection from the local validation interviews the Australian research team met to: decide whether to accept or reject changes to the wording of the term, description, and examples for the lexicon items; consider which current terms might need to be excluded from the lexicon; and consider the list of additional terms that were generated for inclusion. Examples of such decisions are illustrated in Table 2.3.

TABLE 2.3 A sample of results emerging from local validation of the terms of the Australian Lexicon.

Term(s)	Supporting Evidence (from local validation interviews) M = Mathematics Education Researcher E = Education Researcher	Resulting Decision (after consultation by Australian research team)
active listening	Unfamiliar to: E1 M3 M4 M7	Removed from the list of "Terms" and moved to the list of "Phrases"
active reading	Unfamiliar to: E1 E2 E8 E9 M3 M4 M6 M7 M8	
collaborative group work, cooperative group work	Comments included: *distinction not well-understood *terms used synonymously *not natural parlance M1 M5 E2 E3 E5	
group work	Comments included: *consider using this term instead of distinguishing among three types of group work M2 M5 E2 E6	Descriptions refined
guiding	Unfamiliar to: M1 E6 E10 Comments included: *consider the definition in light of similar terms *consider the definition in comparison with the terms **scaffolding**, **prompting**, and **feedback** M2 E1 E6	
discipline (n/v)	Comments included: *distinction between the two terms discipline (n) (the teacher's established standards) and to discipline (v) (expression of disapproval) is cumbersome * not representative of teacher usage E4	Term(s) changed to: **disciplining**
hook	Unfamiliar to: M5 M7 E11 Comments included: *should it be "a hook" or "the hook" *consider the definition in comparison with the term **engagement** M4 M5 E3 E7	Term changed to: **use of a (hook)**

This revised lexicon, in response to the process of local validation, now included 63 Terms, 25 Phrases, and 25 Familiar Activities. Given that the two additional types – Phrases and Familiar Activities – did not represent practice that was named consistently, a decision was made to include only the terms that the research team felt were indicative of a high degree of normative understanding of practice. At this point, the artefact that represented teacher professional vocabulary, the Australian Lexicon, included 63 terms. This version of the lexicon was subsequently subjected to a national validation process.

National validation

A survey was designed to collect the opinions of the broadest relevant community (teachers, teacher educators, researchers in education) regarding the extent to which the draft lexicon was seen as *reasonable*. One hundred and fifty-five participants across Australia responded to the survey involving questions in which they indicated familiarity with the 63 terms and the accompanying operational definitions of the draft lexicon.

To determine whether the lexicon was a reasonable representation of teachers' professional vocabulary the survey included the question, *How familiar are you with the term?* Responses were collected with the following 5-point scale: *Extremely familiar; Very familiar; Somewhat familiar; Not so familiar;* and *Not at all familiar*. A threshold of two-thirds' familiarity was set for validation of the lexical items. If the term was found to be unfamiliar to more than a third of respondents, that is, respondents selected either *not so familiar* or *not at all familiar*, the term was considered not validated for inclusion in the lexicon.

The participants' responses are recorded in Figure 2.1. The red line in the chart indicates the threshold set by the research team for inclusion at 67%. As long as the first three points combine to meet or surpass that value, the term was recognised as familiar to the national community and considered as belonging to their professional lexicon. The results indicated that all the terms met this criterion to be considered validated for inclusion in the lexicon.

In addition to functioning as a tool for validating the terms, the survey also provided an insight into the differing profiles of terms.

High familiarity

By grouping the points *Extremely Familiar* and *Very Familiar*, **High Familiarity** was indicated for 49 of the 63 terms by more than 90% of the respondents. Those terms that had high familiarity for all, that is, all survey respondents chose either *Extremely Familiar* or *Very Familiar*, were the following: **answering questions**, **assessment**, **encouraging**, **giving feedback**, **practising** and **test/testing**.

Less familiarity and mixed profile

By grouping the points of *Not So Familiar* and *Not At All Familiar* **Less Familiarity** was determined for terms for which more than 10% of respondents chose either of these two points. Five terms were identified (% of respondents that indicated less familiarity shown in brackets): **re-teaching** (18%), **peer support** (14%), **elicit understanding** (13%), **reciting** (12%) and **(use of a) hook** (11%).

Three of these terms, as well as indicating **Less Familiarity** proffered a **Mixed Profile**, that is, four of the five points resulted in responses totalling more than 10% each (% of respondents shown in brackets in this order: *Extremely familiar; Very familiar; Somewhat familiar; Not so familiar;* and *Not at all familiar*): **peer support** (50% 19% 17% 12% 2%); **reciting** (51% 23% 15% 11% 1%); and **re-teaching** (48% 18% 18% 14% 4%).

Analysis of this survey question allows us to reflect critically on our professional vocabulary and identify those practices that are familiar and those that are less so. Those terms that are less familiar might be indicative of practices that could be targeted in pre- and in-service teacher education programs if considered activities that promote learning.

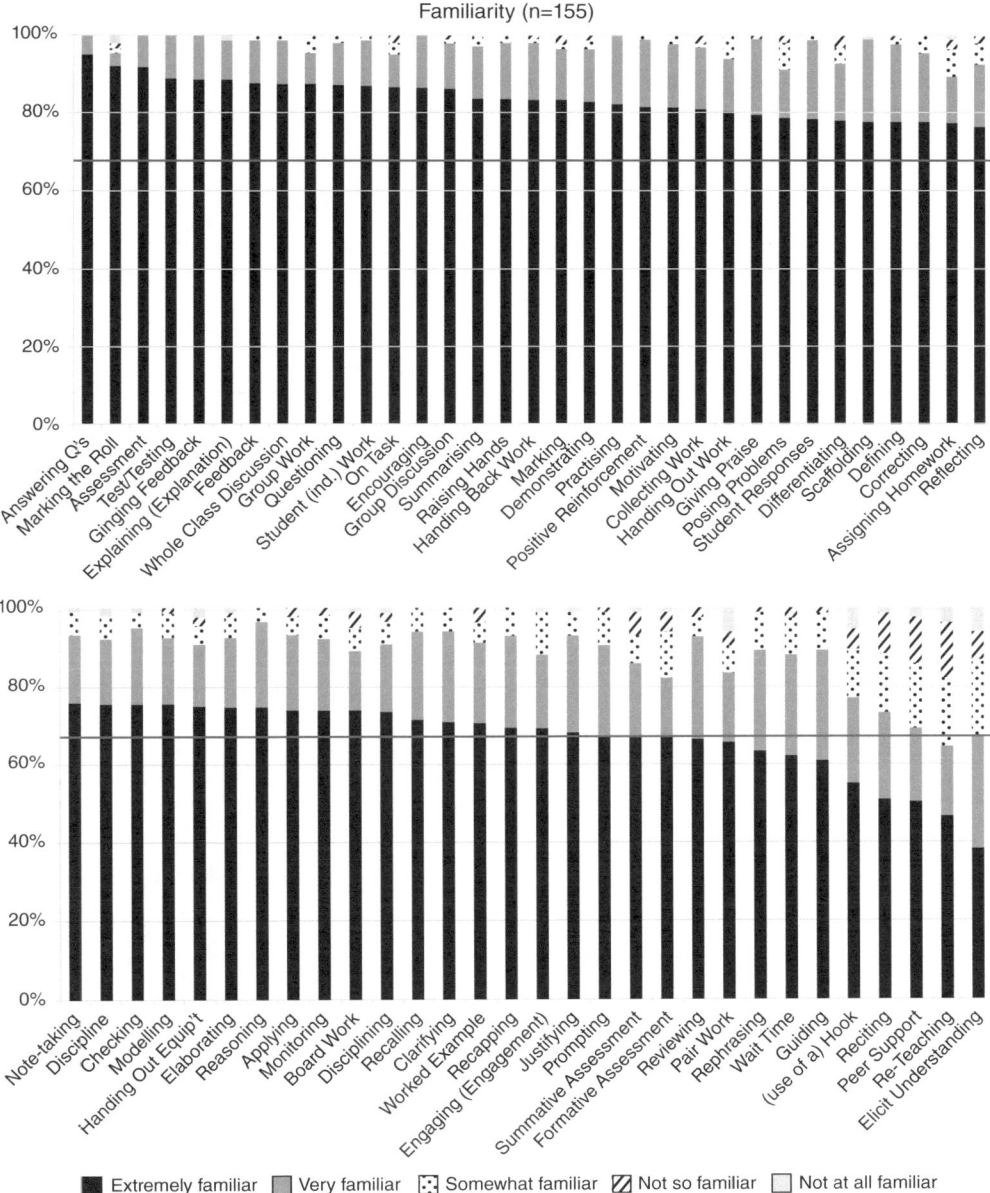

FIGURE 2.1 Percentage responses by 5-point scale to the survey question "How familiar are you with this term?" for the terms in the Australian Lexicon.

Clarity check

The Australian Lexicon was assigned to the Chinese research team for the purpose of checking the clarity of the description and examples. Two expert members of the Chinese research team identified terms whose descriptions they considered problematic, either for reasons of conceptual clarity or clarity of expression. Four terms were revisited, these included **feedback, giving feedback, discipline,** and **disciplining**. After additional discussion with local practitioners it was decided that of these four terms only two would remain in the lexicon and their definitions would be altered to absorb earlier distinctions in meaning. The two terms that remained in the lexicon were **feedback** and **disciplining.**

Phases of development

The Australian Lexicon was documented over a number of years and underwent various stages of development and refinement as shown in Table 2.4. The sixth version of the

TABLE 2.4 Phases of development of the Australian Lexicon.

Phase of Development	Comments	Resultant Lexicon
Initial Lexicon Documentation (August to December 2015)	The initial lexicon was generated by the Australian team after viewing the video material. A distinction was made between "terms," "phrases," and "familiar activities"	*Version One* 69 Terms 17 Phrases 25 Familiar Activities
	Distinction indicated between Teacher terms and Academic terms. Operational definitions for the 69 terms were developed in preparation for validation.	*Version Two* 69 Terms *(inclusive of 6 Academic Terms)* 17 Phrases 25 Familiar Activities
Local Validation (January to September 2016)	Sixty-nine terms were offered for local validation. In response to this subsequent data collection, some of the terms were removed.	*Version Three* 63 Terms 25 Phrases 25 Familiar Activities
	The Australian team made a decision to limit the lexicon artefact to include only the 63 lexical items from the "Terms" category.	*Version Four* 63 Terms
National Validation (October to May 2017)	Following analysis of the national questionnaire all terms were considered validated for inclusion in the lexicon.	*Version Five* 63 Terms
Clarity Check (June to August 2017)	The clarity check was completed in collaboration with two experts from the Chinese national team. Following additional discussions with teacher practitioners the lexicon was reduced further by two terms.	*Version Six* 61 Terms
Final Version		*Final Version* 61 Terms with operational definitions

lexicon, resulting from incremental and iterative negotiation with the mathematics teaching community, is the one presented in *Chapter 3, Australian Lexicon.* We recognise that by its very nature this sixth iteration of the lexicon, similar to a dictionary, will also change over time to reflect the ever-evolving nature of words and practices, and how and when they are used.

Structure

The research team developed a set of categories intended to best organise and communicate the lexicon. These categories were informed by names suggested by a university class of practising teachers undertaking a Master of Education subject – Mathematics: Quality Teaching. In teams of five, these teachers were invited to arrange the lexical items in clusters and give these clusters a name. The teacher-suggested category names are summarised in Table 2.5 and vary in specificity, but less so in number: Team 1 (9 category names), Team 2 (9 category names), Team 3 (6 category names), Team 4 (9 category names), and Team 5 (10 category names).

Almost all teams mentioned assessment, administration, and management in some form and additionally identified both student and teacher activity, strategies, structures

TABLE 2.5 Category names for the Australian Lexicon suggested by teacher teams.

Team	Team 1	Team 2	
Category Names	*Assessment* *Building Self Confidence (Intrinsic)* *Behavioural – Classroom Management* *Classroom Administration* *Metacognitive Skills (Learning Strategies)* *Peer Learning and Teaching* *Setting the Scene/Structuring the Learning Environment* *Teacher Directed throughout Lesson*	*Assessment and Feedback-related* *Classroom Behaviour Management Strategies* *Differentiation Strategies* *Direct Instruction* *Relationships Responses* *Setting Up Classroom/Administration* *Student Directed* *Students Working Together* *Teaching Strategies*	
Team	Team 3	Team 4	Team 5
Category Names	*Assessment* *Group Work* *Management* *Student Output* *Teaching and Learning Strategies* *Tone of Environment*	*Administration* *Assessment* *Classroom Management* *Discussion* *Questioning* *Showing* *Understanding* *Student Activity* *Teacher Movement* *Teaching Strategies*	*Assessment* *Behaviour* *Cognition and Metacognition* *Directing* *Grouping* *Representation* *Speech* *Student Mechanics* *Support* *Teacher Mechanics*

and behaviours. After much reflection and discussion, the research team settled on the following five categories: **Administration**, **Assessment**, **Classroom Management, Learning Strategies,** and **Teaching Strategies.** The research team felt that these five categories captured the essence of those identified by the teacher teams and also captured the activity-based nature of the Australian Lexicon. The 61 terms of the Australian Lexicon were thus organised as follows: Administration (eight terms); Assessment (11 terms); Classroom Management (seven terms); Learning Strategies (27 terms); and Teaching Strategies (49 terms). Terms are presented in more than one category when researchers agreed there was a strong association of the term with each of the categories; for example, twenty-four terms are found in both the Learning Strategies and Teaching Strategies categories. A sample of terms is used to illustrate this in Table 2.6 (see Chapter 3, Table 3.3, for the entire lexicon organised by category).

This collection of sample terms illustrates the variety of context in which a term might apply. For example, *feedback*, could be considered in the context of Assessment as well as in the context of Teaching Strategies. This also indicates that the practice of providing *feedback,* although identified by the same name, may be enacted differently if the function of that feedback differs. Alternatively, the same instance involving the provision of feedback could also serve two different purposes.

The number of terms that are unique to an organisational category and the number of terms that are in more than one category are summarised in Table 2.7.

One of the significant features of the Australian Lexicon is the flexibility of meaning and intention within context, as evidenced by the number of terms that belong to more than one category. This lack of singular intention of most of the terms is discussed in more detail in the following section.

Features of the Australian Lexicon

Upon analysis of the lexicon, a number of features of the Australian Lexicon were identified. These included: a level of flexibility with respect to the development and function of terms; a mix of terms representing actions of teachers and students; and a variety of ways that teachers use terms in conversation.

Flexibility

Within the Australian Lexicon, the terms documented indicate a level of flexibility with respect to their development and function. Three attributes of the lexicon terms that signal this are:

- a prevalence of gerunds and participles
- terms which identify general pedagogical practices
- a lack of terms with a singular pedagogical intention.

Gerunds and participles

A noteworthy feature of the Australian Lexicon is that over half of the lexical items end in -ing, for example, *scaffolding* and *recapping* (see also Table 2.2 and Chapter 3). This

TABLE 2.6 A sample of terms within the organisational categories of the Australian Lexicon.

Sample of terms in one category only

Category	Administration	Assessment	Classroom Management	Learning Strategies	Teaching Strategies
Term	checking handing back work marking the roll	summative assessment test/testing	monitoring positive reinforcement raising hands	practicing reasoning	differentiating posing problems worked example

Sample of terms in more than one category

Category	Assessment	Teaching Strategies		Learning Strategies
Term	elicit understanding feedback reviewing	applying justifying questioning		

TABLE 2.7 Number of terms in the Australian Lexicon unique to the category, and in more than one category by organisational category.

Organisational Category	Number of terms unique to that category	Number of terms in more than one category
Administration	3	5
Assessment	2	9
Classroom Management	1	6
Learning Strategies	2	25
Teaching Strategies	13	36

linguistic presentation (ing) identifies **gerunds** and **participles** that function as nouns or adjectives. An example, with respect to the use of the lexical term *questioning,* follows:

> *Questioning is an essential teacher practice.*
> (gerund/acts as a **noun** but retains its verb qualities)
> Her *questioning* technique engaged all students.
> (participle/acts as an **adjective)**

This use of gerunds and participles lends a flexibility and dynamic quality to the Australian Lexicon not necessarily present in other lexicons, which may have grammatical structures different from those used in English.

General pedagogical practices

Supplementary data collection from local validation, from educators other than those involved in mathematics, confirmed that all 61 terms in the Australian Lexicon identify general pedagogical practices. That is, these terms are not unique to the mathematics classroom.

While the description and pedagogical intent of the terms are familiar to various teaching communities, the practice of the term in other subject domains may be enacted quite differently. For example, in the mathematics classroom, the term *demonstrating* might identify the practice of the teacher working through a task at the board, while in a science classroom, the same term *demonstrating* may identify the practice of the teacher illustrating a sequence of steps required in a practical experiment.

Lack of singular pedagogical intention

Another distinguishing feature of the lexicon is that few terms reveal a singular pedagogical intention. For example, *feedback* might be used to evaluate a student response, appreciate a student's attempt with a difficult problem, or guide their learning. As indicated earlier in the section on Structure, the term *feedback* additionally belongs to more than one organisational category. Forty of the 61 terms in the Australian Lexicon have been identified as belonging to more than one category and only 21 terms belong to a single category (see Table 2.7).

By way of contrast, other terms are specific to position and purpose in an instructional sequence. For example, the lexical terms *recapping* and *reflecting,* are activities that assume experience with a prior educational episode and provide some degree of specificity with respect to intention and location in an instructional sequence. The majority of terms in the Australian Lexicon, however, lacked this level of specificity.

These three attributes of the terms of the Australian Lexicon combine to indicate a lexicon that is somewhat imprecise and ambiguous but also flexible. The lexicon terms are applicable in different domains and contexts. One possible explanation is that Australian teachers have the freedom to develop a highly personalised pedagogical style with an associated personal vocabulary, and in this sense the context in which a term is applied might reflect the teacher's personal educational history. The lexicon appears to have captured a teacher vocabulary that reflects this tolerance of idiosyncrasy in the Australian teaching community.

Whose actions do terms refer to?

The prevalence of certain terms within the Australian Lexicon reflect the importance attached to teacher and student actions. Terms were analysed according to who performs the action: teacher only; student only; teacher or student; teacher and student. Table 2.8 illustrates findings from this analysis.

TABLE 2.8 Number of terms in the Australian Lexicon by classification, including illustrative examples.

Who performs the action?	Terms (number, percentage)	Illustrative Example from the Australian Lexicon
teacher only	*25 of the 61 terms* **41%**	*wait time* A deliberate pause before or after a question is posed or a response given.
student only	*11 of the 61 terms* **18%**	*raising hands* Students put up their hand to attract teacher attention.
teacher or student	*23 of the 61 terms* **38%**	*rephrasing* The teacher (or students) expresses an idea or comment in an alternative way usually for purposes of clarification. The statement requiring rephrasing may be a teacher's comment or a student's comment.
teacher and student	*2 of the 61 terms* **3%**	*whole class discussion* An activity in which the teacher and students have the opportunity to engage in dialogue.
Teacher-performed actions	*50 of the 61 terms;* **82%**	
Student-performed actions	*36 of the 61 terms;* **59%**	
Requires both teacher and student involvement for the action to be performed	*2 of the 61 terms;* **3%**	

The first column in the table indicates the various classifications, while the second and third columns indicate number of terms and examples.

Of the 61 terms in the Australian Lexicon, students' voice and participation in the classroom is a significant characteristic as 36 of the 61 terms (59%) relate to student action. Eleven of these 36 (31%) are wholly student-performed (for example, *on task, practising*). However, the teacher's role is more prominent with 50 of the 61 terms (82%) relating to some teacher action (for example, *differentiating, scaffolding*). Exactly half of these (25 of 50, 50%) are wholly teacher-performed.

Of all the 61 terms there are only two terms for which participation is required from both the teacher and student/s. These examples from the lexicon are *answering questions* and *whole-class discussion*.

This question of "Whose action?" combined with the practices and activities for which we have a name, offers insight into the institutionalised power differentials in our classrooms. The lexicon includes more terms that are specific to naming teacher actions and practice than to naming student actions. However, students are significant contributors, with 36 practices and activities named in the lexicon, in comparison with 50 for teachers, suggesting that students have potential to act, contribute, and be heard.

Use of terms in conversation

The national survey, although primarily designed to validate the lexicon as a reasonable representation of teacher professional vocabulary, also collected responses to the survey question, "Do you use this term in conversation with colleagues?" Responses form 155 participants, to a subset of the lexical items, were collected with the following 5-point scale: *Used extremely often; Used very often; Used moderately often; Used slightly often;* and *Not at all used.*

In order to understand associations between teachers' usage of lexical terms in conversation with colleagues, we mapped on a Cartesian Plane (Figure 2.2) the response to this question against the aggregated results of the question related to familiarity discussed earlier in this chapter (see Figure 2.1). A term was assigned two scores; each of these indicating the average response when a value of five is assigned to the "used extremely often" point, and a value of one is assigned to the "not at all used" point. For example, the score of (3.1, 4.8) for the term *giving praise* indicates that on average the term is *used in conversation moderately often* (x-coordinate) and is *extremely familiar* (y-coordinate). Figure 2.2 displays the resultant scatterplot when all 61 terms are mapped onto the x-y plane as well as detail related to two highlighted sections.

Inspection of the scatterplot confirms our earlier observation that familiarity with these 61 practices is, on average, very high. The most familiar term, with a score of 4.95 is *answering questions*. The range of points across all the terms is [3.86, 4.95] which equates to about one point out of five. *Elicit understanding* is the least familiar of all the terms with an average score of 3.86; nonetheless, it's worth noting that this score equates to a point closer to "very familiar" than "somewhat familiar."

The distribution of terms with respect to "use in conversation" presents a more nuanced picture. With points ranging from 1.92 (*re-teaching*) to 4.73 (*assessment*) a distribution with a clearly populated middle of 44 terms can be seen. These terms are not only highly familiar but also used *moderately often* to *very often* in conversation with colleagues. In Figure 2.2, a collection of seven terms can be seen on the upper end of

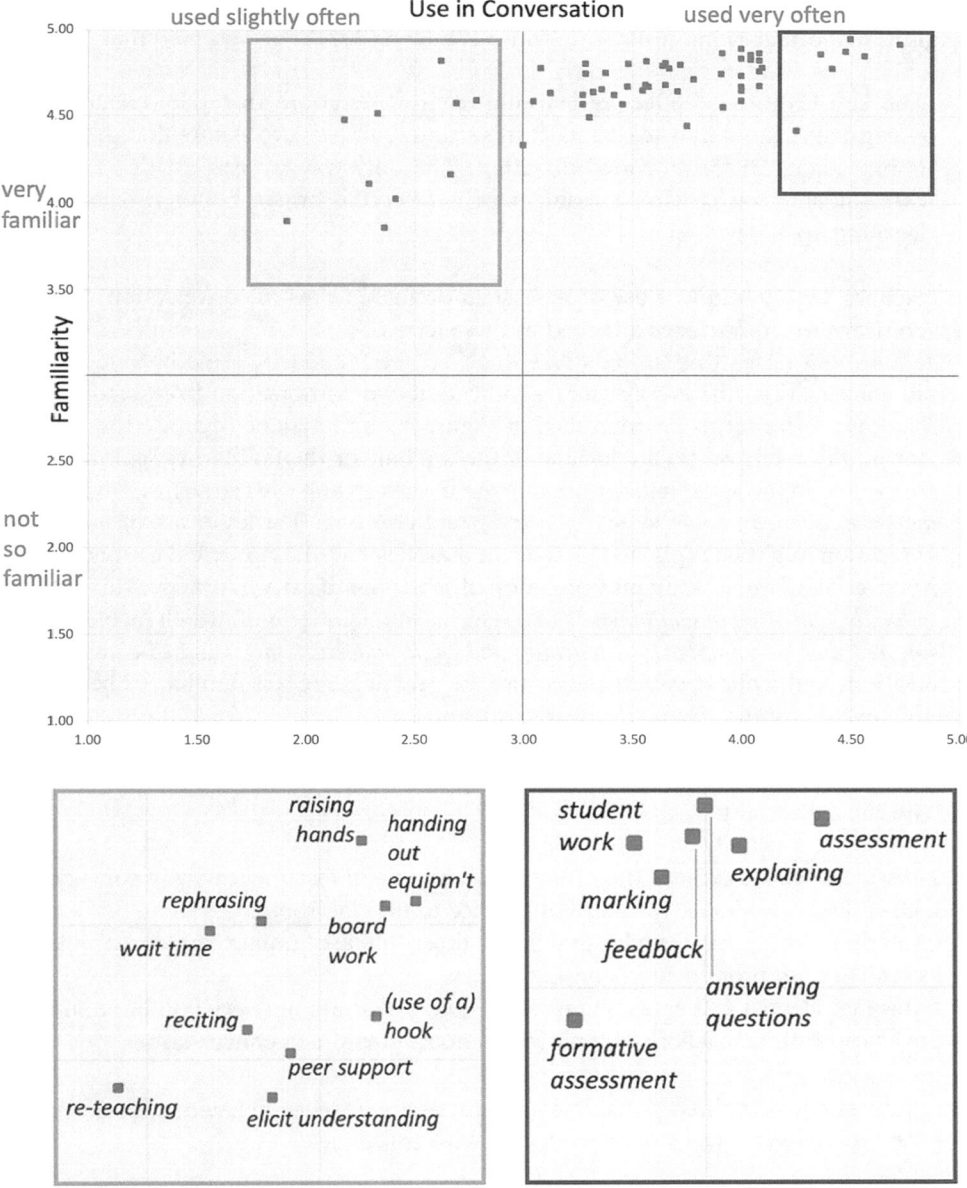

FIGURE 2.2 "Familiarity" of terms in the Australian Lexicon mapped against "Use in Conversation."

use (upper right corner framed with a square) and a collection of ten terms populating the lower end of use (entirely within the upper left quadrant of the graph, framed with a square).

There are seven terms that are not only highly familiar but also used in conversation *very* to *extremely* often. These terms are identified as *formative assessment, student work, marking, feedback, answering questions, explaining,* and *assessment* (see also Figure 2.2).

It is noticeable how many of these terms relate to Assessment and were grouped within this organisational category in our lexicon. Indeed Wiliam (2011) has observed that,

> The key features of effective learning environments are that they create student engagement, and allow teachers, learners, and their peers to ensure that the learning is proceeding in the intended direction. The only way we can do this is through assessment. That is why assessment is, indeed, the bridge between teaching and learning. (p. 81)

The teachers' highly frequent use of assessment-related terms in conversation with their peers confirms the importance attached to these terms.

There are ten terms that are relatively high on the familiarity scale but are used less often in conversation; the average scores indicate use in conversation from *moderately* to *slightly* often. These terms are identified in Figure 2.2 and may be organised further into two groups; those terms that are located in the top half of the familiarity scale and those that are located in the lower half. Those that are in the top half are the terms: *raising hands, handing out equipment, board work, rephrasing,* and *wait time.* The lesser use of these terms in conversation might be explained by the fact that they have less practical use in conversation (such as *board work*) or an interpretation of lesser complexity in practice (for example, *raising hands, handing out equipment*). Those terms in the lower half, which include *(use of a) hook, reciting, peer support, re-teaching and elicit understanding,* might be absent from conversations with colleagues either because the practices are less familiar or because the terms are entirely absent from a teacher's repertoire.

Examining the survey responses to the questions about "familiarity" and "use" in combination has indicated the possibility of three explanatory frames for the location of these terms on the scatterplot:

1. **assessment-related terms** – those that are familiar and feature heavily in conversations with colleagues because of their importance to the classroom;
2. **terms that identify less complex practices** – those that are familiar but less complex and as such are not prominent in conversations;
3. **terms that identify less common practices** – those that are not very familiar and hence, are absent from a teacher's repertoire and do not feature in conversations.

These frames provide three possible explanations for teacher behavior associated with using the lexicon terms when in conversation with colleagues.

Concluding remarks

The Australian Lexicon includes 61 classroom-related terms identifying events, activities and interactions, organised in five teacher-informed categories. The terms of the lexicon have been operationally defined with descriptions and classroom illustrations. In this chapter, we have shared the processes involved in developing this empirical lexicon, and presented some of its characteristic features.

The Australian Lexicon includes terms expressed in the most part as gerunds and participles. This grammatical structure lends to the lexicon a distinct flexible and dynamic quality. The level of flexibility in the lexicon is revealed further when recognising that the terms not only name general pedagogical practices (found both inside and outside of the mathematics classroom), but also reflect more than one pedagogical intention. The lexicon names more classroom activities involving teacher actions than student actions, nevertheless, student voice and participation remain significant characteristics. Almost all of the lexicon terms are highly familiar to teachers; however, the frequency of their use in conversation varies.

Australian teachers of mathematics have the freedom to develop a personal pedagogical style, which includes an associated personal teaching vocabulary. The lexicon has captured a teacher professional vocabulary that reflects breadth and variety in classroom practice, as well as flexibility in describing that practice.

The purpose of this project was to document an Australian Lexicon of teachers of lower secondary mathematics. The lexicon provides a starting point for stimulating and advancing discussion about the practices of the mathematics classroom for the education community. It has the capacity to inform, and be a resource for, pre-service and in-service teacher education. It also has the capacity to support education researchers and practitioners to theorise about classroom instruction and learning both in Australia and internationally.

Acknowledgements

This project has been funded by a Discovery Grant from the Research Council of the Australian Government (ARC-DP140101361) and supported through an Australian Government Research Training Scholarship.

References

Australian Bureau of Statistics (ABS) (2016). *The 2016 census.* https://www.abs.gov.au/
Australian Bureau of Statistics (ABS) (6 February, 2020). *Media release – Schools, Australia 2019* (cat no 4221.0). https://www.abs.gov.au/ausstats
Australian Institute for Teaching and School Leadership (AITSL). (2018). *Australian professional standards for teachers.* https://www.aitsl.edu.au/teach/standards
Boroditsky, L. (2001). Does language shape thought?: Mandarin and English speakers' conception of time. *Cognitive Psychology*, 43 (1), 1–22.
Connell, R. (2009). Good teachers on dangerous ground: Towards a new view of teacher quality and professionalism. *Critical Studies in Education*, 50(3), 213–229.
Darling-Hammond, L., Bransford, J., LePage, P., Hammerness, K., & Duffy, H. (Eds.). (2005). *Preparing teachers for a changing world: What teachers should learn and be able to do.* San Francisco, CA: Jossey-Bass.
Grossman, P., Compton, C., Igra, D., Ronfeldt, M., Shahan, E., & Williamson, P. W. (2009). Teaching practice: A cross-professional perspective. *Teachers College Record*, 111(9), 2055–2100.
Lampert, M. (2000). Knowing teaching: The intersection of research on teaching and qualitative research. *Harvard Educational Review*, 70(1), 86–99.
Lave, J., & Wenger, E. (1991). *Situated learning.* New York, NY: Cambridge University Press.
Levinson, S. C. (2003). *Space in language and cognition: Explorations in cognitive diversity.* Cambridge, MA: Cambridge University Press.
Lortie, D. C. (1975). *Schoolteacher; A sociological study.* Chicago, IL: University of Chicago Press.

Marton, F., Runesson, U., & Tsui, A. B. M. (2004). The space of learning. In F. Marton & A. B. M. Tsui (Eds.), *Classroom discourse and the space of learning* (pp. 3–42). Mahwah, NJ: Lawrence Erlbaum.

Masters, G. N. (2016). Five challenges in Australian school education. *Policy Insights*, 5. Retrieved from: https://research.acer.edu.au/policyinsights/5/

Sapir, E. (1949). *Selected writings on language, culture and personality*. Berkeley, CA: University of California Press.

Shulman, L. (2005). Signature pedagogies in the profession. *Daedalus*, 134(3), 52–59.

Teacher Education Ministerial Advisory Group (TEMAG). (2014). *Action now: Classroom ready teachers*. Canberra, Australia: Australian Government Department of Education.

Weldon, P. (2015). The teacher workforce in Australia: Supply, demand and data issues. *Policy Insights*, 2. Retrieved from: https://research.acer.edu.au/policyinsights/2/

Weldon, P. (2016). Out-of-field teaching in Australian secondary schools. *Policy Insights*, 6. Retrieved from: https://research.acer.edu.au/policyinsights/6/

Wernert, N., & Thomson, S. (2016). Australia. In I. V.S. Mullis, M. O. Martin, S. Goh, & K. Cotter (Eds.), *TIMSS 2015 encyclopedia: Education policy and curriculum in mathematics and science.* Retrieved from: http://timssandpirls.bc.edu/timss2015/encyclopedia/countries/australia

Whorf, B. L. (1956). *Language, thought, and reality: Selected writings of Benjamin Lee Whorf*. Cambridge, MA: MIT Press.

Wiliam, D. (2011). *Embedded formative assessment – practical strategies and tools for K-12 teachers*. Bloomington, IN: Solution Tree Press.

3

AUSTRALIAN LEXICON

Carmel Mesiti, Hilary Hollingsworth, David Clarke,
Amanda Sfindilis Reed and Katherine Roan

An Australian Lexicon

The lexicon presented in this chapter consists of 61 terms considered familiar by Australian middle years teachers in the mathematics education community. Operational definitions were developed for each of the terms in the lexicon and include a general description of the classroom practice (or activity). These operational definitions also include one or two examples from the classroom. A non-example, something that might be thought of as illustrating the practice but is actually not indicative of the term being documented, is also included to help provide a clearer and fuller definition. The words (in English) are presented in the standard order of letters of the Roman alphabet. The spelling convention adopted for this lexicon is British English.

The first lexicon table (see Table 3.1) lists the terms of the lexicons. The second lexicon table (see Table 3.2) includes the operational definition of each term arranged into three columns. The first column lists the **Term** itself, the second column gives its **Description** and the third column presents **Examples** and **Non-examples** from the classroom.

The structure of a lexicon offers insights into the way in which the teaching community conceptualises and organises its practice (see also Chapter 2). The 61 terms of the Australian Lexicon have been organised across five organisational categories informed by a university class of practising teachers (see Table 3.3). These categories are: Administration (eight terms); Assessment (11 terms); Classroom Management (seven terms); Learning Strategies (27 terms); and Teaching Strategies (49 terms). Terms may belong to more than one category; 24 terms are found in both the Learning Strategies and Teaching Strategies categories.

The lexicon documented in this chapter was arrived at through structured negotiations between researchers and the teaching community. The lexicon was progressively consolidated as members of the local and national mathematics education community examined it. The Australian Lexicon, operationalised with descriptions and examples from the classroom, provides a lens through which teachers, policy makers, curriculum developers, teacher educators, and researchers can become aware of Australian mathematics teachers' perspectives on the classroom. This collection of 61 terms has significant practical value, including its use for the study and promotion of teachers' reflective practice.

TABLE 3.1 Australian Lexicon – Terms.

answering questions	applying	assessment	assigning homework
board work	checking	clarifying	collecting work
correcting	defining (giving a definition)	demonstrating	differentiating (differentiation)
disciplining	elaborating	elicit understanding	encouraging
engaging (engagement)	explaining (explanation)	feedback	formative assessment
giving praise	group discussion	group work	guiding
handing back work	handing out equipment	handing out work	(use of a) hook
justifying	marking	marking the roll	modelling
monitoring	motivating	note-taking	on task
pair work	peer support	posing problems	positive reinforcement
practising	prompting	questioning	raising hands
reasoning	recalling	recapping	reciting
reflecting	rephrasing	re-teaching	reviewing
scaffolding	student responses	student (individual) work	summarising
summative assessment	test/testing	wait time	whole class discussion
worked example			

TABLE 3.2 Australian Lexicon – Terms with operational definitions.

Term	Description	Examples and Non-examples
answering questions	The activity of responding to a question or a series of questions from the teacher to students or students to the teacher.	Example: The teacher asks: "What's the area of a triangle with a base of 10 centimetres height of 8 centimetres?" A student responds: "Eighty square centimetres." Another student responds: "Forty square centimetres."
applying	An activity in which a taught procedure or concept is used to solve a problem that is different from the type of problem by which the procedure or concept was initially taught/learned.	Examples: Having been taught Pythagoras theorem using two-dimensional examples, a student makes use of the theorem to solve a problem in three dimensions. Having been taught strategies for solving simultaneous equations, a student employs the taught method to solve word problems in real-world contexts. *Non-example:* *Solving a pair of simultaneous equations after previously solving several similar pairs of equations.*
assessment	An activity undertaken by teacher or students with the primary purpose of generating information about student learning or achievement.	Examples: The teacher administers a test. The teacher observes students while they work, making notes on each student's progress. *Non-example:* *Assigning homework, unless the teacher explicitly indicates that the purpose is assessment.*
assigning homework	The teacher assigns tasks to be completed outside of the lesson typically within a specified time frame. Assigned tasks vary in nature and may include: exercises from a textbook; worksheets; project work; tasks involving computer applications; and finishing off work that was meant to be completed in class.	Example: The teacher writes the homework on the board and invites students to record it in their diaries. The teacher makes a verbal statement about homework that needs to be completed. *Non-example:* *Assigning work for completion during the lesson.*

(Continued)

TABLE 3.2 (*Continued*)

Term	*Description*	*Examples and Non-examples*
board work	The teacher or students record workings, diagrams, illustrations or notes on a board visible to the class.	Examples: A student solves a problem on the board. The teacher writes notes about the properties of a quadrilateral on the board for students to copy. A teacher uses a computer program to illustrate different representations of "one-half" and projects these on the board. *Non-example:* *The teacher records organisational notes, such as the room location for the next lesson, on the board.*
checking	The process by which a teacher or student checks answers, by determining the exactness or correctness of solutions; or checks progress, by determining whether the expected amount of work has been completed.	Examples: The teacher makes notes in her chronicle to indicate the timely completion of homework. The teacher makes a mental note or observation. *Non-example:* *Students annotate their workbook solutions.*
clarifying	The process by which a teacher or student makes an idea, concept, statement or instruction free from ambiguity.	Example: The teacher comments on each written line of a worked solution on the board. *Non-example:* *Repeating a statement made previously.*
collecting work	The teacher gathers responses to assigned tasks students were asked to complete either during class or as homework.	Example: At the end of a given task (that may or may not coincide with the end of the class), the teacher moves around the room to collect work done by students. *Non-example:* *The teacher collects equipment.*
correcting	The process by which a teacher or student identifies correct answers, and possibly identifies and rectifies errors.	Examples: Students annotate their solutions with correct lines of working. Students swap work and annotate each other's. *Non-example:* *Copying solutions from the board into a workbook.*
defining (giving a definition)	The teacher or student gives a clear meaning, identifies essential characteristics, or provides descriptions of a concept or process.	Example: The teacher identifies the characteristics of a quadrilateral that are essential in order for it to be classified as a square. *Non-example:* *A general discussion about the characteristics of two-dimensional objects.*

Term	Description	Examples and Non-examples
demonstrating	An activity undertaken by the teacher or students for the purpose of displaying or illustrating a concept, idea or skill.	Examples: The teacher draws a diagram of a circle on the board to illustrate diameter and circumference. The teacher uses a computer program to display how variables impact particular graphs. *Non-example:* *The teacher provides a definition of similarity of triangles.*
differentiating (differentia-tion)	Any action in which instruction is modified, adapted or varied to cater for student differences.	Examples: The teacher groups students for instruction according to their achievement on a pre-test of the content to be taught. The teacher provides different feedback to students according to her judgement of their emotional or motivational state. *Non-example:* *Students complete the same set of questions regardless of experience with the topic.*
disciplining	The teacher identifies undesirable behaviour that interferes with the learning of students and indicates ways to rectify this behaviour.	Example: A teacher stops her activity and asks students to pay attention to what is being said. *Non-example:* *The teacher asks students to sit down forming table groups of three.*
elaborating	A teacher or student provides additional information (in the form of examples, representations or more detailed descriptions) of ideas, concepts or processes.	Example: The teacher shades the fraction one-half in many alternative representations. *Non-example:* *The teacher has more than one worked example of adding fractions.*
elicit understanding	An activity undertaken by the teacher or students for the purpose of drawing out students' understandings of mathematical ideas, concepts or processes.	Examples: The teacher asks a student to demonstrate a mathematical idea on the board and articulate their understanding of that idea. The teacher asks students to explain how a specific solution relates to a general case. *Non-example:* *Students give answers to multiple-choice questions without elaborating or explaining their choice.*

(Continued)

TABLE 3.2 (*Continued*)

Term	Description	Examples and Non-examples
encouraging	An action undertaken by the teacher for the purpose of promoting students' efforts or willingness to persevere.	Examples: A teacher comments: "Keep trying." "Good effort." The teacher makes clear to a student the progress they have made in a specific area of work. *Non-example:* *The teacher says: "Let's get on with it!"*
engaging (engagement)	A student is actively involved with an educational experience, whereby he/she acts to maintain or extend their contact with the stimulus (typically, in order to increase their knowledge of it).	Examples: A student keeps working on solving a problem after the bell has gone signalling the end of class. A student independently researches an idea or concept that was presented in class. *Non-example:* *A student looks at his watch in anticipation of the end of the lesson.*
explaining (explanation)	Make an idea or situation clear to someone by describing it in more detail or by revealing certain facts.	Example: The teacher gives reasons why a particular mathematical process works in all cases. *Non-example:* *The teacher only provides a solution.*
feedback	The provision of information (typically evaluative) provided in response to the actions of a student or teacher.	Examples: The teacher provides corrective annotations on a student's written test or assignment. The teacher makes an evaluative comment on a student's solution to a problem. *Non-examples:* *The teacher comments on how difficult the test was for her students the year before.* *The teacher gives a student a direction that is clearly not in response to any action by the student.*
formative assessment	Information is collected to identify actions to support further learning.	Examples: A teacher observes students working and makes notes on common misunderstandings in order to better orchestrate the subsequent whole class discussion. Students are asked to describe a difficulty they are currently experiencing in their mathematics work so these can be addressed by the teacher. *Non-example:* *Students complete a test at the end of a topic that measures their understanding of the topic just taught.*

Term	Description	Examples and Non-examples
giving praise	The teacher expresses appreciation of an action or work product of a student.	Examples: The teacher says: "Well done, Anthony, I see you have been working hard on your factorising skills." The teacher says: "I commend you all for your good behaviour during our excursion at the museum." *Non-example:* *The teacher says: "I will give a reward to all who hand in their homework on time."*
group discussion	An activity in which individuals engage in purposeful dialogue about what they are learning.	Example: Four students talk about a learning task. *Non-example:* *A group of students who are sitting together talk about their plans for after school.*
group work	Students work together to complete a given activity.	Examples: The teacher assigns a task to be completed in pairs. A group of four students interact, sharing ideas, for the purpose of solving a problem for which only a group answer is required. Three students, having been assigned an exercise to complete, agree to each answer one-third of the questions and then share all their answers. *Non-example:* *A student asks a friend for the solution to a task assigned as individual work.*
guiding	The teacher offers advice or suggestions in the form of questions or comments to assist students in completing and solving tasks.	Example: A teacher asks: "What else might you do? *Non-examples:* *A teacher says: "Try multiplying."* *The teacher gives step by step instructions.*
handing back work	The teacher returns student work with corrections or other feedback.	Example: The teacher hands back a corrected Algebra test. *Non-example:* *Asking students to correct their own work.*
handing out equipment	The teacher provides students with resources to be used in the planned activities of the lesson.	Example: At the beginning of the lesson the teacher distributes calculators to each student. *Non-example:* *Students who were absent are given the work from the previous lesson.*

(Continued)

TABLE 3.2 (*Continued*)

Term	Description	Examples and Non-examples
handing out work	The teacher assigns tasks by distributing a prepared worksheet.	Example: The teacher moves around the room and distributes the work to be completed as she clarifies what is intended by the task. *Non-example:* *The teacher says: "Do the activity on page 62."*
(use of a) hook	The engaging introduction of a topic or sub-topic that captivates students' attention.	Example: The teacher introduces polyhedra with a video detailing how a soccer ball, a truncated icosahedron, is made. *Non-example:* *The teacher hands out the assignment and invites students to research polyhedra on the internet.*
justifying	An activity undertaken by the teacher or students that involves expressing why particular mathematical processes, solutions or theories work, and providing evidence.	Example: The teacher encourages students to explain why a mathematical generalisation holds true. *Non-example:* *The teacher makes a statement about how a particular mathematical idea is true in all cases but gives no explanation related to how or why this is so.*
marking	Teacher corrects work handed in for assessment by students.	Examples: The teacher marks a multiple-choice test by indicating which responses are correct. The teacher corrects student test papers at his desk while students are working independently. *Non-example:* *The teacher scans student work for indicators of confusion.*
marking the roll	The teacher records student attendance. This is usually done at the beginning of a lesson.	Example: The teacher calls individual names alphabetically and waits for the student to respond accordingly. *Non-example:* *The teacher calls on a student to give his/her solution to a given question.*
modelling	A social performance intended for replication by students.	Examples: The teacher verbalises her thoughts when solving a problem. The teacher elicits an exemplary student's explanation. *Non-example:* *The teacher explains each line of working of a problem.*

Term	Description	Examples and Non-examples
monitoring	The teacher observes the students for various reasons (e.g. assessing progress).	Example: The teacher moves around the room to observe whether students are on task. *Non-examples:* While moving around the room the teacher gives hints to assist students in solving problems. The teacher asks a student about how their team fared on the weekend.
motivating	The teacher explicitly encourages student participation in, and enthusiasm for, learning activities.	Example: The teacher incorporates students' interest in a local sporting team to contextualise a mathematics problem. *Non-example:* *The teacher outlines classroom behaviour expectations.*
note-taking	The written recording of spoken or written information.	Example: Students record in their books a worked example completed on the whiteboard. *Non-example:* *A student writes a written solution to a problem (without reproducing either a verbal or text version produced by another source).*
on task	Students are completing the mathematical work indicated by the teacher.	Example: Students are working quietly on assigned exercises.
pair work	An activity in which two students work together.	Example: Two students work together on assigned learning tasks. *Non-example:* *Two students share a textbook while working independently.*
peer support	An activity undertaken by a student to assist another student with some aspect of a learning task.	Example: A student shows a classmate how to do a particular learning task. *Non-example:* *Two students sit together to complete a task but one does all the work.*
posing problems	Tasks, questions or situations are presented (usually by the teacher) for consideration in class.	Example: The teacher, to promote discussion about possible solution approaches, presents a mathematical problem to the class. *Non-example:* *The teacher assigns questions from a textbook for homework.*

(Continued)

TABLE 3.2 (*Continued*)

Term	Description	Examples and Non-examples
positive reinforcement	The provision of information (most commonly by the teacher) intended to convey approval of another's action for the purpose (or with the effect) of encouraging further engagement in that action.	<u>Examples:</u> The teacher comments on a student's neat work The teacher praises a normally disruptive student for raising her hand. *<u>Non-example:</u>* *Teacher criticises a student for inattentiveness.*
practising	The activity of repeating a procedure for the purpose of improving efficiency or accuracy in its use.	<u>Example:</u> A student solves ten consecutive tasks all involving the addition of fractions. *<u>Non-example:</u>* *A student attempts to make use of the property of similar triangles in a real-world context for the first time.*
prompting	The teacher guides the student (usually with a verbal comment) towards a more appropriate or effective response or solution method.	<u>Example:</u> A teacher comments: "Check your working." "Try multiplying." *<u>Non-example:</u>* *The teacher provides the next step in the solution.* *The teacher directs: "Open your books to page three."*
questioning	The activity of asking a question or questions.	<u>Examples:</u> A student asks the teacher why a particular calculation was appropriate The teacher makes a statement ending in an upward inflection, and clearly calling for a response from a student or students. For example: "So, the angles in a triangle sum to 180 degrees, Michael?" *<u>Non-example:</u>* *The teacher makes a statement that is not phrased as a question and does not require any response. For example: "I doubt that approach will work."*
raising hands	Students put up their hand to attract teacher attention.	<u>Example:</u> A student raises her hand to answer a question. *<u>Non-example:</u>* *A student points to another student.*

Term	Description	Examples and Non-examples
reasoning	The process of consciously making sense of things; the thought process by which one thing is connected to another..	Examples: A student recognises that a particular problem resembles a problem completed earlier and employs the same method to solve the second problem. A student correctly identifies the logical flaw in a mathematical proof. _Non-example:_ _A student remembers the meaning of a mathematical term._
recalling	An activity that involves bringing a fact or event back to mind.	Examples: The teacher reminds the students of previously learned conditions for congruence of triangles. The students share their knowledge of the topic to be taught. _Non-example:_ _The students apply the recently learned technique of "solving simultaneous equations by substitution" with a set of new problems._
recapping	A summary of points already taught.	Example: The teacher summarises the steps involved in solving equations. _Non-example:_ _The teacher summarises the topics to be covered._
reciting	The repeating aloud of a rehearsed response.	Example: The class and teacher repeat the four times multiplication tables aloud in time with each other. _Non-example:_ _The teacher asks, "What is four times four?" and many of the students call out, "Sixteen."_
reflecting	An activity in which students consider the effectiveness or progress of their learning (i.e. their developing knowledge, skills and understandings).	Examples: The teacher asks students to identify and describe three new skills they have learnt during a unit of work. The teacher asks students to identify and describe an aspect of their current study that they do not understand. _Non-example:_ _A student recounts what they did during the lesson._

(_Continued_)

TABLE 3.2 (*Continued*)

Term	Description	Examples and Non-examples
rephrasing	The teacher (or students) expresses an idea or comment in an alternative way usually for purposes of clarification. The statement requiring rephrasing may be a teacher's comment or a student's comment.	Example: The teacher repeats a student answer with minor improvements. *Non-example:* *The teacher uses her tone and expression to emphasise certain words or phrases in a student's comment.*
re-teaching	An activity that addresses skills and ideas previously covered that have not been well understood.	Example: The teacher goes over common student errors in a completed exercise. *Non-example:* *The teacher says: "Remember you did this in year seven."*
reviewing	An activity that addresses skills and ideas previously covered.	Examples: The teacher says: "Let's go over the solution to Exercise 6." The teacher asks students to design three questions (and provide solutions) that demonstrate their understanding of the topic. *Non-example:* *The teacher says: "Today we covered the addition of fractions."*
scaffolding	A form of (typically verbal) guidance intended to influence a student's thinking in order to assist them in the achievement of some cognitive task (e.g. solving a problem or learning a new skill).	Examples: The teacher (while moving around the room) asks a student whether the student's current approach is effective (stimulating the student to reflect on their approach). The teacher asks: "Is there a diagram you could draw?" *Non-example:* *The teacher suggests that the student use the method just taught.*
student responses	A student's statement or action triggered by a teacher's direction or request.	Examples: Students record their solutions on the board. Students offer alternate definitions of the term "quadrilateral." *Non-example:* *A student asks for extra equipment.*
student (individual) work	An activity in which students work by themselves.	Example: Students work on their own and at their own pace on assigned learning tasks. *Non-example:* *A student copies a classmate's work.*

Term	Description	Examples and Non-examples
summarising	Recording or reporting the main points of a discussion or text material.	Examples: Students record three key points about a mathematical concept investigated. At the end of the topic on simultaneous equations, the teacher records the different approaches for solving these. *Non-example:* *The teacher says: "That concludes our unit on simultaneous equations."*
summative assessment	Information is collected for the purpose of summarising students' understanding or assessing their competence in a skill or procedure.	Examples: Students complete a test that measures their understanding of a topic just taught. Students represent their achievements over the course of a year by constructing a portfolio of work samples. *Non-example:* *The teacher questions a class for the purpose of identifying persistent misconceptions to be addressed in the next lesson.*
test/testing	A situation in which individuals are required to attempt a task or tasks for the purpose of demonstrating their competence in specific targeted types of performance or their understanding of targeted concepts or procedures.	Examples: Students complete a complex problem in a given time limit to assess their ability to apply a taught procedure. Students complete a stratified set of tasks intended to locate their level of competence or understanding. *Non-example:* *Students spontaneously investigate an emergent mathematical issue or dilemma.*
wait time	A deliberate pause before or after a question is posed or a response given.	Example: The teacher says: "What is the area of a rectangle that is 9 cm by 7 cm? (pause) Frank? (pause) Susan?" A student responds: "63 square centimetres" The teacher replies: "Very good, Susan. *Non-example:* *The teacher asks a question and answers it without pausing (that is, with no time provided for a student response).*
whole class discussion	An activity in which the teacher and students have the opportunity to engage in dialogue about the learning.	Example: The teacher invites students to share their knowledge of different number systems with the class. *Non-example:* *All students respond to a teacher question in unison.*

(Continued)

TABLE 3.2 (*Continued*)

Term	Description	Examples and Non-examples
worked example	The teacher (or student) writes out the steps involved in order to illustrate the type of solution expected to a problem or task with or without student involvement.	Example: The teacher writes out the solution to a problem on the board, providing oral explanations and clarifications along the way. *Non-example:* *Students record their solutions at the board.*

TABLE 3.3 Australian Lexicon – Terms by category.

Administration	*assessment, assigning homework, checking,* collecting work, handing back work, handing out equipment, handing out work, marking the roll
Assessment	*assessment, correcting, elicit understanding, feedback, formative assessment, marking, monitoring, reviewing, student responses,* summative assessment, test/testing
Classroom Management	disciplining, *encouraging, giving praise, monitoring, motivating, positive reinforcement, raising hands*
Learning Strategies	*answering questions, applying, board work, checking, clarifying, correcting, defining, engaging, group discussion, group work, justifying, marking, note-taking,* on task, *pair work, peer support,* practising, *questioning, raising hands, reasoning, recalling, reciting, reflecting, rephrasing, reviewing, student (individual) work, summarising*
Teaching Strategies	*answering questions, applying, assigning homework, board work, checking, clarifying, defining,* demonstrating, differentiating, elaborating, *elicit understanding, encouraging, engaging,* explaining, *feedback, formative assessment, giving praise, group discussion, group work,* guiding, (use of a) hook, *justifying, marking,* modelling, *monitoring, motivating, note-taking, pair work, peer support,* posing problems, *positive reinforcement,* prompting, *questioning, raising hands, reasoning, recalling,* recapping, *reciting, reflecting, rephrasing,* re-teaching, *reviewing,* scaffolding, *student responses, student (individual) work, summarising,* wait time, whole-class discussion, worked example

Note: Terms in italics are in more than one category.

Acknowledgements

This work was funded by a Discovery Grant from the Research Council of the Australian Government (ARC-DP140101361) and supported through an Australian Government Research Training Scholarship. We would like to express immense gratitude to our colleagues, Annette Amos and Caroline Bardini, as well as teachers and teacher educators from around Australia who participated in the negotiation of this lexicon. A special note of gratitude to the talented teacher who invited us into her classroom to videotape a single lesson of year 7 Mathematics.

Bibliography

Presentations and Papers

Presentations of the work of the Australian team to interested research and practitioner audiences has contributed to the enactment of the negotiative methodology in documenting and refining this lexicon of teacher practice (see also Chapter 2). The work of the Australian national team has been presented locally and internationally at the following meetings and conferences:

- Australian Association for Research in Education (AARE) Annual Conference 2019
- American Educational Research Association (AERA) Annual Meeting 2017, 2018
- Colloquium at Pontificia Universidad Católica de Chile 2019
- Contemporary Approaches to Research in Mathematics, Science, Health and Environmental Education (CAR) Symposium 2017, 2018, 2019
- Congress of the European Society for Research in Mathematics Education (CERME) 2017
- European Association for Learning and Instruction (EARLI) Biennial Conference 2015, 2017, 2019
- European Conference on Educational Research (ECER) 2016, 2017, 2018
- International Congress on Mathematical Education (ICME) 2016
- Mathematical Association of Victoria (MAV) Annual Conference 2017
- Mathematics Education Research Group of Australasia (MERGA) Annual Conference 2017, 2019
- Oceania Comparative and International Education Society (OCIES) Annual Conference 2016, 2017
- International Group for the Psychology of Mathematics Education (PME) Annual Conference 2016, 2017, 2018, 2019
- International Symposium Elementary Mathematics Teaching (SEMT) 2019

A selection of peer-reviewed conference publications from the Australian research team include:

Clarke, D. J., Mesiti, C., Cao, Y., & Novotna, J. (2017). The lexicon project: examining the consequences for international comparative research of pedagogical naming systems from different cultures. In T. Dooley, & G. Gueudet (Eds.), *Proceedings of the Tenth Congress of the European Society for Research in Mathematics Education* (pp. 1610–1617). Dublin, Ireland: ERME.

Mesiti, C., & Clarke, D. J. (2018). The professional, pedagogical language of mathematics teachers: A cultural artefact of significant value to the mathematics community. In E. Bergqvist, M. Österholm, C. Granberg, & L. Sumpter (Eds.), *Proceedings of the 42nd Conference of the International Group for the Psychology of Mathematics Education* (Vol. 3, pp. 379–386). Umeå, Sweden: PME.

Mesiti, C., & Clarke, D. J. (2017a). The international lexicon project: Giving a name to what we do. In R. Seah, M. Horne, J. Ocean, & C. Orellana (Eds.), *MAV17 Achieving excellence in M.A.T.H.S* (pp. 31–38). Melbourne, Australia: The Mathematical Association of Victoria.

Mesiti, C. & Clarke, D. J. (2017b). Structure in the professional vocabulary of middle school Mathematics teachers in Australia. In A. Downton, S. Livy, & J. Hall (Eds.), *40 years on: We are still learning! (Proceedings of the 40th Annual Conference of the Mathematics Education Research Group of Australasia)*, (pp. 373–380). Melbourne, Australia: MERGA.

Mesiti, C., Clarke, D. J., Dobie, T., White, S., & Sherin, M. (2017). "What do you see that you can name?" Documenting the language teachers use to describe phenomena in middle school mathematics classrooms in Australia and the USA. In B. Kaur, W.K. Ho, T. L. Toh, & B. H. Choy (Eds.), *Proceedings of the 41st Annual Meeting of the International Group for the Psychology of Mathematics Education* (Vol. 2, pp. 241–248). Singapore: PME.

Mesiti, C., Clarke, D. J., Roan, K., Hollingsworth, H., Cao, Y., Yu, G., Novotna, J., Zlabkova, I., & Dobie, T. (2016). Discourse about the mathematics classroom. In C. Csikos, A. Rausch, & J. Szitányi (Eds.), *Mathematics Education: How to Solve It? PME40* (pp. 357–363). Szeged, Hungary: PME.

Mesiti, C., Clarke, D. J., & van Driel, J. (2019). Describing and prescribing classroom practice: Do we have a common language? In G. Hine, S. Blackley, & A. Cooke (Eds.), *Mathematics Education Research: Impacting Practice (Proceedings of the 42nd annual conference of the Mathematics Education Research Group of Australasia)*, (pp. 492–499). Perth, Australia: MERGA.

Mesiti, C., Novotná, J., Clarke, D. J., Hošpesová, A., & Hollingsworth, H. (2019). *Speaking about the mathematics classroom: A comparison of the professional lexicons of teachers in Australia and the Czech Republic*. In M. Graven, H. Venkat, A. Essien, & P. Vale (Eds.), *Proceedings of the 43rd Conference of the International Group for the Psychology of Mathematics Education* (Vol. 3, pp. 89–96). Pretoria, South Africa: PME.

4

WHAT WE CAN NAME IN THE CLASSROOM

A Chilean Lexicon of middle-school mathematics teachers

Elisa Calcagni, Valeska Grau, Mónica Cortez and Daniela Gómez

Introduction

Communities of practice, and teachers in particular, share a discourse that allows them to learn from each other (Kelly, 2006). It is thus reasonable to expect differences in the language that researchers and teachers employ to describe the classroom. Nonetheless, teachers' language does not often feature in academic writing, becoming less visible than that of researchers. As researchers, we more commonly publish for academic journals with a similar readership and not necessarily for the practitioner audience. On the other hand, practitioners are not requested to document their knowledge about their practice or to subject their pedagogical practices to scientific scrutiny. Bridges across these two types of knowledge have to be created to increase synergy and improve practices and research (McIntyre, 2005). This book, and the Chilean Lexicon in particular, represent an attempt to build these much-needed connections.

Background

The national context

Chile is a South American country with 18 million inhabitants and a relatively high income in the region (OECD, 2017). It is a rather new nation, independence from Spain having been declared two centuries ago. The country's democratic tradition was interrupted in the 1970s, when a military coup seized power from the socialist Salvador Allende. The military dictatorship lasted for 17 years, profoundly shaping the country's political and economic landscape by violently imposing a market-driven system. This led to privatisation and deregulation of the provision of social services such as healthcare and education. The country has a presidential democratic system since 1989, but the tenets of the free-market model have remained largely unaltered to this date.

Economically, Chile is usually labelled a success story in the region concerning economic growth, income and reduction of poverty, and is considered to be on the verge of development (OECD, 2017). Notwithstanding, high-income inequality and issues with the quality of education, that is, its inability to adapt to a "knowledge-based" economy, have been signalled as serious obstacles for the country's development (OECD, 2017).

Chile's official language is Spanish, which is spoken by virtually the entire population. Four main indigenous languages are spoken by indigenous communities in the country and have to a limited degree been incorporated in the educational system in the past decade. Spanish is the official language in the majority of Latin America a noteworthy feature of the language is that it is overseen by the Royal Spanish Academy of Language (RAE). This is a Madrid-based organisation, whose role is to ensure linguistic unity among Spanish-speaking territories, and it has procedures in place for admitting words to their dictionary publications (RAE, 2014). These publications have acknowledged and included some degree of regional variations in vocabulary and word usage. Indeed, these national and regional variations make us think that despite sharing a language, there is great diversity across the continent, posing the question reagrding applicability of our findings to other Latin American countries.

The educational system

The Chilean educational system predates the country's independence, although at that time it was mainly attended by the elites. The system underwent several stages of expansion in the 20th century to the point that, nowadays, primary and secondary school enrolment is practically universal (Centro de Estudios MINEDUC, 2015). Overall, the country lags behind the OECD in terms of completion of educational degrees at secondary and tertiary levels attainment, although it has starkly improved in the last decades (OECD, 2017). Compulsory schooling comprises two years of preschool, eight years of primary, and four years of high school (15–18-year-olds). Class sizes are on average between 25 and 31 students per class, around 10 students larger than the OECD, with the smallest classes found in private schools (OECD, 2017). Distinctive features of the educational system are its mixed-provision structure and, interestingly, its centralised nature. Regarding the latter, the Ministry of Education oversees a compulsory national curriculum, periodic universal standardised testing [SIMCE], a quality of education school-level assessment, and standards for "good teaching." These standards are assessed through a centralised and high-stakes National Teacher Evaluation compulsory for teachers of municipal schools (Manzi, González & Sun, 2011).

The current structure of the educational system was shaped during the military dictatorship. Three main educational providers run schools in the country: municipal schools, voucher schools, and private schools. Under the market-driven logic, state subvention is allocated according to student enrolment in municipal and voucher schools, and parents' "freedom" to choose their children's schools is prioritised (Raczynski & Salinas, 2008). Roughly 90% of children attend municipal or voucher schools. Private schools in turn cater for the richest 10%, and are fully privately funded (Centro de Estudios MINEDUC, 2012). Until recently, voucher schools were allowed to charge school fees, to make profits, and to select their students, all of which municipal schools could not do. This resulted in a segregated system according to parents' socio-economic status, and net educational achievement results that are strongly predicted by the socio-economic composite of schools. Now, once this factor is accounted for, municipal schools perform similarly to the other two providers (Mizala & Torche, 2012). Recent educational reforms have aimed to address a decade of grassroots movements demanding equal and quality education for all.

The measures include the aforementioned changes to voucher schools, as well as expanding public education enrolment (currently around 40%) and relocating its administration from the municipal level to larger geographical regions with specialised administrators. These changes are too recent to estimate their impact in educational enrolment and results.

Mathematics teachers and teaching in the country

Regarding mathematics teachers, primary-school teachers – who are qualified to teach from 1st to 8th grades – require a generalist teaching professional undergraduate degree that lasts 4–5 years. In the first primary sub-cycle, teachers usually teach all subjects, whereas in the second sub-cycle they often focus on 2–3 subjects, sometimes holding a subject-matter diploma. The percentage of courses on the didactic of mathematics in generalist pre-service teaching programmes is marginal (Varas et al., 2008), and only reaches 7% to 15% in high-school "specialist" teaching degrees (Felmer & Perdomo-Díaz, 2014). A recent study on teachers' perceptions showed that teachers point to disciplinary didactics as a weak aspect of their Initial Teacher Education programs.

With regards to the mathematics curriculum, the country has a tradition of compulsory curricula created by the Ministry, with efforts to de-centralise planning starting in 1990 but falling flat. The curriculum was heavily focused on contents and procedural learning until in 1996, a curricular reform was introduced, outlining for the first time "fundamental goals." These went beyond the mastery of traditional algorithmic mathematics knowledge to include problem-solving skills as part of mathematics education goals. This applied first to the first sub-cycle of primary school and was included up to high school in 1998. It was not until 2012, however, that a substantial curricular update formulated the curriculum in terms of skills development rather than focusing on contents as such (Ministerio de Educación, 2012). These skills function as axes that should appear across content units. They include problem-solving, understood as resolving a situation without knowing a given procedure, and is seen as a means and an end in itself; modelling, which consists in manipulating, creating, and evaluating mathematical models; representing, which the Ministry relates to metaphorising by moving from the concrete or familiar to the more abstract; and arguing and communicating which relates to convincing others and also taking part in mathematical discussions.

Studies focusing on mathematics teaching have primarily analysed lesson videos produced in the context of the national teacher evaluation. They have consistently shown that lessons feature teacher-led instruction at the whiteboard as well as question-and-answer sequences presenting concepts and procedures, followed by individual student work to practise skills (Preiss, Calcagni, Espinoza, & Grau, 2016). The identified issues in mathematics teaching could be linked with students' performance in achievement tests such as PISA, where Chilean students have a 2-schooling-years attainment lag when compared to the OECD average (OECD, 2017). The high socio-economic inequality in the country has a strong effect on student attainment, so that students' socio-economic background impacts more on their performance than other countries participating in PISA (OECD, 2017).

The Chilean educational system is having important accomplishments in terms of coverage, while still having a long way to go to achieve high-quality education for all. The coming decade should see major reforms unfold with regards to school provision and the establishment of a new system for professional development. Looking at mathematics education in particular, results in the subject lag behind other educational areas, and public efforts and research are needed to help promote high-quality teaching and learning in mathematics.

Construction of the lexicon

Our national team is formed by four researchers with a background in educational psychology, Valeska Grau, Elisa Calcagni, Carolina Araya and David Preiss, and three experienced and highly regarded mathematics teachers, Mónica Quezada, Gladys Díaz, and Daniela Gómez. In constructing the Chilean Lexicon, we followed the overall project methodology. Nevertheless, this section will describe aspects of the methodology that were unique to the Chilean team in four phases.

Phase 1: Constructing the Chilean Lexicon

At this stage, our team included the four researchers and the teachers Mónica Quezada and Gladys Díaz. It is noteworthy that an initial challenge was to find teachers who were experts in middle-school mathematics and were also proficient in English. This turned out to be a rare combination, and the two teachers worked in a private school with a very strong English program.

After observing the Chilean video together, each team member observed up to two other videos and generated terms individually, to share them at our meetings. In building our list of commonly used terms, the main challenge was to distinguish what constituted a lexical item, enabling us to move away from long descriptive phrases that provided video accounts but were not commonly used. Another issue we faced was that some of the terms proposed by researchers were more representative of an educational researcher's lexicon, rather than a teacher's (e.g. *andamiaje* or "scaffolding"). Thus, teachers' views in the team were prioritised when making decisions in order to capture the practitioners' vocabulary.

Phase 2: Local (team) validation

In phase 2 we refined our lexicon in reaction to the Lexicon team meeting. The two mathematics teachers checked the terms and their descriptions, seeking to avoid circularity and gain clarity, and they completed some of the missing examples. The resulting list was further validated by Daniela Gómez, a third mathematics teacher with experience in peer mentoring and research. This phase resulted in 74 terms with descriptions and examples.

Phase 3: Refining the lexicon

Review of lexical items with the local community

The third phase involved a validation process with members of the national mathematics education community – teachers, teacher educators, educational researchers, and educational policymakers – employing surveys and in consultation with leading local experts. The goals were to assess whether the terms and descriptions were endorsed by the local community, and to obtain a collated list of well-known terms with a reasonable degree of consensus.

TABLE 4.1 Survey structure, content, and administration.

Survey structure and content		
	Round 1	*Round 2*
Number of forms	5	4
Section 1 – Demographics	Age range, positions held, gender, levels at which they taught mathematics	Age range, positions held, gender, levels at which they taught mathematics, academic degrees
Section 2 – Questions about full lexical items	14–15 items each Knowledge of the terms (yes/no format) Match between term, description, and examples	16–17 items each Familiarity with the terms (very familiar/ somewhat familiar/not familiar) Match between term, description, and examples Term usage
Section 3 – Questions about terms-only	-	50
Administration		
	Round 1	*Round 2*
Format and distribution	Google forms emailed to contacts	Google forms emailed to contacts and posted on relevant Facebook groups
Time of administration	June–July 2016	June–September 2017
Participants	28 participants 16 women–12 men 3–8 responses per form	46 participants 29 women–17 men 6–18 responses per form

The surveys were conducted using Google forms in two rounds, Round 1 taking place in 2016 and Round 2 in 2017 (details about each survey can be found in Table 4.1). Regarding the construction of the surveys, we divided the terms into different forms therefore participants answered questions about a subset of the whole lexicon. The surveys included a demographics section, followed by questions about the lexical items. In the construction of Round 2 questions, we kept the main characteristics of Round 1, while building on other teams' experience with their surveys, enriching our demographic questions and including questions about term usage and familiarity with the term-only (as opposed to full lexical items). Only 28 responses were obtained in the first round; therefore, we conducted a second round of surveys aiming to increase the number of participants.

After the first round of review, seven of the terms were considered well-known to participants and therefore excluded from the second survey. This allowed us to reduce the number of lexical items to be reviewed and the number of forms in round two.

The analysis of the second round of survey items showed that all terms were known to two thirds or more of the participants, which was the threshold criterion to consider that terms should be included on the lexicon. All 74 terms were considered reasonably familiar.

However, we divided the terms into two levels of familiarity, depending on participants' responses. "Very well-known terms" are those for which two thirds or more of the participants indicated that the term is very familiar (57 terms), and "Fairly well-known terms" are those for which the two-thirds cut-off point is met when combining responses that indicate "very familiar" and "somewhat familiar" (17 terms).

Circular clarity check

We then took part in the "circular clarity check," in which the Finnish team examined our lexicon. This led to the update of some of the examples, and to minor changes to ensure a more uniform formulation. As an additional source of validation, at this point we consulted with two leading scholars in the field of mathematics and mathematics education in Chile, who read the full lexicon and made some comments and suggestions. One of them suggested some additions, which the three teachers in the team thought were too academic and were thus not included in our lexicon.

Categorising the lexical items

After the revision, two of our team's teachers organised the terms. The exercise resulted in five categories, with a handful of terms belonging to two categories and most of them being part of one. Two of the researchers checked this categorisation and suggested minor adjustments, resulting in the current lexicon structure.

The Chilean Lexicon

Structure of the lexicon

The Chilean Lexicon comprises 74 lexical items, most of which are short phrases. These are organised in five categories: Subject-matter didactics, general pedagogy, teaching interactions, structure/routine, and classroom climate. Most of the terms belong only in one category, while five of them have been assigned to two categories.

One of the key features of the Chilean Lexicon is its relatively general character in terms of subject matter, given that only one category (with 22 terms) is specific to mathematics, whereas the remaining categories refer to more general teaching topics. Another feature that stands out is that in their phrasing, almost all the terms are of a descriptive nature, meaning they do not contain evaluative words or references to quality, for example, *tareas matemáticas* (mathematical task), or *revisión de resultados* (checking results or answers'). Only three exceptions can be found in the Classroom climate category: *clima apropiado para el aprendizaje* (adequate learning climate), *curso normalizado* (normalised class'), and *interrupción de la clase* (lesson disruption).

"Subject-matter didactics" (didáctica de la disciplina) – 22 terms

This refers to elements that in the team's view belong specifically to mathematics teaching or have a special character in the mathematics classroom. 17 terms in this category are "very well-known terms" and the remaining five terms are "fairly well-known terms." Among the more familiar ones are a series of terms that refer to kinds of problems or tasks, like *problema desafiante* (challenging problem), or *problema no rutinario* (non-routine problem).

Other terms refer to aspects of the mathematical work and are applicable to different tasks, such as *valorar procedimientos* (valuing procedures), *sistematización* (systematisation), or *fundamentar* (providing justification). Two terms that refer to the use of resources (use of technology and use of manipulatives) are also very well-known. Well-known only terms include a few with didactical flavour (didactic variables; mathematical translation of a student's reasoning, and "developing content within a problem"), as well as *ayudame-morias* (memory aide) and *ejercicio metacognitivo* (metacognitive exercise). Regarding the match between term, description, and examples, in general participants consider them a good fit (over 70%), with the exception of two of the less familiar terms ("mathematical translation of a student's reasoning," "memory aide").

"General pedagogy" (metodologías generales) – 18 terms

General pedagogy includes elements of teaching, as the subject matter didactics do, but these can be found across subjects. It comprises 18 terms, some of which refer to how students' work is set up, for example, *trabajo grupal* (group work), or *trabajo individual* (individual work). Other terms refer to forms of evaluation such as *autoevaluación* (self-evaluation), *evaluación entre pares* (peer assessment) and yet other terms refer to actions centred around boardwork: *hacer pasar a la pizarra* (asking someone to come to the board), or *copiar materia del pizarrón* (copying contents from the board). Two terms refer to the use of resources (technology and textbooks). Finally, two terms belong in this category and in "pedagogical interaction": *revisar resultados/respuestas* (checking answers) and *puesta en común* (putting in common). It is worth noting that all the terms in this category fall in the "very well-known term" category. Thus, this category contains what appear to be well-established lexical items that participants easily recognise. Perhaps this relates to the fact that there are teaching elements that can be found across subjects and are therefore more frequently discussed.

"Teaching interactions" (interacción pedagógica) – 17 terms

This category comprises 17 terms that relate to behaviours and verbal aspects of interactions in teaching. Some terms refer to dynamics around questioning, such as kinds of questions or question-and-answer sequence. Other terms point to broader elements in a lesson where interactions are key, such as *monitorear el aprendizaje* (monitoring learning), *facilitar el aprendizaje* (facilitating learning), and *revisión de resultados y respuestas* (checking results and answers). Aside from the two terms that belong in this category and "general pedagogy," three terms overlap between interactions and subject matter didactics: *argumentar* (to argue), *profesor explicita su razonamiento* (the teacher makes his/her reasoning explicit), and *ejercicio metacognitivo* (metacognitive exercise). Ten terms in this category appear to be "very well-known terms," whereas the other seven terms are "fairly well-known terms."

"Structure/routine" (estructura/rutina) – 13 terms

The 13 terms in this category refer to elements that give the lesson a structure, or actions that are performed regularly across lessons, such as *enunciar el objetivo de la clase* (stating the lesson goals), *enunciar el objetivo/tema/plan de la clase* (stating the lesson's theme and objective), and the main phases of a lesson (such as beginning, development, and closure of a lesson). Yet other terms refer to setting up activities and organising participants and materials. In general, terms in this category fall in the "very well-known term" category. Only two terms are "fairly well-known," namely "stating the lesson plan" and "connecting with the curriculum." Regarding the lesson plan, this might be because the other lexical items referred to similar actions – stating goals and stating theme – seemed like more familiar alternatives.

"Classroom climate" (Clima de aula) – 9 terms

This final category comprises relational, emotional, and behavioural aspects of the teaching and learning situation. Unlike other categories where either positive or neutral elements are named, this category includes positive and negative situations (specifically, disruptions and their management). Only four terms in this category are "very well-known terms" for participants: *refuerzo positivo* (positive reinforcement, *clima apropiado para el aprendizaje* (adequate climate for learning), *interrupción de la clase* (lesson disruption), and *petición de ayuda* (asking for help). The remaining five terms are "fairly well-known terms."

Key features of the lexicon

Examining the language of mathematics teachers in relation to the local context

There are several terms related to the way in which teachers organise their work and the materials in the classroom within the category structure/routine. These seem to refer to the key elements of organisation in Chilean classrooms, for example, *enunciar el plan de la clase* (stating the lesson plan) or *enunciar el tema de la clase* (stating the lesson topic), *organización de materiales para la clase* (organising materials for the lesson) or *organización física de los estudiantes* (physical arrangement of the students). There are also terms related to phases of the lesson such as *inicio de la clase* (beginning of the lesson), *desarrollo de la clase* (lesson development) or *cierre de la clase* (lesson closure) and *conectar con plan curricular* (connecting with curricular plan). Therefore, a clear emphasis on the organisation seems to characterise the Chilean Lexicon. This could be related to the National Teacher Evaluation and the good teaching framework, which play a role in defining how a "good lesson" should be structured. One of the assessment instruments consists of a videotaped lesson and the dimensions observed are as follows: Classroom environment, structure of the lesson, and classroom interactions. The second dimension (structure of the lesson) is highly related to the way in which teachers organise their lessons. Considering that this is such a high-stakes evaluation, it is likely to have an important influence on teachers' lexicon. In fact, when reviewing the results of teachers' evaluation since its creation in 2004, all the scores have increased, which is an evidence of participants' appropriation of the way in which these dimensions are assessed (Manzi et al., 2011).

As noted earlier, various terms in the lexicon come from general pedagogy and are applicable across subjects, such as *hacer pasar a la pizarra* (Asking someone to come to the board), or *copiar materia del pizarrón* (Students copy contents from the board), which are all good examples or another characteristic of Chilean classrooms: the importance of the board in teaching practices. There has been research showing that in Chilean lessons, teachers' discourse predominates largely over students discourse or group work (Preiss et al., 2016). However, the lexicon also shows terms for other kinds of interactions such as *trabajo en grupo* (group work) or *trabajo en parejas* (work in pairs). It is common to find students sitting in groups and sometimes doing activities together; however, this does not always reflect real peer collaboration.

Another common feature of Chilean classrooms is reflected on the term *secuencia pregunta-respuesta* (question-and-answer sequence), characterised by a teacher-led interaction, including a series of questions and answers that varies widely in their quality, order, and pedagogical aim. A study conducted by Radovic and Preiss (2010) found that the questions posed by Chilean teachers in mathematics lessons focus on controlling the lesson flow rather than on understanding mathematical concepts. Also, most questions are rote questions, emphasising repeated practice of routine problems.

Despite its general character, there is a group of terms related to the pedagogy of mathematics. Indeed, some of these terms reflect the teacher-led teaching practices portrayed above, such as *ayudamemorias* (memory aide), *problema rutinario* (routine problem), and *formalizar procedimientos* (formalising procedures). Yet other terms focus on other aspects of mathematics teaching. For instance, it can be thought that a few of them require more disciplinary knowledge, such as *traducción matemática del razonamiento de un estudiante* (mathematical translation of a student's reasoning), or *problema no rutinario* (non-routine problem). These pedagogical terms imply not only the teacher translating between student' thinking and mathematical terms, but also to pose problems with non-obvious solutions, that require students to relate various contents and ask them to "do" mathematics. Some terms convey key mathematical activities, for instance, *comprobación matemática* (mathematical check). Others refer to the way in which teachers mediate students' learning such as and *valorar procedimientos* (Valuing procedures).

Analysis of agency

The analysis of agency responds to the question "who performs the action?," which was applicable to 60 of the terms. Examining these terms, it can be asserted the actions in our lexical items have the teacher as the clear protagonist, with 36 terms referring to teachers' actions or activities across the five categories, and 9 terms depicting actions that could be done by either the teacher or the student. Only 11 terms were exclusive to students. All of the latter belong in two categories: "general pedagogy" (e.g. "individual/pair/group work," "self-assessment," and "peer assessment") and "classroom climate" (e.g. "normalised class," "asking for help," "spontaneous student contribution," "lesson disruption," and "student notices teacher's mistake"). Finally, four terms correspond to joint teacher and students' activities. Examples of terms with the teacher as the exclusive actor are presented in Table 4.2.

Upon examining student-performed actions or terms (see Table 4.3), it is noteworthy that even these terms depict students' roles as rather passive. Perhaps the most significant indication of this is the term "spontaneous student contribution," which seems to signal that student participation is otherwise highly scripted.

Form and function

This analysis considers the character of the terms, considering whether they portray forms or functions. These categories were applicable to 72 of the terms. Our lexicon appears to

TABLE 4.2 Terms that refer to teacher-performed actions.

Term	Description	Examples
introducir nuevo procedimiento/contenido (introducing new content/procedure)	The teacher presents topics or procedures that are unknown to students (explicitly flagging the novelty).	After a few lessons about first degree equations with one unknown value, the teacher indicates that a new topic will be addressed: equation systems with two unknown values.
promover distintas formas de resolución de problemas (promoting different ways of solving problems)	The teacher aims for students to use and express different forms of solving mathematical problems. This can be done by guiding them to do so, or asking students to present more than one form of solution.	A group states they solved a problem using two subtractions. The teacher asks them to solve it differently, this time by using at least an addition.

TABLE 4.3 Terms that refer to student-performed actions.

Term	Description	Examples
autoevaluación (self-evaluation)	Written or oral assessment in which a student evaluates their own work.	Students are asked to answer questions such as "I know how to add fractions with the same denominator: yes – no."
intervención espontánea de alumnos (spontaneous student contribution)	The student spontaneously makes a contribution related to the lesson.	The teacher indicates: OK, here we have that the tree diagram is one way of finding the sample space. Let us move on to the next topic. A student adds "but miss, I did it using multiplication."

focus mainly on functions, that is, actions' goals or purposes are explicitly stated in 43 terms (e.g. "promoting diverse problem-solving strategies," "facilitating learning," "introducing new contents"). A further 10 terms refer to both form and function, and only 19 terms refer to forms only (e.g. "use of manipulatives"). Table 4.4 exhibits examples to illustrate the three categories.

The more specific analysis indicates that most of the terms focus on the functions of actions and on the teacher. Interestingly, integrating both results shows that, even though the majority of terms refer to teacher-exclusive actions, only two of these are classified as "form." Meanwhile, six of the students' exclusive terms belong in this category. This reinforces the idea that teachers are the protagonists in the Chilean Lexicon, with a substantial number of terms devoted to describing their actions in terms of their intentions. As noted before, video-based studies of Chilean classrooms relate to this characteristic of our lexicon, given that mathematics lessons across educational levels have been found to emphasise teacher-led activities focused on transmitting information and practicing skills, with scarce opportunities for students' substantive contributions (Preiss et al., 2016).

TABLE 4.4 Lexical items that illustrate "form" and "function."

	Term	Description	Examples
Function	**formalizar procedimientos** (formalising procedures)	The teacher systematises the steps of a given procedure.	The teacher systematises the steps to solve an equation system with two unknown quantities: "To find the unknown values, first we replace one of the terms, y, and we look for the value of the other term, x. Having found the value of the first term x we replace it for its numerical value and we look for the value of the second term y."
Form	**copiar la materia del pizarrón** (students copy contents from the board)	The students write down on their notebooks what is written on the board	
Both (Form and Function)	**puesta en común** (putting in common)	The teacher orchestrates the joint review of a completed task without giving out the right answer. This implies that knowledge is distributed among students and the teacher.	The class is reviewing the results of a worksheet. The teacher asks: "Did anyone get a different result? Why?"

Insights from the lexicon review

In this section, we examine further the results of the survey to provide insights into participants' familiarity and usage of the Chilean Lexicon.

Terms seem to be familiar to the local community

The review surveys allowed us to check if the terms were known to members of the relevant communities. Most of the terms were found to be very well-known to participants, and some terms appeared to be fairly well-known terms. This finding gives an indication that the terms are reasonably familiar to mathematics teachers and educators, which is relevant when assessing the validity of the lexicon in the local context. Unfortunately, we do not have the same degree of information regarding term usage, because data about this was only collected in the second round of surveys regarding full lexical items.

Participants in the survey did not declare the same level of familiarity or usage for all terms. Therefore, examining which terms appear to be best known in the lexicon is of interest. We considered the familiarity data in conjunction with the usage information and identified the ten "most common" terms which fall under percentile 75 or above in terms of familiarity (86,6% of "very familiar" responses) and in frequency of use (80% of very frequent use). These common terms are as follows: *Inicio de la clase* (beginning of the lesson), *desarrollo de la clase* (lesson development) *cierre de la clase* (lesson closure*)*, *trabajo individual/trabajo en parejas/trabajo grupal* (individual work/pair work/group work), *hacer pasar a la pizarra* (asking someone to come to the board), *resolución de ejercicios matemáticos* (solving mathematical exercises), *refuerzo positivo* (positive reinforcement), and *clima apropiado para el aprendizaje* (adequate climate for learning). What is noteworthy about this list is that six of the most common terms refer to lesson phases and common arrangements of participants, which point to elements that constitute key building blocks of every lesson and are common across subjects. On the other hand, only one term belongs in the subject matter didactics category, and it is a fairly generic term (solving mathematical exercises), which reinforces the sense that the Chilean mathematics lexicon is a fairly generic one.

Demographics of familiarity

We examined participants' demographic characteristics (experience in academic-related work and age) and how these interacted with familiarity including 74 respondents. The number of respondents for each item is indicated in parentheses in Figure 4.1. To assess the differences in the distribution of the ordinal variable across groups we performed the non-parametric test Independent samples Mann–Whitney *U* test on SPSS v22.

Work experience in academia. We considered whether participants had reportedly worked in university positions including research, lecturing at university level, and/or being a practicum supervisor. In total, 42 participants had such experience and 32 did not. In most of the cases, participants' experience working in academia did not make a difference in their reported familiarity with the terms. This shows that terms are generally well-established in the communities that are relevant to our lexicon.

There are, nonetheless, a few exceptions depicted in Figure 4.1. In the subject matter didactics category, there are significant differences with regards to two terms referring to types of problems, which are more familiar for participants with experience in academia: *problema no rutinario* (non-routine problem) and *problema rutinario* (routine problems). The same was found for two terms related to argumentation: *fundamentar* (to justify) and *argumentar* (to argue), which belong in subject-matter didactics and pedagogical interaction. In the general pedagogy category, the only difference was found in *registrar respuestas en la pizarra* (writing answers on the board), which is more familiar for people with experience in academia. In the pedagogical interactions category, a significant difference was found for *facilitar el aprendizaje* (facilitating learning), again in favour of participants related to academia. Finally, no differences across these groups were found in the classroom climate and the routine/structure categories. It is noticeable that the bulk of the terms showing differences belong in the subject-matter didactic category, indicating that there might be more dissimilarities in the lexicon specific to mathematics than in other areas of professional vocabulary. Possibly, this could relate to the fact that didactics of mathematics does not feature prominently in teacher undergraduate degrees (Varas et al., 2008) and thus part of its lexicon could be less well-known for teachers who are not linked to the academic context.

Participants' age. We compared familiarity with the terms according to participants' age, which was recorded employing 5-year age ranges from under 25 to 65+. We created two groups to compare participants who were 35 and under (*n* = 36) and 36 and over (*n* = 38). Again, only a few terms were found to differ in familiarity across groups, suggesting that the lexical items are known to participants of different ages. Figure 4.2 shows the

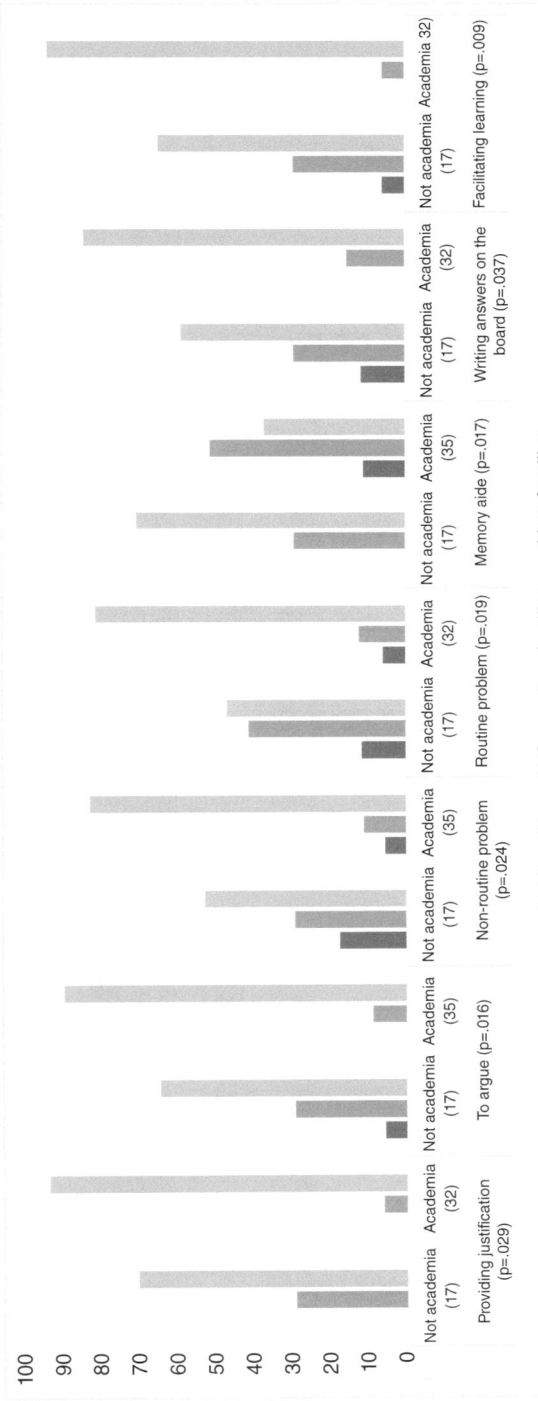

FIGURE 4.1 Terms in which working in academia made a significant difference in familiarity.

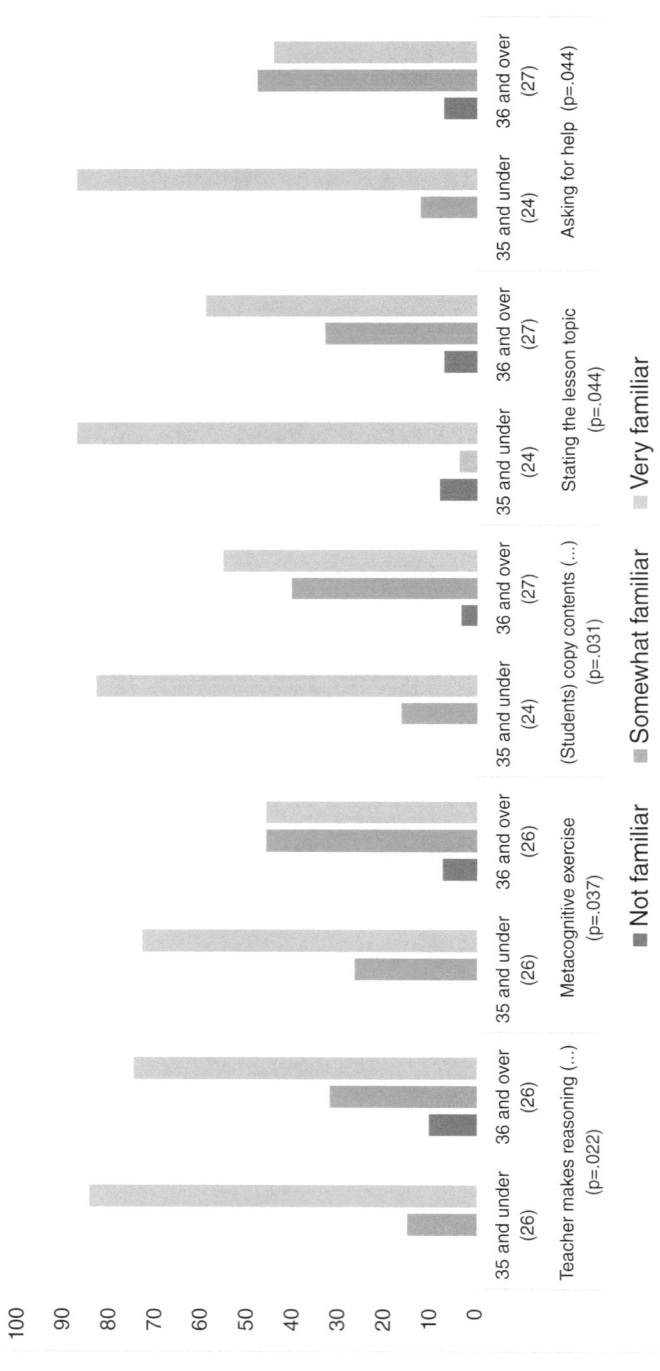

FIGURE 4.2 Terms in which age made a significant difference in familiarity.

percentages of responses in each category, indicating that all of the terms in which differences emerged are more familiar to younger participants. These are interesting findings in that they point to terms that may be newer to the Chilean mathematics teachers' Lexicon, and/or that may be part of more recent teacher education. The fact that these terms do not coincide with the work-related differences points to their novelty. Specifically, differences were found in: *profesor explicita su razonamiento* ('teacher makes his/her reasoning explicit') and *ejercicio metacognitivo* (metacognitive exercise) in the subject matter didactics category; *copiar materia de la pizarra* (Copying content from the board) in the general pedagogy category; *enunciar el tema de la clase* (stating the lesson theme) in the structure/routine category; *petición de ayuda* (asking for help) in the Classroom climate category. No differences were found in the interaction category, which is interesting because the category contains many of the "fairly well-known terms," indicating that the lower level of familiarity is not age-related, nor is it explained by participants' involvement in academia.

Concluding thoughts

When considering the exercise of creating the lexicon, it is worth noting that it was extremely difficult for our academic group to connect with professional teaching communities and academic mathematics communities. This also resonates with the local context, where contacts between teachers from different schools, between researchers and teachers, and even between researchers working in different departments are infrequent. Indeed, collaborations between academia and practitioners usually do not go beyond the latter acting as research participants. Perhaps for these reasons, it was also very difficult to complete a reasonable sample of respondents in our review. Despite these difficulties, the terms in our lexicon were reasonably known to participants of different ages and work experience. After constructing and reviewing the Chilean Lexicon, we can say that there is a level of shared professional terminology that Chilean mathematics teachers recognise and report to employ. A key direction of future work would be to disseminate our findings further among mathematics teachers' communities.

Our Chilean Lexicon features predominantly short phrases as opposed to single-word terms. For the most part, they refer to things in a neutral form, and most of the terms that involve agency are actions performed by teachers. Our analyses reveal that the lexicon of Chilean middle-school mathematics teachers is mainly a generalist one. This raises questions about the depth of mathematics education that the generalist degree required to teach at this level has been able to achieve with respect to professional vocabulary. On the other hand, we think that all the efforts put by the Chilean government in the last 15 years to create a framework for teaching and disseminate the dimensions included in the teacher evaluation, might have helped to develop a reasonably common language regarding what is important to observe and name in the classroom. Exploring the relationship between the terms and the country's "Framework for good teaching" would be an interesting step forward, to examine how these guidelines relate to the identified professional language.

We think that further efforts involving the lexicon should be exploring its potential uses in practice, especially in initial and in-service teacher education, and its function as a resource to engage in productive professional dialogue. Such a tool could help practitioners in sustaining a focus on problems of practice in mathematics teaching. These discussions could be greatly enriched by sharing and observing the elements and practices of teaching we already have labels for, but also by learning from what other cultures can name.

Acknowledgements

This project has been partially funded by Measurement Centre (MIDE UC), Pontificia Universidad Católica de Chile and Proyecto REDES 170141 de CONICYT – Chile.

References

Centro de Estudios MINEDUC. (2012). Realidad educativa en Chile: ¿Qué aprendemos de la Encuesta CASEN 2011?

Centro de Estudios MINEDUC. (2015). Variación de matrícula y tasas de permanencia por sector.

Felmer, P., & Perdomo-Díaz, J. (2014). *La Resolución de Problemas en la Matemática Escolar y en la Formación Inicial Docente*. Retrieved from www.fonide.cl

Kelly, P. (2006). What is teacher learning? A socio-cultural perspective. *Oxford Review of Education*, 32(4), 505–519.

Manzi, J., González, R., & Sun, Y. (Eds.). (2011). *La evaluación docente en Chile*. Santiago de Chile: MIDE UC.

McIntyre, D. (2005). Bridging the gap between research and practice. *Cambridge Journal of Education*, 35(3), 357–382.

Ministerio de Educación. (2012). *Bases curriculares Matemática*. Retrieved from http://www.curriculumenlineamineduc.cl/605/w3-propertyvalue-53,688.html

Mizala, A., & Torche, F. (2012). Bringing the schools back in: The stratification of educational achievement in the Chilean voucher system. *International Journal of Educational Development*, 32(1), 132–144.

OECD. (2017). *Education in Chile*. https://doi.org/10.1787/9789264284425-en.

Preiss, D. D., Calcagni, E., Espinoza, A. M., & Grau, V. (2016). ¿Cómo se enseña el lenguaje y las matemáticas en las salas de primer y segundo ciclo básico en Chile? Principales hallazgos de una serie de estudios observacionales en clases de lenguaje y matemáticas. In J. Manzi & M. R. García (Eds.), *Abriendo las puertas del aula: Transformación de las prácticas docentes* (pp. 153–184). Santiago de Chile: Ediciones UC.

Preiss, D. D., Calcagni, E., & Grau, V. (2015). Classroom Research and Child and Adolescent Development in South America. *New Directions for Child and Adolescent Development*, 2015(147), 85–92. https://doi.org/10.1002/cad.20093

Real Academia Española (RAE), (2014). *El Diccionario de la lengua española. Edición 23ª*. Consultada en https://www.rae.es/sites/default/files/Dossier_Prensa_Drae_2014_5as.pdf.

Raczynski, D., & Salinas, D. (2008). Fortalecer la educación municipal. Evidencia empírica, reflexiones y líneas de propuesta. In C. Belleï, D. Contreras, & J. P. Valenzuela (Eds.), *La agenda pendiente en educación: Profesores, administradores y recursos: Propuesta para la nueva arquitectura de la educación chilena* (pp. 105–133). Santiago de Chile: Universidad de Chile; Fondo de las Naciones Unidas para la Infancia UNICEF.

Radovic, D., & Preiss, D. D. (2010). Patrones de discurso observados en el aula de matemática de segundo ciclo básico en Chile. *Psykhe*, 19(2), 65–79.

Varas, L., Felmer, P., Gálvez, G., Lewin, R., Martínez, C., Navarro, S., Ortiz, A., & Schwarze, G. (2008). Oportunidades de preparación para enseñar matemáticas de futuros profesores de educación general básica en Chile. *Calidad en la Educación*, 29, 64–88.

5

CHILEAN LEXICON

Valeska Grau, Elisa Calcagni, Mónica Cortez, Daniela Gómez,
Gladys Díaz, Carolina Araya and David D. Preiss

A Chilean Lexicon

The lexicon presented in this chapter consists of 74 terms, mostly short phrases and some single-word terms. The original terms are in Spanish, the official language of Chile. Operational definitions accounting for what the term means were developed alongside one or two examples. These are meant to illustrate the way in which the term may occur in the classroom. The words (in Spanish and English) are presented in the standard order of letters of the Roman alphabet and the spelling corresponds to Chilean Spanish and Standard British English, respectively.

The first lexicon table (see Table 5.1) lists all the terms of the lexicons. The second lexicon table (see Table 5.2) includes the operational definition of each term arranged into three columns. The first column lists the term itself, the second column gives its description while the examples are reported in column three.

The 74 terms in the lexicon were organised by the teachers in the team according to their thematic relationships, reflecting different aspects of the teaching-and-learning practice (see Table 5.3). The categories are: Subject-matter didactics (22 terms); General pedagogy (18 terms); Pedagogical interaction (16 terms); Classroom climate (nine terms); and Structure/routine (12 terms). Only one of them has a mathematical focus, and a handful of terms belong to more than one category.

The Chilean Lexicon documented on this paper features predominantly short phrases as opposed to single-word terms, and most of the terms that involve agency are actions performed by teachers. Our analyses reveal that the lexicon is mainly a generalist one. This raises questions about the depth of mathematics education that the generalist degree required to teach at this level has been able to achieve with respect to professional vocabulary. Thus, the lexicon gives some clues to design and implement professional development programmes and initial teacher educators' curricula.

TABLE 5.1 Chilean Lexicon – Terms.

activación de conocimientos previos (activating previous knowledge)	**alumno se da cuenta de un error del profesor** (student notices teacher's mistake)	**ampliar aporte de alumno/a** (building on a student's contribution)	**argumentar** (arguing)
atención a estudiantes con necesidades educativas especiales (adjustments for students with special educational needs)	**autoevaluación** (self-evaluation)	**ayudamemorias** (memory aide)	**cierre de la clase** (lesson closure)
clima apropiado para el aprendizaje (adequate climate for learning)	**comprobación matemática** (mathematical check)	**conectar con plan curricular** (connecting with curricular plan)	**contra-preguntar** (counter-asking)
copiar la materia del pizarrón (students copy contents from the board)	**curso normalizado** (normalised class)	**dar instrucciones para trabajo grupal** (giving instructions for group work)	**desarrollo de la clase** (lesson development)
desarrollo del contenido dentro de un problema (developing contents within a problem)	**descripción de material a utilizar** (description of materials that will be used)	**ejercicio metacognitivo** (metacognitive exercise)	**enunciar del objetivo de la clase** (stating the learning objective)
enunciar el plan de la clase (stating the lesson plan)	**enunciar el tema de la clase** (stating the lesson topic)	**evaluación entre pares** (peer assessment)	**evaluación formativa** (formative assessment)
facilitar el aprendizaje (facilitating learning)	**formalizar procedimientos** (formalising procedures)	**fundamentar** (providing justifications)	**hacer pasar a la pizarra** (asking someone to come to the board)
inicio de la clase (beginning of the lesson)	**interrupción de la clase** (lesson disruption)	**intervención espontánea de alumnos** (spontaneous student contribution)	**introducir nuevo procedimiento/ contenido** (introducing new content/procedure)
manejo conductual de la disrupción (behaviour management)	**monitoreo del aprendizaje** (monitoring learning)	**organización de materiales para la clase** (organising materials for the lesson)	**organización física de los estudiantes** (physical arrangement of the students)
parafrasear lo que dice un/a alumno/a para clarificar (paraphrasing what a student says to clarify)	**petición de ayuda** (asking for help)	**pregunta metacognitiva** (metacognitive question)	**pregunta por elaboración** (elaboration question)

pregunta por información (information question)	**presentar contenidos con contexto de la vida real** (presenting contents in real life contexts)	**problema desafiante** (challenging problem)	**problema no rutinario** (non-routine problem)
problema rutinario (routine problem)	**profesor explicita su razonamiento** (teacher makes his/her reasoning explicit)	**promover apoyo entre pares** (promoting peer support)	**promover distintas formas de resolución de problemas** (promoting different ways of solving problems)
puesta en común (putting in common)	**redondear ideas** (rounding up ideas)	**refuerzo positivo** (positive reinforcement)	**registro de las respuestas en la pizarra** (writing answers on the board)
remediales (remedial)	**resolución de ejercicios matemáticos** (solving mathematical exercises)	**resolución en conjunto de tareas matemáticas** (solving a mathematical task together)	**revisar que los alumnos hayan hecho la tarea para la casa** (checking homework)
revisión de resultados o respuestas (checking results or answers)	**secuencia pregunta-respuesta** (question-and-answer sequence)	**simplificación de un problema** (simplification of a problem)	**sistematización** (systematisation)
tarea para la casa (homework)	**tareas matemáticas** (mathematical tasks)	**técnicas para apoyar la comprensión del enunciado de un problema** (techniques to support understanding of the wording of a problem)	**trabajo en grupo** (group work)
trabajo en parejas (pair work)	**trabajo individual** (individual work)	**traducción matemática del razonamiento de un alumno** (mathematical translation of a student's reasoning)	**uso de material concreto** (using concrete materials)
uso de recursos multimediales (use of multimedia resources)	**uso de recursos tecnológicos -propios de la matemática** (use of technological resources -specific to mathematics)	**uso de texto escolar** (use of textbook)	**uso del humor** (use of humour)
valorar procedimientos (valuing procedures)	**variables didácticas** (didactic variables)		

TABLE 5.2 Chilean Lexicon – Terms with operational definitions.

Term	Description	Examples
activación de conocimientos previos (activating previous knowledge)	The teacher mentions previously addressed contents, that relate to the current lesson's topic, either by lecturing or by asking students questions.	In the initial phase of a lesson focused on adding fractions with different denominators, the teacher reminds the students how to add fractions with the same denominator.
alumno se da cuenta de un error del profesor (student notices teacher's mistake)	A student points out a mistake that the teacher has made in a procedure.	A student notices a mistake in the calculations, and s/he tells the teacher.
ampliar aporte de alumno/a (building on a student's contribution)	The teacher builds on a student's spontaneous contribution, in relation to the learning objective, through lecturing or follow-up questions	In a lesson focused on geometric 3D shapes, a student observes there is a resemblance between the formulae for calculating the volume of the cone and of the cylinder. The teacher answers that this is the case, and there is a relationship between the two. This is followed by an experimental demonstration.
argumentar (arguing)	The student expresses verbally their reasoning about a procedure.	"To solve this problem, I used a division because it was about distributing in equal parts, and we know how many things we want to distribute, and the number of people that will get things."
atención a estudiantes con necesidades educativas especiales (adjustments for students with special educational needs)	Adjustments to the teaching strategies, contents or evaluations aimed for students with diagnosed special educational needs.	Adjusting the level of difficulty of an exercise in a worksheet or test for students with special educational needs.
autoevaluación (self-evaluation)	Written or oral assessment in which a student evaluates their own work.	Students are asked to answer questions such as "I know how to add fractions with the same denominator: yes – no."
ayudamemorias (memory aide)	In order to help students, the teacher flags some aspect of the contents, highlighting its importance. It can include giving tips.	When solving a division on the board the teacher writes the subtractions using a different colour.

Term	Description	Examples
cierre de la clase (lesson closure)	Final phase of a lesson, in which the lesson contents or activities are somehow synthesised.	The lesson focused on whole numbers, and at the end of the lesson, the teacher asks students: what did you understand about adding whole numbers? Can you write a definition showing how it is done? Can you establish different cases? Express with examples and explain how you solved the calculations.
clima apropiado para el aprendizaje (adequate climate for learning)	The teacher and students create adequate, non-threatening conditions (explicitly or implicitly) whereby initiative and mistakes are allowed.	Students often share their questions and doubts; students do not only participate when they think they have the right answer.
comprobación matemática (mathematical check)	Calculations related to validating results obtained in a task.	Applying the inverse operation to prove an arithmetic calculation
conectar con plan curricular (connecting with curricular plan)	Situating students' learning in a wider curricular or temporal context (unit, curriculum).	The teacher says, "This unit focuses on integer numbers, which we will address this year. In the past years, you have seen natural numbers, and in the future, you will learn about rational numbers, and in high school real numbers, and another kind of numbers called imaginary."
contra-preguntar (counter-asking)	After a student expresses doubt or asks something, the teacher does not give an answer and, instead, asks other students to reply instead.	Student: "I don't understand, why is 200 equal to 2 times 10 squared?" Teacher: let's see, step by step. First, how much is 10 squared?"
copiar la materia del pizarrón (students copy contents from the board)	The students write down on their notebooks what is written on the board	
curso normalizado (normalised class)	The class behaves according to the class' rules, without the need for explicit demands. This can be inferred in that the lesson flow is continuous and there are no major disruptions or attempts to manage students' behaviour.	The teacher arrives in the classroom and the class follows his/her instructions step by step; the students listen to the teacher and complete the proposed tasks.

(*Continued*)

TABLE 5.2 (*Continued*)

Term	Description	Examples
dar instrucciones para trabajo grupal (giving instructions for group work)	The teacher gives instructions for students to work in groups. These can refer to forming groups, how many participants to include, time frame, tasks.	The teacher says, "Form groups of three, this is the worksheet you will work on, it has 15 exercises. You can discuss your responses and agree on them."
desarrollo de la clase (lesson development)	This is the longest phase in a lesson, in which the proposed activities or contents are addressed.	After introducing the topic, the teacher proposes activities for the students, to address the lesson contents. The activities can involve from lecturing and explanations to group work focused on problem-solving.
desarrollo del contenido dentro de un problema (developing contents within a problem)	The teacher introduces or expands terms and mathematical concepts in the context of problem-solving activities, in a whole-class interaction.	In a lesson focused on geometric 3D shapes, the teacher asks the students: "How much liquid from this cone can the cylinder hold?"
descripción de material a utilizar (description of materials that will be used)	The teacher describes a learning resource that will be used.	The teacher tells the students that they will work with an algebra book, and they will focus on standard algebraic identities.
ejercicio metacognitivo (metacognitive exercise)	The teacher leads students through a conversation, to construct or re-construct the reasoning behind a procedure.	After a student has shown how she solved a problem involving quantities contained in other quantities, the teacher asks: "Why did you decide to divide? How did you know which number was the divisor? What did you do to identify relevant information?"
enunciar del objetivo de la clase (stating the learning objective)	The teacher states the lesson objectives.	The teacher says, "In this lesson, we will learn how to add and subtract fractions with different denominators."
enunciar el plan de la clase (stating the lesson plan)	The teacher states the activities that will be done in the lesson.	The teacher says, "In this lesson you will work in groups of three on a worksheet for half an hour and then we will check your answers together."
enunciar el tema de la clase (stating the lesson topic)	The teacher states the topic of the lesson, without making the mathematical objective explicit, because the students are expected to discover this on their own.	The teacher says, "In this lesson, we will solve situations of fair sharing."

Term	Description	Examples
evaluación entre pares (peer assessment)	Students privately review and check their peers' answers (in groups or individually).	After solving a problem during group work, the groups exchange the worksheets for reviewing and feedback among peers.
evaluación formativa (formative assessment)	Brief assessment, written or oral, which is not marked and provides information about current achievement and/or how to improve.	At the end of a lesson, the students individually complete a worksheet with two questions similar to the ones addressed during the lesson. The next day, they check the procedures and answers and students can assess their progress.
facilitar el aprendizaje (facilitating learning)	Managing a student's mistake, doubt or confusion. It can involve explaining an exercise again, asking new questions, scaffolding, giving cues.	Working with a worksheet that involves technical language, the teacher helps students by adjusting to the language that students use.
formalizar procedimientos (formalising procedures)	The teacher systematises the steps of a given procedure.	The teacher systematises the steps to solve an equation system with two unknown quantities: "To find the unknown values, first we replace one of the terms, y, and we look for the value of the other term, x. Having found the value of the first term x we replace it for its numerical value and we look for the value of the second term y."
fundamentar (providing justifications)	Student or teacher justifies a procedure using mathematical properties or theorems.	Proving an equality, for instance by explaining why the fourth root of a squared equals the square root of a.
hacer pasar a la pizarra (asking someone to come to the board)	The teacher invites students to come to the board to solve a problem.	The teacher invites students "who wants to come to the board to solve this equation?"
inicio de la clase (beginning of the lesson)	Initial phase in a lesson, in which typically the topic or lesson objectives are introduced.	In the beginning of the lesson, the teacher says hello to the students and states the lesson goals and activities.
interrupción de la clase (lesson disruption)	Different forms of disruption, coming from one or more students, that obstruct the normal flow of the lesson.	Students chatting without paying attention to the teacher impeding others' ability to hear.

(*Continued*)

TABLE 5.2 (*Continued*)

Term	Description	Examples
intervención espontánea de alumnos (spontaneous student contribution)	The student spontaneously makes a contribution related to the lesson.	The teacher indicates: OK, here we have that the tree diagram is one way of finding the sample space. Let us move on to the next topic. A student adds "but miss, I did it using multiplication."
introducir nuevo procedimiento/ contenido (introducing new content/procedure)	The teacher presents topics or procedures that are unknown to students (explicitly flagging the novelty).	After a few lessons about first degree equations with one unknown value, the teacher indicates that a new topic will be addressed: equation systems with two unknown values.
manejo conductual de la disrupción (behaviour management)	The teacher uses some strategy to manage behaviours that disrupt the lesson, in order to restate the lesson flow.	A teacher says "David, please put away your mobile and focus on the task"
monitoreo del aprendizaje (monitoring learning)	The teacher supervises student work and checks their understanding.	There are different forms of monitoring, like walking between the desks and simply observing what students are doing, or asking questions to check for understanding.
organización de materiales para la clase (organising materials for the lesson)	Arranging and distributing materials that will be used.	The teacher hands out a set of material to each group.
organización física de los estudiantes (physical arrangement of the students)	Special arrangement of students in the classroom; this might change during the lesson.	The students are seating in groups of four, with two tables forming a square.
parafrasear lo que dice un/a alumno/a para clarificar (paraphrasing what a student says to clarify)	Teacher's active listening. It aims to check that s/he has understood a student's contribution, or to clarify the contribution for other students to understand.	A student says, "A cylinder resembles a soda." The teacher says, "Philippe, I think you mean the shape of the cylinder is similar to the shape of a can of soda."
petición de ayuda (asking for help)	A student expresses doubt, lack of understanding or uncertainty.	Students ask the teacher to approach their group when they face difficulties with an exercise.
pregunta metacognitiva (metacognitive question)	Questions that focus on the previous learning process or on what will happen next, or on the challenges found along the way. They focus on the learning process.	In an instructional sequence in which the teacher has intentionally selected problems with a similar structure, she then asks, "How did you know this problem could be solved this way?"

Term	Description	Examples
pregunta por elaboración (elaboration question)	Questions that aim for students to articulate extended answers, making explanations or arguments explicit.	After a student states the value of x in an equation, the teacher asks her to explain how she found the value: "Could you explain the steps you followed to find the unknown"?
pregunta por información (information question)	Questions that demand for specific information, checking for understanding, or generating interactions.	A teacher asks, "How many faces does a cube have?"
presentar contenidos con contexto de la vida real (presenting contents in real life contexts)	Relating mathematical topics to everyday life situations, aiming to introduce or apply a new topic.	When working with percentages, relating them to sales and discounts expressed in percentages, mentioning retail sales and build from there.
problema desafiante (challenging problem)	Problem that aims to engage or motivate students.	The teacher presents the first numbers in the Fibonacci series and asks: "Which do you think are the next numbers in this series?"
problema no rutinario (non-routine problem)	Working on a problem for which students do not already know the solving procedures.	The following problem is introduced: "Inside a cube with 12 cm. Of edge there is a cone with 12 cm. Of diameter and 12 cm. Height. How much volume of the cube is not occupied by the cone?"
problema rutinario (routine problem)	Working on a problem for which students apply known procedures. Typically, the level of difficulty is low.	A shop is offering 25% discount on all its products. If you want to buy a shirt that originally costs $12.000, how much would the corresponding discount be?
profesor explicita su razonamiento (teacher makes his/ her reasoning explicit)	The teacher explains why s/he has done something or will do something, stating metacognitive reasons to model argumentation.	A teacher explains, "When I am solving a division I ask myself: how many times does 5 go in 8? Although I could also ask 5 times what is the closest to 8 I can get without going over?"
promover apoyo entre pares (promoting peer support)	The teacher encourages students to support each other in solving mathematical tasks.	The teacher asks the students to explain an exercise to their classmates, or to provide support in the solving process.
promover distintas formas de resolución de problemas (promoting different ways of solving problems)	The teacher aims for students to use and express different forms of solving mathematical problems. This can be done by guiding them to do so, or asking students to present more than one form of solution.	A group states they solved a problem using two subtractions. The teacher asks them to solve it differently, this time by using at least an addition.

(Continued)

TABLE 5.2 (*Continued*)

Term	Description	Examples
puesta en común (putting in common)	The teacher orchestrates the joint review of a completed task without giving out the right answer. This implies that knowledge is distributed among students and the teacher.	The class is reviewing the results of a worksheet. The teacher asks: "Did anyone get a different result? Why?"
redondear ideas (rounding up ideas)	The teacher synthesises ideas that emerged from dialogue.	The teacher states, after a question and answer sequence: "Some of the methods you have explained that are used to determine the sample space are very similar, because they all involve drawing the possible combinations. These drawings are called diagrams."
refuerzo positivo (positive reinforcement)	The teacher or students give motivational messages, verbal or behavioural.	After a student has solved an exercise at the board the teacher tells her, "Congratulations! Keep up the good work."
registro de las respuestas en la pizarra (writing answers on the board)	The right answers are written on the board.	The teacher or a student writes on the board the answers and solutions for tasks on a worksheet that is being reviewed.
remediales (remedial)	Activities that aim to strengthen knowledge or skills that have been found to be weak after an evaluation.	
resolución de ejercicios matemáticos (solving mathematical exercises)	Activity in which students calculate or solve using operations	Finding the value of x in the expression $5x + 3 = -2x + 6 \times 4$ by writing the procedure.
resolución en conjunto de tareas matemáticas (solving a mathematical task together)	The teacher solves a problem or exercise with the students in a public interaction.	The teacher writes down a non-routine problem on the board and they solve it collectively with students.
revisar que los alumnos hayan hecho la tarea para la casa (checking homework)	Checking students' work done at home.	The teacher collects the notebooks and checks the homework that was assigned in the previous lesson.
revisión de resultados o respuestas (checking results or answers)	The teacher states if a task or process students have completed is correct. This implies the knowledge lays with the teacher.	The class is reviewing a worksheet, students read out their answers, and the teacher replies indicating whether they are correct.

Term	Description	Examples
secuencia pregunta-respuesta (question-and-answer sequence)	Dialogue led by the teacher, with a series of questions and answer that vary in terms of their quality, sequence, and teaching intentions.	Teacher: what is the cube of 2? Student: 8; Teacher: and the cube of 3?; Student: 27; Teacher: and the cube of 4?; Student: 64; Teacher: why is it that we call "cube of" the numbers raised to the third power?
simplificación de un problema (simplification of a problem)	Changes to the original formulation of a problem, or questions that decrease the level of difficulty of a problem, when students do not know how to approach it.	Original problem: 3/4 kg of flour is necessary to make a tray of cookies. How much flour is necessary to make half a tray? Simplification: If I want to make two trays of cookies, what do I need to do to find out how much flour I will use?
sistematización (systematisation)	The teacher systematises the abstract mathematical ideas that emerge from an episode of mathematical work (problems, activities, exercises).	After the students have compared different examples of quantitative variables describing their differences and similarities, the teacher asks students for their ideas, and establishes the definition of continuous and discrete quantitative variables.
tarea para la casa (homework)	Work assigned to be done at home.	At the end of the lesson, the teacher indicates which exercises from the workbook students need to complete before the next lesson.
tareas matemáticas (mathematical tasks)	Mathematical activities to be worked on by the students, that are proposed by the teacher.	Solving problems with additions.
técnicas para apoyar la comprensión del enunciado de un problema (techniques to support understanding of the wording of a problem)	The teacher teaches or reminds students of strategies to understand a word problem, for instance, highlighting relevant information.	The teacher asks the students to read a non-routine problem from a worksheet. Then, she asks "what are the first steps we need to take to solve this problem?"
trabajo en grupo (group work)	Students work in groups of three or more participants	
trabajo en parejas (pair work)	Students are arranged in pairs.	
trabajo individual (individual work)	Students solve mathematical tasks individually.	

(*Continued*)

TABLE 5.2 (*Continued*)

Term	Description	Examples
traducción matemática del razonamiento de un alumno (mathematical translation of a student's reasoning)	The teacher uses mathematical terms to explain or paraphrase what a student is doing when writing the solution to a problem.	The teacher says "look, Arthur is trying which of these numbers fits in the equation in order to check his answer on the multiple choice test"
uso de material concreto (using concrete materials)	Use of manipulative materials (dice, coins, etc.) as part of mathematical activities	Using 1 × 1 cm. Cubes to form cubes with different side lengths (2 × 2, 3 × 3) and discover the formula to calculate the volume of cubes.
uso de recursos multimediales (use of multimedia resources)	Use of technological resources, such as projector, video, etc.	The teacher uses the projector to show students some slides in which problems are written.
uso de recursos tecnológicos -propios de la matemática (use of technological resources -specific to mathematics)	Use of mathematical software or calculator.	Use of a specialised software such as GeoGebra, to solve non routine problems on the board.
uso de texto escolar (use of textbook)	Use of texts or worksheets from a textbook that include exercises or problems for students to solve.	The teacher uses routine exercises from a textbook for students to practice during a lesson.
uso del humor (use of humour)	Use of humour during a lesson (with different purposes).	
valorar procedimientos (valuing procedures)	Asking students to write down their procedures while solving problems, and focusing on the logic underlying alternative procedures.	A problem that is solved by multiplying "3 × ½" was solved doing this multiplication, adding 1/2 three times, and dividing 3 into 2. The mathematical knowledge underlying both processes are discussed, and they are assessed in a different context: "what if the multiplication was 23 × ½?" In doing this, the teacher highlights the value of the different procedures as a function of the relationship between the numbers.
variables didácticas (didactic variables)	Elements in a mathematical task that are modifiable during planning and teaching.	The mathematical task involves quantifying a collection: - numeric field; − use of manipulatives; − using the objects [that form the collection].

TABLE 5.3 Chilean Lexicon – Terms by category.

Didáctica de la disciplina (subject-matter didactics)	*arguing*, challenging problem, developing content within a problem, didactic variables, formalising procedures, mathematical check, mathematical task, mathematical translation of a student's reasoning, memory aid, *metacognitive exercise*, non-routine problem, promoting different ways of solving problems, providing justification, routine problem, systematisation, solving a mathematical task together, simplification of a problem, solving mathematical exercises, *teacher makes his/her reasoning explicit*, valuing procedures, use of technology, use of manipulatives
Metodologías generales (general pedagogies)	asking someone to come to the board, *checking results or answers*, formative assessment, giving instructions for group work, group work, homework, individual work, pair work, peer assessment, presenting contents in real-life contexts, promoting peer support, *putting in common*, remedial, self-assessment, students copy contents from the board, use of multimedia resources, use of textbook, writing answers on the board
Interacción pedagógica (teaching interactions)	adjustments for students with special educational needs, *arguing*, *checking results or answers*, counter-asking, elaboration question, facilitating learning, information question, *metacognitive exercise*, metacognitive question, monitoring learning, paraphrasing what a student says to clarify, *putting in common*, question-and-answer sequence, rounding up ideas, *teacher makes his/her reasoning explicit*, techniques to support understanding of the wording of a problem, widening student's contribution
Estructura/ rutina (structure/ routine)	activating previous knowledge, beginning of the lesson, checking that students have done their homework, connecting with curricular plan, description of materials that will be used, introducing new content/ procedure, lesson closure, lesson development, organising materials for the lesson, physical arrangement of the students, stating the lesson objective, stating the lesson plan, stating the lesson topic
Clima de aula (classroom climate)	adequate climate for learning, behavioural management of disruptions, demand for help, lesson disruption, normalised class, positive reinforcement, spontaneous student contribution, student notices teacher's mistake, use of humour

Note: Terms in italics are in more than one category.

Acknowledgements

This work has been partially funded by Measurement Centre (MIDE UC), Pontificia Universidad Católica de Chile and Proyecto REDES 170141 de CONICYT – Chile.

Bibliography

Presentations

The work of the Chilean national team has been presented internationally at the following meetings and conferences:

- American Educational Research Association (AERA) Annual Meeting 2018
- Colloquium at Pontificia Universidad Católica de Chile 2019
- European Association for Learning and Instruction (EARLI) Biennial Conference 2017
- Kaleidoscope Conference for Graduate Students of Education at University of Cambridge 2017

6

EXPLORING THE LEXICON OF MIDDLE-SCHOOL MATHEMATICS TEACHERS IN CHINA

Yiming Cao, Guowen Yu and Lianchun Dong

Introduction

As introduced in Chapter 1, the Lexicon Project was established initially to conquer the obstacles encountered by researchers from different communities using different languages from each other. To identify the pedagogical naming systems employed by different cultures, the larger project involved researchers from ten different communities, including China. The joint aim is to document the professional terms employed by mathematics educators within each of the ten countries, and to identify named practices of middle-school mathematics classrooms. This chapter presents our Chinese results.

Many dialects are used around China, but Mandarin is the official language and the compulsory instructional language in schools. What we research in this paper is teachers' professional Mandarin language in mathematics classrooms, but not dialects. China has a long history of using standard language in teaching, thus some of the Chinese terms which emerged many years ago are still used today. Consider the term 画龙点睛 (hua long dian jing; *finishing touch*) as an example. It is a Chinese idiom, which emerged in the Tang Dynasty, and in the modern era, it has been used in the pedagogical context to refer to some actions or words by teachers or students in a constructive way when solving a certain problem. However, we had not documented a pedagogical lexicon until 1992, when Shanghai Education Press released a massive 12-volume Educational Dictionary, with Gu (1992) as chief editor. The Chinese Lexicon shares some terms with this dictionary, but it still has its own constraints due to the various contexts of Chinese mathematical classrooms.

In the process of developing the Chinese Lexicon, we included the pedagogical terms but excluded the mathematical terms, following the protocols of the larger project. The lexicon results presented in this chapter are used by the Chinese middle-school mathematics teachers in their everyday classroom practice, but the frequency of using these terms differs among teachers.

In addition to producing the Chinese Lexicon, we also conducted some local analyses on this lexicon, established a Chinese Lexicon Structure, and built the critical pedagogical behaviours (CPBs) of Chinese mathematics classrooms. Actually, every "term" in the Chinese Lexicon corresponds to a specific classroom pedagogical behaviour together with a description of this behaviour. We examined the Chinese Lexicon as representative of a fundamental structuring by Chinese middle-school mathematics teachers of the world of their classrooms.

The Chinese education system and mathematics teaching culture

In this section, we present the general information of Chinese mathematics education, including the education system and the cultures of mathematics teaching. In China, compulsory education covers children aged 7–16 years of age. This includes 6 years in primary school (Grades 1–6), and 3 years in middle school (Grades 7–9). We use a uniformed national curriculum program, with a limited number of textbooks issued by the government. Generally, the choice of textbooks is made by the regional education department, meaning that teachers and their schools cannot decide which textbook to use. Based on the national curriculum, there are three levels of curriculum (including mathematics): the national level, the local level (generally on province level or city level), and the school-based curriculum. The local and school-based curricula are influenced by the notion of "mathematics for all" and support the national curriculum to create a mathematics culture which includes a real-world application of mathematics. This is important as China is a multi-nation country in which areas of eastern and western China are at quite different levels of economic development.

China has a long history of mathematics dating from the Zhou Dynasty (1046–256 BC) with the appearance of The Nine Chapters of the Mathematical Art (Chemla & Guo, 2004), but a short history regarding the modern mathematics education system. China's long history of an exam culture remains as a deep influence. Due to historical and political reasons, following the Qing Dynasty (1636–1911) Chinese mathematics education was influenced deeply by two different educational systems: the United States and the Soviet Union. This caused a swing between Soviet and American education ideas, such as emphasising the basic knowledge and basic skill in mathematics, so-called double base (basic skill and basic knowledge) (Zhang, Li, & Tang, 2005, p. 189), exams and exercises, and the teaching procedures in classroom. In Chinese culture, even though teaching has been recognised as a profession since the establishment of The Teacher Act in 1994, it is still believed that the knowledge for teaching can be developed "from examples and by doing" (Li, Huang, Bao, & Fan, 2011, p. 82). This echoes what Wang (2013) has remarked about university being the best place for learning advanced mathematics content, while the core purpose of teacher preparation is to learn subject knowledge because prospective teachers can develop their pedagogical knowledge from their future teaching practice.

Methodology for data collection and analysis

In this section, we present our methods in two parts, the first relates to generating the Chinese Lexicon, which is followed by four detailed steps on how we conducted our analysis of this lexicon.

Methods for generating the Chinese Lexicon

The central lexicon team had developed a protocol for generating each country's lexical terms and conducting the local review, national review as well as the international review (see more in Chapter 1). While we adopted some of the procedures of the central team,

TABLE 6.1 Example list of Chinese Lexicon terms.

Term	Description
按顺序回答 **an shun xu hui da** (answer in turn)	The teacher asks questions to students in turn, for example, the teacher asks students in a row the answers to questions on an examination.
查漏补缺 **cha lou bu que** "check the leak, fill the gap" (fill the gaps)	The teacher helps students to find gaps in the original knowledge system and supplement for what they do not know well before to complete their knowledge structure.

we took a different approach when undertaking local analyses. This approach is different from those in other national research teams.

In the Lexicon Project, there were nine videos of middle-school mathematics classrooms from each country. The Chinese Lexicon central team members and some experienced teachers and mathematics education researchers watched the videos and developed 40 terms after the viewing of these nine videos. Some experts were invited to check, refine and enrich the 40 terms, after which we had 123 terms altogether, including Chinese terms and corresponding descriptions, English translations and video excerpts. Local validation was conducted among the team and national validation involving experts from around China were conducted. At the conclusion, we were able to determine a final Chinese Lexicon which can be used by the teachers, researchers and policy makers.

The procedure for generating the Chinese Lexicon consisted of four main steps: (1) video viewing by the team members and experts, (2) local review, (3) national review, and (4) the clarity check. The final product included for each term: Chinese characters, Pinyin, the term in closest English, a literal translation, and a description in English. Examples of terms from the lexicon are given in Table 6.1.

Step 1: Video viewing by team members and experts

To name what we could see and the middle-school mathematics teachers could experience in their classrooms, the research team members, school teachers, mathematics education researchers and postgraduate students observed the filmed lessons from each of the nine participating countries and identified in Chinese "those things for which you have a name." We did so according to the known Chinese pedagogical language system and recorded the beginning and end time of the associated video moments. Throughout this process, we recorded each single Chinese character or short phrases that we believed to be widely used by our teachers and to be understood in a similar way by our teachers. In addition, the research team constructed a description of the particular named activity. These operational definitions were recorded in the original language (Chinese); the researchers were mindful that in translating the description into English the main features of the activity were maintained and illustrated. In this way, we generated an initial list of 40 terms.

Step 2: Local review

An expert panel of experienced school teachers and university professors was invited to add and delete terms according to their teaching and researching experience. This panel revised the draft lexicon to now include 123 terms. Among the 123 terms, 102 were thought to be widely used and 12 were given provisional status as seldom used, while 9 were judged as seldom used but of sufficient pedagogical significance to be included in the Chinese Lexicon.

All the terms and descriptions were validated locally. Each of the terms was discussed within the whole team to ensure the lexicon was clear. Then we interviewed three experienced experts, an associate professor with more than 25 years' experience in high school and university, an experienced teacher with more than 40 years' experience as a middle-school teacher and a lecturer with nearly 15 years' experience in university as a mathematics education researcher to recheck all the terms and descriptions. We submitted the feedback to our team expert panel for the next step and discussed again with the whole team.

Step 3: National review

We developed a questionnaire in Chinese, taking advantage of a national conference that many teachers and researchers participated in, and invited 165 people from primary schools, secondary schools and universities to take part in the questionnaire process. We kept the same questionnaire structure used with the central team for our national validation. The following table, Table 6.2, indicates the demographic information of the 165 participants.

All the terms and their descriptions offered for national validation have been agreed to as known to at least two-thirds of the 165 respondents.

Step 4: Clarity check

Our Chinese Lexicon was assigned to the German team for a clarity check while the Chinese team checked the Australian Lexicon. We discussed the feedback from the German team and revised our lexicon according to the feedback. Each term was revised by two experts invited by the central Chinese team, and when we received all the modifications, we ran again the expert checking of the revised version of Chinese Lexicon, and each term was re-checked again by two experts. After this process, we had a clearer and more precise lexicon.

TABLE 6.2 Demographic information of the 165 participants of national validation.

Area	Job Category	Teaching Experience	Grade Level
East: 81 Middle: 35 West: 49	Teacher: 132 Researcher and Coach: 23 Student: 10 (4 PhD student and 6 Master Degree student)	Less than 3 years: 37 4–10 years: 28 More than 10 years: 100	Primary school: 7 Secondary school: 125 Tertiary: 23
Age	Gender	Qualification	Years of Tertiary Study
Less than 30: 50 30–49: 97 More than 50: 18	Male: 65 Female: 100	PhD & Master: 66 Bachelor of Science: 97 Junior college: 2	Less than 4 years: 10 4 years: 77 6–7 years: 53 More than 8 years: 25

Methods for local analyses

Taking the lexicon as the starting point, the local analyses aimed at investigating the critical level of the pedagogical behaviours in classrooms listed in the lexicon. Drawing on the results of research on "critical pedagogical incidents," both "critical incidents" and "CPBs" play crucial roles in controlling the efficiency and order of classroom, as well as students' gaining of knowledge. It has also been shown that CPBs, to some extent, play important roles in informing teachers' training (Yang & Wang, 2015). This section is consequently focusing on "critical behaviour" and the importance of specific pedagogical behaviours and the interplay between the two. The word "critical" in the phrase "CPB" seeks to emphasise the importance of particular classroom pedagogical behaviours. That is to say, to name one specific behaviour as "critical" is actually a judgment of value, which is based on the perceived importance of the particular behaviour. This study uses the Chinese Lexicon to identify the CPBs in middle-school mathematics classrooms in China.

As the identification of the CPBs and their levels relies on the number of connections among terms, the first step is to build the connection of the pedagogical behaviours in the whole structure so as to get the Correlation Index in Class (CIC), which has been defined as the number of pedagogical behaviours that respondents associated with each individual pedagogical behaviour (Cao & Yu, 2017). A low-level connection indicates that this specific behaviour is likely to influence the events of the mathematics classroom to a lesser extent. Given that the connection with other pedagogical behaviours is an indication of the likely influence of the behaviour on the other activities of the mathematics classroom, it is also reasonable to infer that such a behaviour is not very "critical."

The three levels, A, B, and C were defined as follow:

- *Level A: The highest rank of critical behaviours*. These pedagogical behaviours are highly used; they play crucial roles in teaching and learning. In addition, their connections with other pedagogical behaviours are also very strong.
- *Level B: The middle rank of critical behaviours*. The frequency of use in practical teaching and learning is less than those in Level A. Their connection with other pedagogical behaviours is also strong.
- *Level C: The bottom rank of critical behaviours*. These pedagogical behaviours have little or no connection with other pedagogical behaviours; in other words, they are, to some extent, relatively independent.

Given the results from the CIC, these "weaker" pedagogical behaviours, of which there were 107 in total, were categorised as level C (the lowest level), while the rest, 16 terms, were left for a further level identification since they were more frequently connected to other terms. To determine the significance of particular terms within the Chinese Lexicon, these 16 pedagogical behaviours were assigned to two additional levels (A and B) according to the level of connectivity of the pedagogical behaviours signified by the lexical terms as well as the results of the questionnaire (see Figure 6.1). In this questionnaire, the respondents are asked to rank the remaining 16 terms for their usage in practice, usage in communication and the educational importance.

At this stage, the internal relationship among the terms was further analysed. We sorted the 16 terms according to four aspects (agency, hierarchy, sequence and coincidence) and

Key Terms in The Middle School Mathematics Classroom

Education Experience (year): _____ Title:_____ Position:_____

Instructions

Thanks for your cooperation! Please note that all your information will not be disclosed. The questionnaire is used to identify the key terms that contribute most to a successful middle school mathematics class in your opinion. Your are asked to sequence the terms separately according to the frequency of use in practice, frequency of use in speech, and the educational importance (1 means the least and 15 means the most).

Note: Write a same number if you think they are the same in frequency of use or of the same educational importance.

Frequency of use in Practice	Frequency of use in Speech	Educational Importance
( ) Teacher presentation	( ) Teacher presentation	( ) Teacher presentation
( ) Teacher explanation	( ) Teacher explanation	( ) Teacher explanation
( ) Teacher assessing	( ) Teacher assessing	( ) Teacher assessing
( ) Teacher feedback	( ) Teacher feedback	( ) Teacher feedback
( ) Teacher questioning	( ) Teacher questioning	( ) Teacher questioning
( ) Teacher tutoring	( ) Teacher tutoring	( ) Teacher tutoring
( ) Classroom management	( ) Classroom management	( ) Classroom management
( ) Summary of the class	( ) Summary of the class	( ) Summary of the class
( ) Student listening	( ) Student listening	( ) Student listening
( ) Student answering	( ) Student answering	( ) Student answering
( ) Student presentation	( ) Student presentation	( ) Student presentation
( ) Student assessing	( ) Student assessing	( ) Student assessing
( ) Student feedback	( ) Student feedback	( ) Student feedback
( ) Do exercise	( ) Do exercise	( ) Do exercise
( ) Self-learning	( ) Self-learning	( ) Self-learning
( ) Cooperative learning	( ) Cooperative learning	( ) Cooperative learning
( ) Teacher-student-interaction	( ) Teacher-student-interaction	( ) Teacher-student-interaction

FIGURE 6.1 Questionnaire related to key terms in the middle-school mathematics classroom.

found the critical levels (Levels A and B) of the 16 terms. Our examination of this internal relationship was guided by four questions:

1. *Who is responsible for the actions implied by the term?*
2. *Is there a possible hierarchical relationship among the terms?*
3. *Can any of the terms be arranged in a chronological sequence according to their likely occurrence in the course of a lesson? Are any of the terms coincident or overlapping in their occurrence or focus?*

And responses to questions of one, two and three helped determine the response to question four:

4. *There is some hierarchical structure among the terms, a kind of critical level (Level A or B), which "level" does each term belong to?*

Of these 16 CPBs, 5 CPBs belong to Level A (the most critical level), 11 CPBs belong to Level B, and the remaining 107 lexical terms were designated as Level C (Cao & Yu, 2017).

Results and discussion

In this section, we present our analysis results in two parts: a general aspect on the whole lexicon and a further analysis on the inner structure of these 123 lexicon terms.

Results about the Chinese Lexicon

After the national review and clarity check, we confirmed all the 123 candidate terms are included in the lexicon, which is detailed in the next chapter. The national review results showed all the terms for this national validation had been agreed to by at least two-thirds of the 165 respondents. In Part D of the questionnaire, we showed the full list of all the terms and found familiarity for each term. The following table, Table 6.3, shows the terms that are not familiar to at least 10% of all the respondents. In the list, we can find that the most unfamiliar term 提出变式 (ti chu bian shi; *propose variation*) is still familiar to more than 3/4 of the respondents (124 out of 165).

TABLE 6.3 Terms that are not familiar to at least 10% of all the respondents.

Term	Respondents unfamiliar with term (%)	Term	Respondents unfamiliar with term (%)
提出变式 Propose variation	25	实验 Trail	13
画龙点睛 Finishing touch	24	暴露问题 Expose errors	12
挑战 Challenging	22	教师自答 Teacher answers herself	12
课堂生成 Teaching in the moment	17	突发事件处理 Organisational matters	12
教师诊断 Teacher diagnosing	16	感悟 Perception	12
变式教学 Teaching with variation	15	矫正 Correction	12
下课仪式 Formal ending of class	15	师生共答 Teacher and students answer together	10
按顺序回答 Answer in turn	15	铺垫 Bridging	10
小组辅导 Group tutoring	13	背诵 Reciting	10
默写 Dictation	13	体验 Experience	10
个人活动 Individual activity	13	操作 Operating	10
衔接 Connecting	13		

Results and discussion on local analysis on lexicon structure

Extracting pedagogical behaviours

We collected CPBs to check whether there were any additional behaviours appearing. According to Tripp (1993), we can stop checking additional events when there are at most two to three new ones appearing among every hundred. The lesson videos did not cover all possible pedagogical behaviours of middle-school mathematics lessons, the research team only extracted 40 (40/123) pedagogical behaviours. In addition, the Chinese national research team invited other experts and experienced teachers to supplement the team of experts. This expert panel increased the number of terms until no further terms were identified.

Expert proof to extract CPBs

To complete the Critical Incident analysis, the activities of the mathematics classrooms were actually organically considered as a whole. Therefore, researchers firstly identified the connections among pedagogical behaviours, based on the extent of connections, further constructed the CIC and then investigated the relatively more critical behaviours in the classroom. The term "critical" here means that one specific pedagogical behaviour is strongly related to many other pedagogical behaviours, thus the changing of classroom teaching and learning leads to the creation of a situation whereby "if one moves, the whole class moves in response" (牵一发而动全身).

The Chinese national research team had an idea of building connections and structure of pedagogical behaviours different from those used by other teams and further proposed that the following four views could be used to construct the CIC:

- *Agency*. Agency differs between the terms from S (Student), T (Teacher) to ST (Student and Teacher together).
- *Hierarchical*. Level One: On the first level, the terms could be regrouped according to whether the term referred to Teachers, Students or to Teacher–student Interactions. Level Two: The category "Teachers" could be divided into: *Classroom Management, Demonstration, Questioning, Feedback, Summarising, Explanation*, and *Tutoring*. The category "Students" included: *Classroom Management, Demonstration, Questioning, Feedback, Summarising, Doing Exercise, Collaborative Studying, Self-learning* and *Listening*. The category "Teacher–student Interaction" had no sub-structure at this level.
- *Coincident*. The terms refer to activities that can happen at the same time. For example, *Group Report* and *Student Listening* – when a group is reporting their findings or answers, the other students should be listening carefully in the class.
- *Sequential. Teacher Questioning* and *Student Answering* are an example of a pair of tasks that are intrinsically sequential – when the teacher asks a question to the whole class, this action is typically followed by an individual answering or the class answering together.

The Chinese Lexicon Structure was formed by the following different steps. The first step was categorising the terms according to the subjects of activities including teachers (T), students (S) and their interactions together (T&S). The second step was dividing teacher

TABLE 6.4 Agreement Rate.

	First level Agreement Rate				Second level Agreement Rate			Time connection Agreement Rate		
	A-B-C*	A-S**	B-S	C-S	A-B	A-S	B-S	A-C	A-S	C-S
Consistent num	121	102	103	99	81	104	100	102	109	115
Agreement Rate	0.98	0.82	0.83	0.80	0.65	0.84	0.81	0.82	0.88	0.93

* A, B and C refer to experts.
** S represents the draft Chinese lexicon.

behaviours (T), student behaviours (S) and their interactive behaviours (T&S) into differ-ent categories (the categories including *Classroom management, Presentation, Questioning, Feedback, Summarising, Teacher explaining, Tutoring, Doing exercise, Student listening, Cooperative learning, Self-learning, and Teacher–student interaction*). Last but not least, all of the 123 pedagogical behaviours had been attributed to the second level and form the diagram of hierarchical structure. In addition, the hierarchical structure detailed the following relationships: sequential, coincident, and no connection, with regard to time.

The research team then invited three experts and researchers to validate the structure, which also provided an endorsement of the validity of this local study. The following table, Table 6.4, indicated the agreement by different experts, with each other and the prelimi-nary model formed by the research team. When reading Table 6.4, A, B, and C refer to the experts while S represents the preliminary model of the Chinese Lexicon. We accepted that agreement had been reached when at least two experts agreed (Khandelwal, 2009).

Table 6.4 shows that both the agreement rate by and among experts and the research team all attained an acceptable level (>0.8). After discussion and negotiation, the experts and researchers continued to validate the structure through a process of negotiated con-sensual revision. The agreement rate kept increasing and eventually was greater than 0.9, which indicated a very stable structure.

According to Perreault Jr. and Leigh (1989), the categorised validity exponent formula helped check the stability and consistency of the structure. The validity exponents of the structure in terms of first level, second level, and time-relationship dimension were 0.99, 0.79, and 0.9, respectively; the average validity was 0.89, which indicated a high level of validity.

We also visualised the connections between one term and another using the network visualisation software Polinode, this allowed us to explore the number of links by term which are captured in Table 6.5. These were considered as "CPBs," and by examining their level of connectivity these terms had been allocated to level A or level B. We can see from the table that the term with the fewest links was *Tutoring*, which is quite important in Chinese middle-school mathematics classrooms, we included it into higher level and excluded the other 107 terms with no more than two links to other terms.

There were three groups of numerical values obtained from the questionnaire in Figure 6.1, indicating:

1. the rank of the frequency of use in the classroom;
2. the frequency of use in communication; and
3. the importance to teaching and learning of the 16 pedagogical behaviours.

TABLE 6.5 Number of links per term.

Pedagogical Behaviours	Links	Pedagogical Behaviours	Links
Student listening	34	Cooperative learning	11
Teacher explanation	32	Teacher questioning	9
Self-study	24	Student explanation	7
Student feedback	23	Student displaying	7
Teacher feedback	17	Teacher displaying	6
Show content	14	Teacher–student interaction	6
Classroom management	13	Student questioning	5
Student doing exercise	12	Tutoring	3

Therefore, sets of three values were identified for each specific teacher respondent. These values helped to identify CPBs in level A and B.

All 124 questionnaires that were distributed were completed (a response rate of 100%). Excluding those that could not be used because of a lack of sufficient information, we were able to include the data from 107 questionnaires. In the questionnaire, the teacher was asked to rank the 16 terms in Figure 6.1 for their usage in practice, usage in communication and educational importance. The ranking 1–16 indicated least (1) to most (16).

Interviews with some teachers had shown that while some pedagogical behaviours could get high scores for their importance, in practical teaching, they might not be used very often, or their importance can actually be shown by reference to other contexts and situations outside classrooms. Consequently, in the classroom, their importance might not be so obvious. Similarly, while some other pedagogical behaviours might appear frequently in the classroom or daily conversations, their importance in teaching and learning are not high. For these latter ones, while they can be seen to be high-frequency behaviours, further investigation proves that they cannot be classified as "CPBs." In short, "CPBs" are indicated with a respective importance of the three factors rated by five experienced experts.

Therefore, a three-variable-array for every behaviour had been given by every interviewee; then by calculating the average value, we obtained 16 three-variable-arrays forms the point pattern in a coordinate system. In the coordinate system, the distance from one specific point to (16,16,16) can be used to imply the extent of critical level (see also Table 6.6).

Table 6.6 shows that five terms had a weight value greater than ten and a distance to the ideal point less than ten. This distinguished these terms from the others in the table. Therefore, these five pedagogical behaviours were designated as constituting level A: *Teacher–student interaction, Teacher displaying, Student listening, Student doing exercise*, and *Teacher questioning*. Another 11 pedagogical behaviours were in level B: *Teacher explanation, Self-study, Student feedback, Teacher feedback, Show content, Classroom management, Cooperative learning, Student explanation, Student displaying, Student questioning*, and *Tutoring*.

TABLE 6.6 Weight value of each critical pedagogical behaviour and its corresponding distance to (16, 16, 16).

Pedagogical Behaviour	Weight Value	Distance to (16, 16, 16)
Teacher–student interaction	11.88	7.26
Teacher displaying	11.27	8.17
Student listening	11.01	8.78
Student doing exercise	10.69	9.05
Teacher questioning	10.62	9.45
Self-study	9.92	10.67
Student feedback	9.56	11.30
Teacher explanation	9.39	11.48
Student displaying	9.37	11.67
Show content	9.29	11.74
Cooperative learning	8.99	12.15
Student explanation	8.72	12.60
Student questioning	8.33	13.25
Teacher feedback	8.18	13.40
Classroom management	8.12	13.95
Tutoring	7.56	14.38

Note: These values in the second column were calculated as weighted average using the rates fixed with five experienced teachers, and the formula is: Weight value = 0.37*value (use in practice) + 0.21*value (use in speech) + 0.42*value (educational importance).

Conclusions

As part of the larger international project, The International Classroom Lexicon Project, this research, using questionnaires and interviews, studied the difference in "critical level" of 123 pedagogical behaviours in Chinese mathematics classrooms. First, the CIC of every pedagogical behaviour indicated the level of connection of one pedagogical behaviour (lexical term) with other pedagogical behaviours. Then those pedagogical behaviours whose CIC was less than three were allocated to level C. Those behaviours with a CIC greater than three were allocated to level A and level B, and included 16 pedagogical behaviours, identified in Table 6.5.

Questionnaires and interviews were used to rank these 16 pedagogical behaviours by their level of connectivity. In addition, the corresponding distance of specific point to (16, 16, 16) also showed a term's "critical level." Combining these two methods, it was shown that, there were five pedagogical behaviours whose powers are bigger than ten and their distances to (16, 16, 16) were less than ten:

Teacher–student interaction, *Teacher displaying*, *Student listening*, *Student doing exercise*, and *Teacher questioning*. These five pedagogical behaviours were designated as constituting level A. Another 11 pedagogical behaviours were in level B: *Teacher explanation*, *Self-study*, *Student feedback*, *Teacher feedback*, *Show content*, *Classroom management*, *Cooperative learning*, *Student explanation*, *Student displaying*, *Student questioning*, and *Tutoring*.

Meanwhile, the extraction of pedagogical behaviour can never ignore the information from experienced teachers. Therefore, further analysis by the 47 teachers who have more than 10 years of teaching experience and relatively high professional qualification confirmed that the "top five" CPBs were: *Teacher–student interaction*, *Teacher questioning*, *Teacher displaying*, *Student listening*, and *Student doing exercise*. This is consistent with what was obtained from the whole sample. Therefore, it is reasonable to identify these pedagogical behaviours with level A and claim that they are CPBs. The role of the experts was to guarantee the reliability and validity of the questionnaire to make sure that the data helped to address the research question.

Implications

Our Chinese team sought to identify an inherent structure in the Chinese Lexicon. The results prioritised connectedness. This has significant implications for the way in which the lexicons generated by other countries might be viewed. The Chinese lexicon also offers insight into the pedagogical principles that historically underlie Chinese classroom practice.

This study showed that in middle-school mathematics classrooms, *Teacher–student interaction*, *Teacher questioning*, *Teacher displaying*, *Student listening* and *Student doing exercise* get higher scores in "critical level" and have the greatest influence in terms of developing teaching and learning, as well as teachers' professional development. The research results showed that *Teacher–student interaction* has the highest critical value, which is not only the focus of practical teaching and learning in classroom but also a highly debated topic among educational researchers and teachers (Cao & He, 2009).

Interviews with four experienced teachers endorsed the "critical level" of the pedagogical behaviours in level A. *Teacher displaying* is related to the logical direction of teaching; in other words, reasonable and suitable displaying tends to show the development of knowledge in a logical way. In addition, there were several crucial questions including how to show the content in a way that is more easily acceptable for students, what kind of examples should be chosen, how is the atmosphere of the lesson, how do teachers instruct students, and, how to display content. These are all crucial questions.

The importance of *Teacher–student interaction* is also very obvious, in the sense that the whole class is actually an organic whole for communication and interaction. Furthermore, communication, interaction and inspiration between teachers and students are all very important, so teachers have to be more careful when picking up questions and methods for explaining so as to construct a more effective basis from which to interact.

Particularly for mathematics, *Student doing exercise* plays an essential role not only in after-class practices but also during the class. That is to say, *Student doing exercise* during class helps to understand the procedures of their acquiring of knowledge and expose their confusion which provides useful information for the subsequent teaching and learning.

Students listening intertwines with many other pedagogical behaviours, so its importance in class can never be underestimated. *Students listening* and other pedagogical behaviours interact in a variety of ways, for instance: when students are listening, how teacher explains, how they (students) think and solve problems at the same time, how teacher instructs them, are important considerations. In order to increase the efficiency of *students listening*, it is more useful to focus on improving the simultaneous efficiency of both *students listening* and other pedagogical behaviours rather than insist on listening and accepting passively, independently of the other behaviours on which it is dependent.

The research of Chinese Lexicon Structure building and CPB in this study could be extended through comparison with similar analyses in other countries. Investigation of the characteristics of CPBs in different cultural contexts, focusing on the role orientation of teachers and students in different CPBs, could contribute to significant insights into the classroom teaching and learning of mathematics internationally.

The Chinese Lexicon will change and the hotspots of mathematics education research and teachers' concerns in their classrooms will also change over time, so we consider it might be possible and useful to renew either these key terms or the entire lexicon every 5–10 years to see the changes of teachers' naming system and bring new insights to the mathematics education research in China.

Acknowledgements

During our work, some experienced middle-school teachers, coaches as well as university professors helped us enormously and we thank them for their support and involvement. This project was supported partially by Beijing Normal University.

References

Cao, Y. M., & He, C. (2009). 初中数学课堂师生互动行为主体类型研究——基于LPS项目课堂录像资料 [Research of the type of teacher-student interaction behaviour subject in junior middle school mathematics classroom in the LPS video data]. 数学教育学报, 18(5), 38–41.

Cao, Y. M. & Yu, G. W. (2017). 中学数学课堂教学行为关键性层级研究 [Research on the critical level of the critical pedagogical behaviours in the middle school mathematics classroom]. 数学教育学报, 26(1), 1–6.

Chemla, K., & Guo, S. (Eds.). (2004). *Les neuf chapitres: le classique mathématique de la Chine ancienne et ses commentaires* [The nine chapters: Classic mathematics of ancient China and its commentaries]. Paris: Dunod.

Gu, M. Y. (1992). 教育大辞典 [*Educational dictionary*]. Shanghai: Shanghai Education Press.

Khandelwal, K. A. (2009). Effective teaching behaviours in the college classroom: A critical incident technique from students' perspective. *International Journal of Teaching and Learning in Higher Education*, 21(3), 299–309.

Li, Y., Huang, R., Bao, J., & Fan, Y. (2011). Facilitating mathematics teachers' professional development through ranking and promotion practices in the Chinese mainland. In N. Bednarz, D. Fiorentini, & R. Huang (Eds.), *The professional development of mathematics teachers: Experiences and approaches developed in different countries* (pp. 82–92). Canada: Ottawa University Press.

Perreault, W. D., Jr., & Leigh, L. E. (1989). Reliability of nominal data based on qualitative judgments. *Journal of Marketing Research*, 26(2), 135–148.

Tripp, D. (1993). *Critical incidents in teaching: Developing professional judgement*. New York, NY: Routledge.

Wang, J. (Ed.). (2013). 中国数学教育:传统与现实 [*Mathematics education in China: Tradition and reality*]. Nanjing: Jiangsu Education Publishing House.

Yang, Y. D., & Wang X. (2015). 运用关键性教学事件分析支撑中国式数学课例研究 [Employing crucial incidents analysis to support Chinese mathematical Lesson Study]. 数学教育学报, 24(3), 40–47.

Zhang, D., Li, S., & Tang, R. (2005). 中国大陆的"双基"数学教学[The "Two basics": Mathematics teaching and learning in Mainland China]. In L. Fan, N. Wong, J. Cai, & S. Li (Eds.), 华人如何学习数学[*How Chinese learn mathematics*] (pp. 189–207). Nanjing: Jiangsu Educational Press.

7

CHINESE LEXICON

Yiming Cao, Guowen Yu and Lianchun Dong

A Chinese Lexicon

There are 123 terms in the Chinese Lexicon that were documented after collaborative work among researchers, experienced teachers and coaches. Most of the terms (102/123) are widely used by middle-school Mathematics teachers in their practice or communication, 12 terms are less frequently used, and nine others are seldom used but of sufficient pedagogical significance to be included in the lexicon (see also *Chapter 6*).

In this chapter, we present our lexical terms with Chinese characters, Pinyin, the closest English translation and a description in English. The first table (see Table 7.1) lists each of the terms organised alphabetically by Pinyin. The second table (see Table 7.2) includes two columns, the first column lists the term and the second column gives its description.

The Chinese Lexicon was collaboratively developed by the researchers and teachers from different areas of China, making it a nationally-accepted lexicon. It's the first time that Chinese mathematics researchers give attention to teachers' structural language and practice in their classes, this offers insights into the pedagogical principles that historically underlie Chinese classroom practice. In addition, the Chinese Lexicon will help assist: teachers in their professional development; teacher educators' in their course development; and policymakers' decisions about curriculum and classroom teaching.

TABLE 7.1 Chinese Lexicon – Terms.

按顺序回答 **an shun xu hui da** (answer in turn)	板书 **ban shu** (blackboard writing)	暴露问题 **bao lu wen ti** (expose errors)	背诵 **bei song** (reciting)
变式教学 **bian shi jiao xue** (teaching with variation)	变式训练 **bian shi xun lian** (training with variation)	布置家庭作业 **bu zhi jia ting zuo ye** (assign homework)	布置课堂练习题 **bu zhi ke tang lian xi ti** (assign class exercise)
布置任务 **bu zhi ren wu** (assign a task)	猜想 **cai xiang** (guess)	操作 **cao zuo** (operating)	查漏补缺 **cha lou bu que** (fill the gaps)
出示例题 **chu shi li ti** (show exemplary tasks)	出示练习 **chu shi lian xi** (show exercises)	创设情境 **chuang she qing jing** (creating situations)	订正作业 **ding zheng zuo ye** (revise work)
独立探究 **du li tan jiu** (independent exploration)	发现问题 **fa xian wen ti** (generate problems)	反思 **fan si** (reflection)	分析问题 **fen xi wen ti** (analyse problems)
分组 **fen zu** (grouping)	复习 **fu xi** (review)	概括 **gai kuo** (outline)	概念讲解 **gai nian jiang jie** (concept explanation)
感悟 **gan wu** (perception)	个别辅导 **ge bie fu dao** (individual tutoring)	个别回答 **ge bie hui da** (individual student answers the question)	个别提问 **ge bie ti wen** (individual questioning)
个人活动 **ge ren huo dong** (individual activity)	观察 **guan cha** (observation)	合作探究 **he zuo tan jiu** (cooperative exploration)	画龙点睛 **hua long dian jing** "draw dragon dot eye" (finishing touch)
活动前准备 **huo dong qian zhun bei** (preparation before the task)	记笔记 **ji bi ji** (take notes)	记忆 **ji yi** (memorising)	检查 **jian cha** (checking)
检查作业 **jian cha zuo ye** (check homework)	矫正 **jiao zheng** (correction)	教师表扬 **jiao shi biao yang** (teacher's praise)	教师答疑 **jiao shi da yi** (teacher answers (student's confusion))
教师读题 **jiao shi du ti** (teacher reading questions)	教师反馈 **jiao shi fan kui** (teacher's feedback)	教师反问 **jiao shi fan wen** (teacher inquires)	教师鼓励 **jiao shi gu li** (teacher encouraging)

教师讲解 **jiao shi jiang jie** (teacher's explanation)	教师讲评 **jiao shi jiang ping** (teacher's comment)	教师肯定 **jiao shi ken ding** (teacher affirms)	教师批评 **jiao shi pi ping** (teacher's criticism)
教师评价 **jiao shi ping jia** (teacher's assessing)	教师启发 **jiao shi qi fa** (teacher's inspiration)	教师提问 **jiao shi ti wen** (teacher questioning)	教师提醒 **jiao shi ti xing** (teacher reminding/ refocus/ focus)
教师小结 **jiao shi xiao jie** (teacher's brief summary)	教师巡视 **jiao shi xun shi** (teacher walks and sees)	教师演示 **jiao shi yan shi** (teacher's presentation)	教师要求 **jiao shi yao qiu** (teacher's requirement)
教师诊断 **jiao shi zhen duan** (teacher diagnosing)	教师追问 **jiao shi zhui wen** (teacher makes a detailed inquiry)	教师自答 **jiao shi zi da** (teacher answers herself)	解决问题 **jie jue wen ti** (solve problems)
课堂管理 **ke tang guan li** (classroom management)	课堂练习 **ke tang lian xi** (seat work)	课堂生成 **ke tang sheng cheng** (lesson happens)	课堂总结 **ke tang zong jie** (class summarising)
理解 **li jie** (understanding)	例题讲解 **li ti jiang jie** (examples explanation)	默写 **mo xie** (dictation)	判断 **pan duan** (judgement)
铺垫 **pu dian** (bridging)	强调易错点 **qiang diao yi cuo dian** (stressing common mistakes)	求解 **qiu jie** (solving)	全班回答 **quan ban hui da** (all the class answer)
全班提问 **quan ban ti wen** (questioning the class)	上课仪式/组织教学 **shang ke yi shi/zu zhi jiao xue** (class beginning ceremony)	生生互动 **sheng sheng hu dong** (student–student interaction)	生生互评 **sheng sheng hu ping** (student-and-student assessment)
师生共答 **shi sheng gong da** (teacher and students answer together)	师生共同探讨 **shi sheng gong tong tan tao** (teacher and students together explore)	师生互动 **shi sheng hu dong** (teacher–student interaction)	实验 **shi yan** (trail)
梳理知识体系 **shu li zhi shi ti xi** (sort out the knowledge system)	随堂测试 **sui tang ce shi** (class test)	提出变式 **ti chu bian shi** (propose variation)	提出教学目标 **ti chu jiao xue mu biao** (stating teaching objectives)

(Continued)

TABLE 7.1 (*Continued*)

提出问题 **ti chu wen ti** (propose problems)	体验 **ti yan** (experience)	挑战 **tiao zhan** (challenging)	调动积极性 **diao dong ji ji xing** (arouse enthusiasm)
突发事件处理 **tu fa shi jian chu li** (emergency treatment)	拓展 **tuo zhan** (extension)	维持秩序 **wei chi zhi xu** (maintain order)	习题讲解 **xi ti jiang jie** (exercises explanation)
下课仪式 **xia ke yi shi** (class ending ceremony)	衔接 **xian jie** (connecting)	小组辅导 **xiao zu fu dao** (group tutoring)	小组汇报 **xiao zu hui bao** (group report)
小组活动 **xiao zu huo dong** (group work)	小组讨论 **xiao zu tao lun** (group discussion)	新课导入 **xin ke dao ru** (introduction of new lesson)	学法指导 **xue fa zhi dao** (learning methods guiding)
学生读题 **xue sheng du ti** (student reading questions)	学生反馈 **xue sheng fan kui** (student response)	学生回答 **xue sheng hui da** (student answer)	学生讲题 **xue sheng jiang ti** (student's explanation)
学生纠错 **xue sheng jiu cuo** (student correcting)	学生评价 **xue sheng ping jia** (student assessing)	学生上黑板/学生板演 **xue sheng shang hei ban/xue sheng ban yan** (student writes on the blackboard)	学生思考 **xue sheng si kao** (student's thinking)
学生探究 **xue sheng tan jiu** (student's exploration)	学生提出疑问 **xue sheng ti chu yi wen** (students proposing confusion)	学生提问 **xue sheng ti wen** (student questioning)	学生听讲 **xue sheng ting jiang** (students listening)
学生展示 **xue sheng zhan shi** (student show)	学生自评 **xue sheng zi ping** (student self-assessment)	学生做题 **xue sheng zuo ti** (students do exercise)	引起学生注意 **yin qi xue sheng zhu yi** (attract students' attention)
应用 **ying yong** (application)	预习 **yu xi** (preview)	证明 **zheng ming** (proof)	知识点讲解 **zhi shi dian jiang jie** (knowledge points explanation)
自学 **zi xue** (self-learning)	作图 **zuo tu** (drawing)	作业展示 **zuo ye zhan shi** (work show)	

TABLE 7.2 Chinese Lexicon – Terms with descriptions.

Term	Description
按顺序回答 **an shun xu hui da** (answer in turn)	The teacher asks questions to students in turn, for example, the teacher asks students in a row the answers to questions on an examination.
板书 **ban shu** (blackboard writing)	Teacher writes a title, draws graphs, writes the process of solving exemplary tasks or key mathematics facts, procedures, rules, or theorems on the blackboard. It can represent both, the teacher's behavior of writing on the blackboard and the information the teacher writes on the blackboard.
暴露问题 **bao lu wen ti** (expose errors)	The students are asked or invited to show or present their incomplete thoughts.
背诵 **bei song** (reciting)	It is one of the ways to test results of memorising; students present orally the concepts, theorems, formulas, and so on they have memorised.
变式教学 **bian shi jiao xue** (teaching with variation)	A teaching method where the teacher transforms the conditions of the original example or varies the questions asked of the students. This is undertaken with the purpose of strengthening and deepening students' understanding of theory and the execution of examples.
变式训练 **bian shi xun lian** (training with variation)	The teacher transforms the conditions or questions of the original problem and asks the students to do the same in order to strengthen their understanding of the content, their grasp of the problem, and expand their knowledge.
布置家庭作业 **bu zhi jia ting zuo ye** (assign homework)	The teacher assigns some tasks that students should do after class, which include tasks for reviewing what they have just learned and sometimes tasks for previewing the new content.
布置课堂练习题 **bu zhi ke tang lian xi ti** (assign class exercise)	The teacher gives some problems to students to finish in class.
布置任务 **bu zhi ren wu** (assign a task)	The teacher shows the prepared exemplary tasks, exercises or explorations to the whole class through oral language, blackboard, multimedia or the learning guide and raises problems that need to be settled.
猜想 **cai xiang** (guess)	The teacher or students predict the possible results before obtaining the conclusion.
操作 **cao zuo** (operating)	Dealing with the elements of knowledge information (such as the key elements of the concept, the conditions of the problem, the elements of the graphic and physical objects, and so on), removing and piecing and trying to find out the related structure.

(Continued)

TABLE 7.2 (*Continued*)

Term	Description
查漏补缺 **cha lou bu que** (fill the gaps)	The teacher helps students find gaps in the original knowledge system and supplement what they do not know well to complete their knowledge structure.
出示例题 **chu shi li ti** (show exemplary tasks)	The teacher shows the prepared exemplary tasks on the blackboard, in the teaching material, or through multimedia.
出示练习 **chu shi lian xi** (show exercises)	The teacher shows the prepared exercises on the blackboard, in the teaching material, or through multimedia.
创设情境 **chuang she qing jing** (creating situations)	The teacher creates learning environment or scenes for students with the help of real life, teaching aids, language description, and so on.
订正作业 **ding zheng zuo ye** (revise work)	After the teacher checks or marks students' homework or seatwork, the students correct their mistakes.
独立探究 **du li tan jiu** (independent exploration)	The individual student explores some particular learning content on his own without the teacher's explanation.
发现问题 **fa xian wen ti** (generate problems)	Problems emerge in the process of student learning and identified by the students or teacher.
反思 **fan si** (reflection)	The teacher or students give further consideration to the knowledge, the methods of problem-solving and the teaching or learning methods in order to summarise experience and lessons as well as to promote deeper understanding.
分析问题 **fen xi wen ti** (analyse problems)	Examination of the problems that students meet in their process of learning to prompt the problem-solving.
分组 **fen zu** (grouping)	The whole class is divided into groups according to some criteria (e.g. a group may be formed from students who are already sitting in proximity to each other) and each group is assigned a learning/exploratory activity.
复习 **fu xi** (review)	Students re-learn what they have learnt through answering questions or doing exercises under the teacher's guidance in order to consolidate what they have learnt before and have the learnt content included in their knowledge structures.
概括 **gai kuo** (outline)	Refine and associate the knowledge, the thinking, and methods of solving problems which have been learned in a period of time to construct a refined knowledge structure.
概念讲解 **gai nian jiang jie** (concept explanation)	The teacher shows and explains the concept to the students.

Term	Description
感悟 **gan wu** (perception)	Students experience and finally make sense of the knowledge in the process of learning.
个别辅导 **ge bie fu dao** (individual tutoring)	The teacher gives tutoring and guidance to a particular student on knowledge, psychology, and behavior.
个别回答 **ge bie hui da** (individual student answers the question)	An individual student answers a question the teacher asks.
个别提问 **ge bie ti wen** (individual questioning)	Teacher asks an individual student a particular question. The question can be asked before or after the student is named.
个人活动 **ge ren huo dong** (individual activity)	The student does some operations individually, for example, paper cutting and coin tossing.
观察 **guan cha** (observation)	The students examine the information (text or graphics information) in learning materials, and try to find the information they need.
合作探究 **he zuo tan jiu** (cooperative exploration)	Students explore in groups or in pairs some particular learning content without the teacher's prior explanation.
画龙点睛 **hua long dian jing** "draw dragon dot eye" (finishing touch)	The teacher or students give some constructive views on a certain problem or knowledge, and it makes the teaching more successful.
活动前准备 **huo dong qian zhun bei** (preparation before the task)	Students organise their learning materials (e.g. textbook and note book), learning instruments (e.g. ruler compass) as well as mental preparation (e.g. mental focus, mental preparedness) for the learning tasks ahead.
记笔记 **ji bi ji** (take notes)	On the basis of the teacher's explanation and students' own thinking, the students write what the teacher teaches or the conclusions and key mathematics facts, procedures, rules, or theorems they have explored under the teacher's guidance in their notebooks in the process of learning.
记忆 **ji yi** (memorising)	Students strengthen concepts, theorems, and formulas, in their minds.
检查 **jian cha** (checking)	The teacher or students check students' progress in solving the tasks or what the students show, in order to know if the students have finished and evaluate the quality of them.

(*Continued*)

TABLE 7.2 (*Continued*)

Term	Description
检查作业 **jian cha zuo ye** (check homework)	The teacher checks students' homework in order to know whether they have finished it and whether they had problems with the homework.
矫正 **jiao zheng** (correction)	The teacher corrects the mistakes in problem-solving or content understanding.
教师表扬 **jiao shi biao yang** (teacher's praise)	The teacher speaks highly of the students' good behaviour or of the valuable results they have produced through exploratory work.
教师答疑 **jiao shi da yi** (teacher answers (student's confusion)	The teacher answers the questions or confusion students have.
教师读题 **jiao shi du ti** (teacher reading questions)	The teacher reads the problem, and at the same time, the teacher can write the problem on the blackboard or show the problem with the aid of multimedia.
教师反馈 **jiao shi fan kui** (teacher's feedback)	Teacher reacts to or gives some comments on a student's behavior or performance, helping students to be clear about their success or failure, or provides some guidance to them and encourages them to think more.
教师反问 **jiao shi fan wen** (teacher inquires)	The teacher has some doubts about the students' answer, and asks the students to rethink and answer again.
教师鼓励 **jiao shi gu li** (teacher encouraging)	The teacher inspires students to persevere when their confidence is shaken or the teacher praises them and encourages them to continue to make effort when they have finished the difficult learning tasks well.
教师讲解 **jiao shi jiang jie** (teacher's explanation)	The teacher explains particular problems, concepts, theorems, or principles systematically in order to reveal the relationships, structures, and essence of the problems.
教师讲评 **jiao shi jiang ping** (teacher's comment)	The teacher explains and gives some comments on students' finished tasks such as their homework and their answers to teacher's questions.
教师肯定 **jiao shi ken ding** (teacher affirms)	The teacher agrees with students' behaviors such as their response to a question.
教师批评 **jiao shi pi ping** (teacher's criticism)	The teacher identifies the student's misconduct in relation to poor behaviour (e.g. playing on a mobile phone, not paying attention).

Term	Description
教师评价 **jiao shi ping jia** (teacher's assessing)	The teacher comments on students' learning approaches and their written work.
教师启发 **jiao shi qi fa** (teacher's inspiration)	The teacher sparks and motivates students' thinking through coaching or hinting, in order to encourage the students to find out the nature and structure of the problems.
教师提问 **jiao shi ti wen** (teacher questioning)	The teacher asks a question and an individual student or the whole class can be asked to answer the question.
教师提醒 **jiao shi ti xing** (teacher reminding/ refocus/focus)	The teacher tries to focus students' attention, when finding students distracted, while he is teaching the key part of the lesson.
教师小结 **jiao shi xiao jie** (teacher's brief summary)	The teacher summarises skills, methods, principles of problem-solving, key points that need more attention as well as the mathematics facts, procedures, rules or theorems, and their relations. It usually happens after an explanation or problem-solving.
教师巡视 **jiao shi xun shi** (teacher walks and sees)	The teacher walks into the classroom and observes the students to record students' problems, get some information about students' learning and activities when they are working on their learning tasks.
教师演示 **jiao shi yan shi** (teacher's presentation)	The teacher shows the content, such as graphs, to the class through teaching equipment, for example, a projector, Geometer's Sketchpad, graphic calculators.
教师要求 **jiao shi yao qiu** (teacher's requirement)	The teacher sets some detailed objectives for the students' learning activities, such as a limitation of the time for completing the task.
教师诊断 **jiao shi zhen duan** (teacher diagnosing)	The teacher makes a judgement about students' psychological state and learning effects through the process and results of their learning.
教师追问 **jiao shi zhui wen** (teacher makes a detailed inquiry)	The teacher has the impression that the students have not fully understood or lacked key information in their answers and asks more follow-up questions.
教师自答 **jiao shi zi da** (teacher answers herself)	The teacher answers her own question.
解决问题 **jie jue wen ti** (solve problems)	Analysis of the problems that students meet in the process of learning and then solving them.

(Continued)

TABLE 7.2 (*Continued*)

Term	Description
课堂管理 **ke tang guan li** (classroom management)	The teacher or students' pedagogical behaviour to support the teaching aims, regulate the interpersonal relationships, improve the teaching environment as well as guide students learning.
课堂练习 **ke tang lian xi** (seat work)	The teacher gives students some exercises to consolidate the exemplary tasks they have just learnt or to review the previous knowledge. The students try to complete the exercises independently in their notebooks.
课堂生成 **ke tang sheng cheng** (lesson happens)	The teacher adjusts his/her teaching plans by making instructional use of an unexpected event, which is beyond the intended plan for the lesson.
课堂总结 **ke tang zong jie** (class summarising)	The teacher or students lead the summary of the content for the whole class. The generalisation will help the students integrate new knowledge, skills, and thinking methods into their existing knowledge structures. This will help students to enhance their understanding and form a personal structured knowledge system.
理解 **li jie** (understanding)	Students identify the generation of knowledge and its internal logic and grasp the connotation of knowledge in the process of learning.
例题讲解 **li ti jiang jie** (examples explanation)	The teacher shows the exemplary task and explains it to the whole class.
默写 **mo xie** (dictation)	It is one of the ways to test results of memorising that students write the concepts, theorems, formulas they have memorised in their notebooks.
判断 **pan duan** (judgement)	Choosing whether the solution of the problem is right or wrong, or true or false.
铺垫 **pu dian** (bridging)	The teacher teaches some content, for the purpose of introducing the subsequent content which is usually the key content of one lesson.
强调易错点 **qiang diao yi cuo dian** (stressing common mistakes)	The teacher reminds the students of the points where they usually make mistakes on understanding or performing, to help them to avoid mistakes.
求解 **qiu jie** (solving)	The teacher or the students finish the process of computing the final answer of the problem.
全班回答 **quan ban hui da** (all the class answer)	The class gives a choral response to the teacher's question after the teacher raises it.

Term	Description
全班提问 **quan ban ti wen** (questioning the class)	Teacher asks a particular question to the whole class.
上课仪式/组织教学 **shang ke yi shi/zu zhi jiao xue** (class beginning ceremony)	This is the beginning of the class and is indicated by a regular pattern of behaviour. The teacher says, "The class begins." The class captain says, "Stand up." Then the teacher states, "Good morning everyone." The students respond, "Good morning teacher," in unison. The act of teachers and students bowing to each other accompanies these interactions, and it signals the formal beginning of class.
生生互动 **sheng sheng hu dong** (student–student interaction)	Students explore and communicate with each other on a certain problem.
生生互评 **sheng sheng hu ping** (student-and-student assessment)	Students assess each other.
师生共答 **shi sheng gong da** (teacher and students answer together)	The teacher and the whole class answer a question together, which is usually asked by the teacher but sometimes by a student.
师生共同探讨 **shi sheng gong tong tan tao** (teacher and students together explore)	The teacher and students study an inquiry question together.
师生互动 **shi sheng hu dong** (teacher–student interaction)	The teacher and students investigate, make discoveries and communicate together about a certain problem.
实验 **shi yan** (trail)	The teacher or students verify the conclusion or obtain a conclusion through activities such as paper cutting and coin tossing, or through virtual operations such as using the Geometer's Sketchpad and the graphic calculator as well as through other constructions.
梳理知识体系 **shu li zhi shi ti xi** (sort out the knowledge system)	After having finished a section or chapter of a textbook, the teacher or the students make a summary of the learnt knowledge in order to form an integral knowledge structure.
随堂测试 **sui tang ce shi** (class test)	Teacher organises a short quiz in class to test how the students master the knowledge they have learnt. The students do it orally or with pen and paper.

(Continued)

TABLE 7.2 (*Continued*)

Term	*Description*
提出变式 **ti chu bian shi** (propose variation)	The teacher or students make changes to the conditions of the original question in order to form new problems or proposes additional questions. These questions/problems are then solved in order to cultivate exploration-minded learners.
提出教学目标 **ti chu jiao xue mu biao** (stating teaching objectives)	The teacher outlines the learning content of the lesson and sets detailed objectives for the class, which includes detailed information about the following three dimensions: knowledge and content; process and methods; and emotional attitudes and values. These dimensions are identified in the curriculum standards.
提出问题 **ti chu wen ti** (propose problems)	The teacher puts forward his/her prepared problems that need to be solved, or clarify the problems students find in their process of learning.
体验 **ti yan** (experience)	Students experience the acquisition of knowledge in the process of learning.
挑战 **tiao zhan** (challenging)	The teacher puts forward more difficult tasks and asks the students to solve them.
调动积极性 **diao dong ji ji xing** (arouse enthusiasm)	The teacher induces students to have a desire to learn certain content by stimulating their interest, arousing, or strengthening their motivation.
突发事件处理 **tu fa shi jian chu li** (emergency treatment)	The teacher copes with the unexpected non-teaching affairs that happen in the class.
拓展 **tuo zhan** (extension)	The teacher adds more exercises, exemplary tasks or other teaching resources on the basis of the original teaching schedule in order to help students to deepen understanding, consolidate the knowledge, or widen the knowledge background.
维持秩序 **wei chi zhi xu** (maintain order)	The teacher excludes adverse factors such as students' misbehaviour in the class in order to complete the teaching.
习题讲解 **xi ti jiang jie** (exercises explanation)	Students solve the problem, and the teacher explains to other students on the basis of the students' thinking.
下课仪式 **xia ke yi shi** (class ending ceremony)	This is the ending of the class and is indicated by a regular pattern of behavior. The teacher says, "The class is over." The class captain says, "Stand up." Then the teacher and students say goodbye to each other and it signals the formal the ending of class.
衔接 **xian jie** (connecting)	The teacher and students establish the connection between the current knowledge and the knowledge that the students have learned or will learn soon.

Term	Description
小组辅导 **xiao zu fu dao** (group tutoring)	The teacher gives some tutoring to a group or answers a question raised by a student from that group.
小组汇报 **xiao zu hui bao** (group report)	An individual student or several ones from a group present their results, which they worked out in group activities or explorations, to the whole class.
小组活动 **xiao zu huo dong** (group work)	Students complete some operations together within a group, for example, paper cutting and coin tossing.
小组讨论 **xiao zu tao lun** (group discussion)	All the students are divided into different groups according to a certain standard and the students in one group discuss or debate together the tasks or explorations under the guidance of the teacher.
新课导入 **xin ke dao ru** (introduction of new lesson)	The teacher introduces the content of the lesson with a selection of the following activities: reviewing previous work; questioning students on previous work; using physical manipulatives; playing games; or proposing a scene.
学法指导 **xue fa zhi dao** (learning methods guiding)	The teacher helps students owith learning methods, such as telling students to pay attention to summarising the rules.
学生读题 **xue sheng du ti** (student reading questions)	Students read the problem and get the key information or the details.
学生反馈 **xue sheng fan kui** (student response)	A student corrects other's mistakes or makes a positive or negative evaluation for others' approaches, or students react to the teacher's words or actions.
学生回答 **xue sheng hui da** (student answer)	The teacher asks a question and students answer orally. Sometimes an individual student answers and then other students can add some comments, sometimes the whole class answers together.
学生讲题 **xue sheng jiang ti** (student's explanation)	An individual student explains his/her findings or the thinking involved in solving the problems, at the front of the classroom or from their seat.
学生纠错 **xue sheng jiu cuo** (student correcting)	Students point out the mistakes made by the teacher during the blackboard writing or explanation as well as the mistakes made by other students.
学生评价 **xue sheng ping jia** (student assessing)	Students make a positive or negative judgment in relation to the teacher (seldom) or other students.

(Continued)

TABLE 7.2 (*Continued*)

Term	Description
学生上黑板/学生板演 **xue sheng shang hei ban/ xue sheng ban yan** (student writes on the blackboard)	An individual student solves the problem and writes it on the blackboard.
学生思考 **xue sheng si kao** (student's thinking)	Students ponder on what they are going to learn in order to get some ideas or conclusions or to understand something.
学生探究 **xue sheng tan jiu** (student's exploration)	Students do some explorations on some particular learning contents without the teacher's explanation. It can be an independent exploration or group work or desk mate work.
学生提出疑问 **xue sheng ti chu yi wen** (students proposing confusion)	Students raise questions about the teacher's explanation, and the questions request the teacher to rethink and to explain again.
学生提问 **xue sheng ti wen** (student questioning)	The student asks questions he comes across in the learning process to the teacher or to other students in order to get answers.
学生听讲 **xue sheng ting jiang** (students listening)	Students pay attention to the teacher's oral presentation, blackboard writing or multimedia demonstration during the process of teacher explanation.
学生展示 **xue sheng zhan shi** (student show)	Students explain problems or a knowledge point to all the class at the front, with the help of multimedia, or in their seats, sometimes the show is in the form of group reporting.
学生自评 **xue sheng zi ping** (student self-assessment)	Students score or judge their own answers true or false according to the answer.
学生做题 **xue sheng zuo ti** (students do exercise)	Students try to complete the exemplary tasks or exercises assigned by the teacher on their own.
引起学生注意 **yin qi xue sheng zhu yi** (attract students' attention)	The teacher tries to attract students attention, for example, by raising his/her voice or quieting down suddenly.
应用 **ying yong** (application)	The concepts, theorems and formulas are further applied and consolidated in the problems.
预习 **yu xi** (preview)	Students learn the knowledge to be learnt in the next lesson when they have finished the objectives of the class and there's still some time left. It often happens out of class.

Term	Description
证明 **zheng ming** (proof)	The teacher or the students give evidence by deduction or reasoning for their correct conclusion or guess.
知识点讲解 **zhi shi dian jiang jie** (knowledge points explanation)	The teacher explains mathematics facts, procedures, rules or theorems that students should learn.
自学 **zi xue** (self-learning)	Students learn a knowledge point or a problem-solving strategy fully by themselves before the teacher explains.
作图 **zuo tu** (drawing)	The drawing of graphics that meet the requirements with a ruler, set square or compass usually.
作业展示 **zuo ye zhan shi** (work show)	The teacher or students present students' work (problem-solving models and questions) and thinking process.

Acknowledgements

During our work, middle-school experienced teachers, coaches as well as university professors helped us enormously and we thank them for their support and involvement. This project was supported by a Discovery Grant from the Research Council of the Australian Government (ARC-DP140101361) and partially by the International Joint Research Project of the Faculty of Education, Beijing Normal University.

Bibliography

Presentations and Papers

The work of the Chinese national team has been presented locally and internationally at the following meetings and conferences:

- Chinese Mathematical Education Academic Conference 2018
- European Conference on Educational Research (ECER) 2016
- International Group for the Psychology of Mathematics Education (PME) Annual Conference 2016
- The Second Chinese Mathematics Education Conference and Doctoral Forum on Mathematics Education 2016

A selection of peer-reviewed conference and journal publications from the Chinese research team include:

Clarke, D.J., Mesiti, C., Cao, Y., & Novotna, J. (2017). The lexicon project: examining the consequences for international comparative research of pedagogical naming systems from different cultures. In T. Dooley, & G. Gueudet (Eds.), *Proceedings of the Tenth Congress of the European Society for Research in Mathematics Education* (pp. 1610–1617). Dublin, Ireland: ERME.

Cao, Y. M. & Yu, G. W. (2017). 中学数学课堂教学行为关键性层级研究 [Research on the critical level of the critical pedagogical behaviours in the middle school mathematics classroom]. *数学教育学报*, 26(1), 1–6.

Mesiti, C., Clarke, D.J., Roan, K., Hollingsworth, H., Cao, Y., Yu, G., Novotna, J., Zlabkova, I., & Dobie, T. (2016). Discourse about the mathematics classroom. In C. Csikos, A. Rausch, & J. Szitányi (Eds.), *Mathematics Education: How to Solve It? PME40* (pp. 357–363). Szeged, Hungary: PME.

8

UNDERSTANDING EACH OTHER WHEN SPEAKING ABOUT THE MATHEMATICS LESSON

The professional Czech Lexicon

Alena Hošpesová, Hana Moraová, Jarmila Novotná and Iva Žlábková

Studying the terminology of the didactics of mathematics

Both in theory and practice, we often need to reflect on pedagogical reality. Teachers, principals, inspectors, researchers, and pupils often try to describe "what is really going on in a classroom," that is, what was observed as going on at a given time and place. Janík and Slavík (2007) emphasise that to grasp and describe "what is actually happening," that is, identifying didactical facts in a lesson in a meaningful way, is the outcome of intellectual work based on observer's/actor's prior experience. Using terms from pedagogy allows us to interpret facts in a meaningful way and also to communicate them. It can be said that the used terminology forms a common interpretative framework. This implies we should ask how the terminology a person is able to use actively affects their perception of pedagogical reality. Rezat and Sträßer (2012) relate different perception to different actors in education and discuss conventions about what it means to be doing mathematics, to be a mathematics teacher or a pupil studying mathematics, about school as an institution, about textbooks and their structure and form, public image of mathematics in a society, and so on. All these have an impact not only on mathematics lessons but also on discourse about these lessons. This may result in a situation when researchers and teachers from different countries may fail to understand each other, since the terminology they use refers to very different situations (compare with Clarke, 2009).

All the above-mentioned arguments are reasons why the Czech Republic entered the International Classroom Lexicon Project. Our chapter starts with brief characteristics of the Czech system of education and shows the process of creation of the Czech Lexicon in conditions of the Czech Republic and its system of education.

The Czech system of education and its influence on the Czech Lexicon

The process of creation of the Czech Lexicon was supported with the analysis of video recordings from lessons of mathematics at lower secondary school level. Thus, to be able to understand the whole process of creation of the Czech Lexicon, it is essential to have some knowledge of the Czech system of education and thus the conditions in which the

Czech Lexicon came to exist. That is why we start with a brief description of the system of education and teacher training in the Czech Republic.

In the Czech Republic, compulsory education (named "basic school") lasts nine years and is divided into two levels. Primary level covers the 1st to 5th year of schooling, lower secondary education 6th to 9th year. Children enter compulsory schooling usually at the age of six. Education in the Czech Republic conforms to the *Framework Educational Programme for Basic Education* (2017, in Czech Rámcový vzdělávací program pro základní vzdělávání[1]). Based on this reference, schools create their own school educational programmes (in Czech Školní vzdělávací program). Even though each school has its own unique curricular document and meets all the prescribed educational outcomes there is also freedom for schools to specialise in their own way. These additional lessons are used for different subjects in different educational areas at each school. Schools may, for example, focus on extending foreign language, mathematics, art or science education.

Mathematics is considered to be a major subject. It involves at least 16% of all lessons in primary school and 12% in lower secondary. The number of lessons at each level can be and often is increased with use of the additional lessons in the timetable.

Teachers and teacher education

To be fully qualified a teacher must finish five years of university studies. The content of the study programmes is considerably different at different universities. Universities have a lot of autonomy in construction of their study programmes. However, all universities respect a binding framework given by the Ministry of Education. Teacher education includes study of the selected subjects, study of subject didactics as well as common core subjects (pedagogical and psychological disciplines). An important part of university studies is teaching practice, which has various forms at different stages of undergraduate studies. Teaching practice includes observing lessons of excellent teachers and students' own teaching.

Pre-service primary-school teachers are educated in such a way that they will be able to teach all subjects that are compulsory at primary-school level. The common practice in Czech schools is that primary-school classes have their class teacher for the majority of lessons. However, the number of lessons of classes do not necessarily correspond to the number of lessons a teacher teaches in one week and thus in practice most teachers specialise somehow when they start their work at school. Some of them will teach more art lessons in more classes, other English, music, even at lower secondary school level.

Teacher preparation for lower and upper secondary schools was divided into two stages: three-year Bachelor degree and a follow-up two-year Master degree study programme in the teaching profession. Mathematics is traditionally studied with biology, physics, chemistry or informatics. Some universities nowadays allow its students to decide on the combination of studied subjects, other have a set of given possible combinations. Graduates of Bachelor degrees are not qualified teachers. Graduates of the Master level study are fully qualified teachers for lower and/or upper secondary schools. They are expected to have sufficient expertise in the studied disciplines and have developed pedagogical-psychological and subject-didactical competences.

The profession of the teacher is regarded as important by society (Walterová, 2001; Vašutová, 2007). Teachers are assessed primarily by head teachers and the institute Czech School Inspection. Influence is also exerted by school boards which is constituted of members of the parent community, of pedagogical staff of the school and representatives of the establisher of the school (municipalities or regional governments). However, the average income of a teacher is only very slightly above the average salary in the country and teachers are the worst paid university graduates. In consequence, schools (especially in the capital Prague) are short of teachers, especially primary teachers, teachers of mathematics and sciences.

Planning for the construction of Czech Lexicon

When constructing the Czech Lexicon, we faced similar obstacles as all subject didactics including didactics of mathematics in the Czech Republic. The language used in the subject didactics should form the basis for descriptions of lessons from both a researcher's and a teacher's perspective, which is, however, not the case. Currently, subject didactics in the Czech Republic are in the process of reconstitution. When creating new theories, they focus on developing onto-didactical and psycho-didactical dimensions of the domain and of research in the subject didactics as well as in confronting different directions of didactical thinking. These efforts can be expected to precise the used terminology (Stuchlíková & Janík, 2015).

Project team

The first step when we entered the project was to make the decision about who should be in the project team to balance the everyday school practice with the world of research. The lexicon must be built on solid terminology used in mathematics education but must at the same time be comprehensible to practising teachers. This led to the decision of building a team made from researcher and teachers (mathematics educators from University of South Bohemia České Budějovice and Charles University, a researcher in pedagogy from University of South Bohemia, a mathematics teacher and a language teacher). The researcher in pedagogy brings her knowledge of terminology used in pedagogy in general. The presence of the language teacher was decided upon for two reasons: to focus on types of discourse and communication in mathematics lessons from the linguistic point of view and also to translate Czech lexicon into English for international use. The mathematics teacher was a part of the team because of his contact with the everyday reality of schools.

Aim and stages of creation of the Czech Lexicon

The aim of our work and effort was creation and validation of a list of comprehensible terms that are used by the community of mathematics educators and mathematics teachers to describe what is happening in a mathematics lesson. This corresponds to the aim of The International Classroom Lexicon Project (Clarke & Mesiti, 2010). From the very beginning, we were planning gradual creation, validation, and modification of a list of terms.

The Czech Lexicon was created in six stages, in which methodological procedures were continually adapted to the situation. The transition from one stage to another was always initiated by confrontation of the proposed term with the observed school practices and with experience from the use of lexicon for communication on pedagogical reality. Table 8.1 presents a brief characterisation of the stages.

TABLE 8.1 Overview of lexicon creation stages.

Stage	Data	Description
1	Video recording of the Czech lesson	Search for common language
2	Video recordings and transcript of the Czech lesson and lessons of other project partners	Search for structure
3	Video recordings and transcript of the Czech lesson and lessons of other project partners	Condensing structure
4	Questionnaires, interviews with educators	Validation of usability and comprehensibility
5	Lexicon structure and terms	Change of structure
6	Lexicon terms	International validation

Constructing the Czech Lexicon

A search for common language

In the first stage the work began with an analysis of a "national" lesson. At the beginning the team members coded the transcript of the Czech lesson individually using open coding (Strauss & Corbin, 1998). The team members divided the lesson into segments that were meaningful to them personally and were looking for suitable terms to describe these segments. Five descriptions of the same lesson were the outcome of this stage of open coding. As was expected, there were huge differences in approaches to description of various phenomena.

These descriptions were of two different types:

- Detailed and very lesson specific terms (Example of terms used for the description of a classroom dialogue in Table 8.2).
- Generalised and grouped into structures to be more universally usable terms (example of the same segment in Table 8.3).

In general, it can be said that detailed terms (Table 8.2, column (a)) were more likely to be used by teachers who were observing the lessons very closely. More general terms were used by researchers, especially the researcher in the field of pedagogy (Table 8.2, column (b)).

Having finished individual work on this stage, all team members were shown coding made by other team members. The joint discussion on the point and possible use of the lexicon resulted in the decision to accept more general terms (example in Table 8.2 (b)) and create a structure. We considered the creation of a structure necessary, especially to make orientation in the lexicon accessible and thus the lexicon usable.

A search for structure

The team faced the task of having to create superordinate terms related to segments within the episode that would also be usable more universally and not tied to the one particular analysed teaching episode. The team members tried to group their terms into larger and

TABLE 8.2 Suggested lexicon terms given by the teacher (a) and by the researcher (b).

Speaker	Transcript of the lesson	Lexicon terms (detailed) (a)	Lexicon terms (more general) (b)
T	*Well, OK now, as some of you have mastered it already and some of you are already working on your homework to have nothing to do at home, I like that, obviously we're getting better again, so we now have it checked. Now I believe all of you are confident you have the calculations right. OK.*	Teacher addresses the whole class, a positive evaluation comment that many students are immediately working on the task	The Teacher (T) assesses the work of students
T	*Let's now stop the discussion, let's listen to each other and show the right solution on the whiteboard. If you're sure you understand, you can continue solving the assigned problems. Who needs some explanations watches the whiteboard, OK? It's up to you. But, listen to each other.*	Attracting attention and a new instruction on how to proceed in other activities (once again the opportunity to work independently on assignments and exercises in addition)	T assigns a new task
T	*So, problem a), 78, 2a. Who will show us? Klárka? Klárka, please, come here*	An instruction for the concrete student to write a solution of the task on the blackboard	
S	*Because the minus two upsilon reverses.*	The student writes the solution on the blackboard and comments	The student solves the task and writes on the blackboard
T	*Now, it does not reverse, you move it and use the inverse sign. Excellent!*	Positive assessment	T clarifies the pupil's explanation and assesses work

more universal categories. During this activity, as well as while comparing the created descriptions with video recordings of other lessons, more items had to be added to the list. The individual items, at this time, could more accurately be described as long descriptions of the teacher's or the pupils' activity as opposed to single words or short phrases. The result of this process was a list of descriptions that worked as codes – the first draft of Czech Lexicon.

An example of the outcome of this stage is presented in Table 8.3. It illustrates how different teacher's activities were classified under the superordinate term "revision."

Condensation of the lexicon structure

Having created a rough structure, the Czech Lexicon was verified. This was done by coding foreign lessons. Each project team member coded two lessons and each lesson was coded

TABLE 8.3 Item "revision" in the initial version of Czech Lexicon.

Opakování a procvičování vědomostí a dovedností			Revision	
Příklady opakování a procvičování vědomostí a dovedností	Učitel (U) klade otázky vyžadující vysvětlení	Examples of revision and practising knowledge and skills	Teacher (T) asks questions requiring clarification or explanation
	U klade otázky vyžadující jednoslovnou odpověď (jde mu o uderženi pozornosti)		T asks questions requiring one-word answer (to keep pupils´ attention)
	U klade řečnické otázky		T asks rhetorical questions
	Formou rozhovoru s více žáky		In the form of dialogue with more pupils
	U řeší úlohu, zapisuje na tabuli a komentuje své řešení		T solves a problem, records it on the board and comments on his/her solution
	U řeší úlohu, zapisuje na tabuli a vyzývá žáky ke komentářům.		T solves a problem, records it on the board and encourages pupils to comment.

by two members of the research team. The results of their coding were compared and the list of terms was extended and restructured. Lexicon items were supplemented by some events that we did not come across in the Czech lesson (e.g. choral response present in the Chinese lesson). We found illustrative examples to explain what is meant by the given term in different educational contexts and conditions.

At the same time, the list of items was compared with national research publications dealing with classroom research. In the case of the Czech Lexicon, we worked with Czech dictionaries (e.g. Průcha, Walterová, & Mareš, 2013).

In the end, we had 97 items in 11 categories that more or less corresponded to stages of a lesson. They terms were organised against the following categories:

- Classroom management;
- Introductory communication;
- Explanation of new topic;
- Revision of a previously taught topic;
- Solving of a problem, Checking individual work;
- Institutionalisation;
- Summary;
- Non-mathematical social interaction;
- Assessment;
- Concluding the lesson; and
- Individual consultation with a pupil.

In each category, we identified several activities of both teachers and/or pupils. Some of them were not terms but descriptions of activities. The use of them indicated a problem of unsettled terminology in the Czech mathematics education theory; many of the words in the draft lexicon are used in everyday vocabulary, and where no exact term existed, the phenomenon is identified with a description.

Validation at national level

Once we had established a draft lexicon the process of validation at a national level could begin. The testing focused on whether teachers and other experts understand the terms from the lexicon and whether they use them actively. Two questionnaires were created and administered among primary and lower secondary school teachers in order to find out:

- How respondents understand proposed terms or their description (1st questionnaire).
- How respondents subjectively perceive their familiarity with chosen terms (2nd questionnaire).

In addition, interviews with experts were carried informally at meetings as well as after presentations at conferences.

First questionnaire – findings

Four different groups of pre-service and in-service teachers were shown a video recording of a Czech lesson and given a worksheet with eight terms to which they were asked to create a description of an activity that would illustrate the given term (*Group work*, *Monitoring pupils' work*, *Feedback*, *Peer assessment*, *Recapitulation of a solution*, *Summative assessment*, *Formative assessment*, *Individual consultation with a pupil*, *Institutionalisation and Non-mathematical interaction*), and a list of ten descriptions of activities to which they were asked to write an appropriate and fitting term in Czech.

The four different groups were: a group of 11 part-time students; prospective teachers of mathematics (Group 1), a group of 5 in-service teachers and mathematics educators taking part in a conference for teachers of mathematics (Group 2); 7 pre-service mathematics teachers at the Faculty of Education, University of South Bohemia (Group 3); and, 28 in-service teachers studying to extend their qualification for teaching mathematics at the Faculty of Education, University of South Bohemia (Group 4).

The goal of this form of national validation was to determine:

- If there are differences among how pre-service and in-service teachers of mathematics speak about mathematics;
- If there are any differences in how they understand particular terms used in the lexicon and what terms they propose to use for description of specific activities typical for a lesson of mathematics.

The analysis of this part focuses on the differences in understanding of the set of given terms.

Group work

A majority of respondents illustrates this term with similar activity – pupils work together, cooperate to find a solution. Five respondents also use the illustration that the teacher divides pupils into groups (two from Group 1, one from Group 3, and two from Group 4). One respondent from Group 2 mentions that the pupils organise work on their own. The highest number of non-central illustrations of the term were come across in Group 4 – two respondents describe group work as a situation when the teacher monitors how pupils cooperate, one says it is a form of teaching, one says the teacher works with all pupils and two respondents say group work is a situation when pupils learn from each other and their communication skills are developed.

Monitoring pupils' work

The majority of respondents understands monitoring as walking around the classroom, observing and checking progress of solution or other pupils' activity. One respondent in Group 1 stresses that the teacher does not intervene and does not correct mistakes. One respondent in Group 1 also states the teacher motivates their pupils in this activity. For another respondent in Group 4 it is important that the teacher evaluates while monitoring, yet another stresses out that monitoring means also giving advice (in contrast to the respondent in Group 1 for whom monitoring means no intervention). Two respondents describe monitoring as feedback for the teacher on how pupils understand, another two respondents describe this activity as asking questions by a teacher to pupils who are working. One respondent gave no illustration and another described the term as keeping work of pupils over a period of time and assessing if pupils have made any progress.

Feedback

The validation shows that the term feedback is understood in a number of very different ways. For some respondents, feedback is information for teachers on how pupils have understood, for other respondents it is information for pupils on their strengths and weaknesses. Respondents rarely see it as information for both parties involved in the teaching-learning process. The more central understanding of the term is checking understanding, asking questions, giving information about a correct solution, but also marking tests and evaluation of pupils' work. One respondent from Group 3 claims that feedback can also have the form of self-assessment, and one in Group 2 says it can also have the form of peer-assessment. One respondent stresses that it can be both verbal and non-verbal. Group 4 is again the group that gives more unusual illustrations, some of which can hardly be seen as feedback. One respondent understands feedback as asking questions to a longer written text, another as questions to the topic at the end of the lesson. One respondent writes that feedback is when the teacher is happy with the pupils' knowledge and work, yet another respondent claims that feedback is when pupils use the solving procedure from one problem in a new context, two respondents cannot explain the term.

Peer assessment

This term appears to be better understood. Respondents describe peer-assessment as an activity in which pupils are explaining to each other, assessing each other, or checking correctness of their solutions and results. Some speak of the opportunity to discuss and develop cooperation. One respondent from Group 2 speaks of an opportunity to learn from each other. Especially in Group 4 respondents describe the form of this activity – swapping notebooks and correcting other pupil's work, one pupil working on the whiteboard, other pupils controlling them and commenting. One respondent understands peer assessment as observing others while working, which is not the central meaning. Another respondent from Group 4 writes that peer evaluation is connected to mathematical games. And yet another respondent writes that it must be fair and that it saves teacher's time and energy. Four respondents (two from Group 3 and two from Group 4 cannot explain the term).

Recapitulation of a solution

This term is most often understood as an activity conducted by the teacher resulting from pupils' solution and strategies. For some respondents, it means writing one correct solution on the whiteboard for everybody to see, for some it is summarising, evaluation of various strategies, precision and further explanation, reformulation of pupils' solutions and arguments, revision or emphasising the most important steps in a solution. Two respondents write that it is the pupil who recapitulates; five write that it can be both the teacher and the pupils who sum up, evaluate or recapitulate.

Summative assessment

A surprising moment connected with this term is the number of respondents who gave no description or illustration (12 respondents). In general, summative assessment is understood as assessment of some whole: end-of-term evaluation, evaluation after a teaching unit, giving grades, giving points, having a norm or working with some scale. Two respondents from Group 1 stress it is assessment of a pupil for somebody else, it is comparative in nature. Some respondents relate summative assessment to assessment at the end of an activity or lesson. One respondent in Group 4 states that summative assessment means giving no grades, just ticking whether the student "knows" or "does not know." One respondent in Group 3 confuses the meaning of summative and formative assessment claiming that formative means giving grades while summative means giving verbal information. One respondent in Group 4 thinks summative assessment means summarising a topic at the end of the lesson. Another respondent writes summative assessment means that the teacher assesses understanding of the objective by the whole class.

Formative assessment

Similar to summative assessment, in the case of formative assessment there was a high number of no descriptions (12 respondents). Also, the number of respondents who do not understand the term is alarming, namely in Group 4 (seven respondents) and one respondent in Group 1. The incorrect descriptions refer to: grading (whereas formative is giving points); giving numbers; or, assessment written in the credit book. In meaningful descriptions teachers identify: feedback to pupils; stimulating assessment; information on progress an individual pupil has made; assessment while doing an activity; as well as, information on where to improve and how to improve.

Individual consultation with a pupil

Description of the term consultation with a pupil does not cause problems. The term seems to be generally well understood. Respondents only stress different aspects of such a situation – some see it as an opportunity to learn more about the pupil's understanding, for explaining what was not understood, some see it as a possibility to support the needs of individual learners (both talented and special needs). Some of them also refer to the need to discuss the pupil's behaviour in the lessons. One respondent also speaks of preparation for entrance exams and mathematical contests. Only one respondent states individual consultation is possible during the lesson. However, the majority of respondents speak of private communication outside of a lesson.

Institutionalisation

This term is misunderstood by a vast majority of respondents. 21 respondents give no description or explanation. Only four respondents give an explanation that is not entirely accurate but seem to be connected with the term (e.g. anchoring a concept in some context, activities become stable patterns). Most other descriptions come out of the belief that the term comes from the word institution – the respondents refer to rules, statuses, interaction in an institution, and so on.

Non-mathematical interaction

The last of the terms was easier for the respondents but still largely misunderstood especially in Group 4. 16 respondents (14 from Group 4) give no explanation or description. In Group 4 those who answer speak of mathematical problems from real life, interdisciplinary tasks, presentation of results, and numeracy, which are not the same as how the researchers understood the term. Only five of the respondents from Group 4 clearly refer to non-mathematical interaction, for example, when a teacher is commenting on what is happening outside. Also in Group 1 two respondents refer to mathematical problems based on real-life situations and one respondent speaks of interdisciplinarity. One misunderstands the term when stating that non-mathematical interaction is a reaction to a non-mathematical problem.

The other part of the validation (i.e. survey into what terms respondents use to name a situation) only confirms that teachers use very different language to describe the same situation. For example, the description *The teacher has a discussion with pupils in which they guide them to formulation of new knowledge* is given many very different terms: deduction, self-knowledge, monitoring of understanding, development of critical thinking, inquiry-based education, Socrates method, mind map, motivation, results of solution of problems, and discussion.

The situation *Pupils discover something new while solving a problem* is labelled as cognition, problem-based learning, gaining new information, deduction, knowledge acquisition, group work, learning through experience, inquiry-based learning, problem-solving, independence, extension, creativity, heuristic method, construction of knowledge, a-didactical situation, objective project solution.

Some terms (monitoring, feedback) are understood differently by different people, which confirms our premise that terminology is not settled. This becomes really visible in the terms summative/formative assessment and institutionalization that are used in literature and do not penetrate into use in practice. Significant differences could be seen in Group 4 because they are teachers who are not qualified for teaching mathematics and have worked for a long time on the first grade of primary school. Their illustrations focus much more often (compared to other groups) on what the teacher is doing than what the pupils are doing.

Second questionnaire – findings

In the second questionnaire, we tried to verify whether teachers are familiar with the terms and use them actively. The form of the questionnaire was inspired by the questionnaire used by the Australian team (Mesiti & Clarke, 2018).

The questionnaire was filled in by 43 teachers from different parts of the country. This is a significant attribute of the respondents of the survey as these teachers graduated from different universities. The frequency of individual answers was processed into tables and graphs. Table 8.4 and Figure 8.1 illustrate the responses to terms that we expected to be harder to understand.

TABLE 8.4 Table with results in selected items from the questionnaire for Czech teachers.

Term	I know the term…				
	..very well	..well	I know it	I have met it	I do not know it
Teacher (T) gives instructions to organize pupils' work	24	14	3	1	1
T gives instructions to disciple pupils	23	10	9	1	0
T communicates information to pupils through an interview with them	23	12	7	1	0
T conducts an interview with pupils – gradually leads to the formulation of new knowledge	18	16	8	0	0
T points out the relationship with the knowledge previously learned	29	12	1	1	0
The pupils solve the task independently	21	18	4	0	0
T monitors the solution of pupils passing through the class	25	13	5	0	0
Mutual pupils' control	9	9	22	3	1
The pupil is at the board and T invites others to comment on what he/she does	10	15	10	7	0
T controls the activity of all pupils, all pupils perform the same activity	19	13	11	0	0

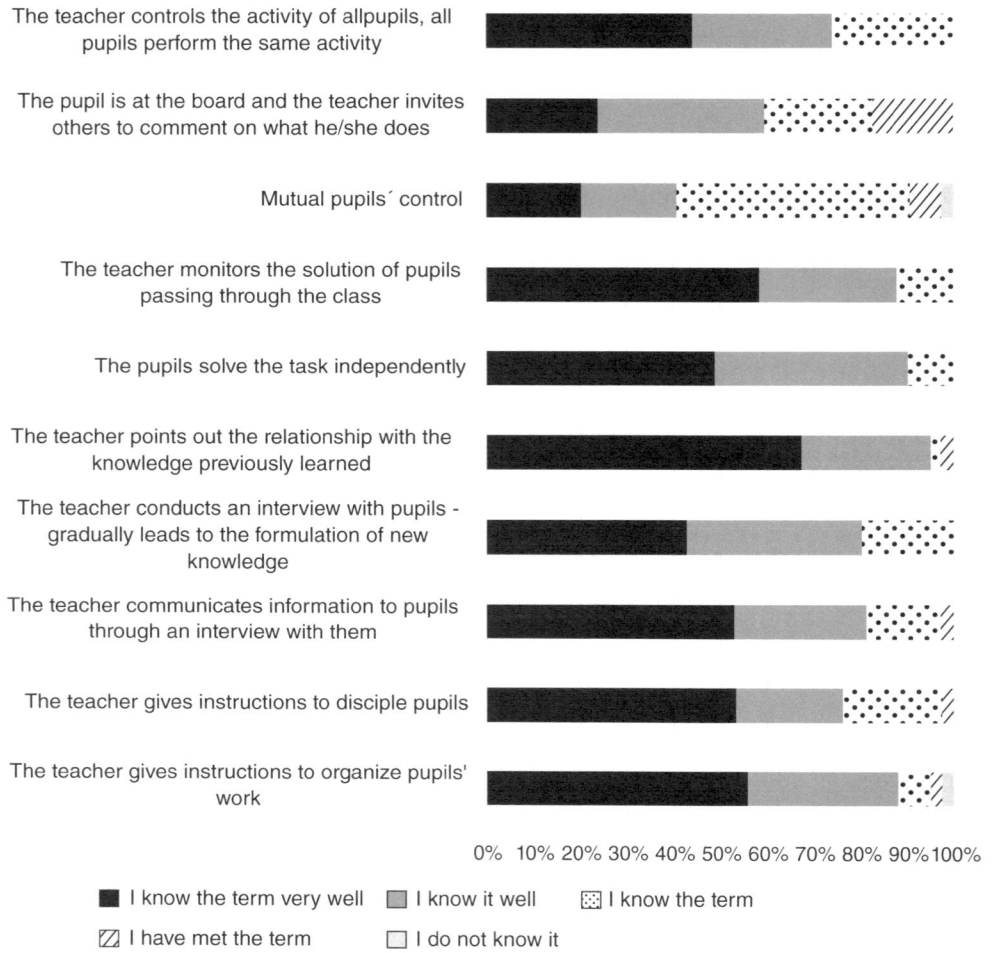

FIGURE 8.1 How familiar Czech teachers are with selected terms?

To summarise the findings of our national validation, the Czech teachers mostly felt confident that they understand terms describing their activity or activity of their pupils in lessons (Figure 8.1). In fact, their understanding of a number of terms does not align with the understanding of these terms by the research team.

Another source of data were interviews with mathematics educators and other experts in the area of education. Comparison of results indicated that there are considerable differences between terminology used by researchers and practising teachers, mainly in the fields that have been developing over the past few years (e.g. in Figure 8.2, the terms scaffolding, institutionalisation and linking). Thus, majority of in-service teachers have not come across them during their university studies. It only confirms the important role of the Czech Lexicon for future teacher education (Novotná, Moraová, Hošpesová, Žlábková, & Bureš, 2016).

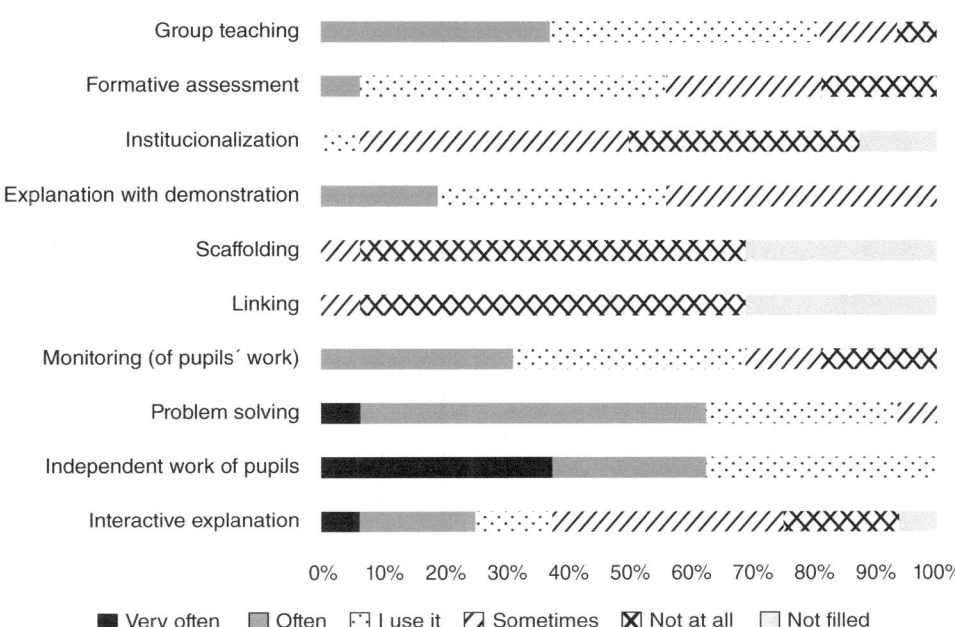

FIGURE 8.2 Relation of terms to the reality of the mathematical classroom.

Change of structure

The process of national validation revealed an important characteristic of the draft lexicon. We realised that there were too many detailed items described in everyday language without a corresponding single term or short phrase, which resulted in difficulties when comparing different lessons and when looking for illustrative examples.

We came to the decision to restructure the lexicon into a new form of fewer terms from didactics of mathematics and/or general pedagogy all of which would be illustrated by several examples. The lexicon was transformed into an open system which could be continuously supplemented with new examples. The comparison of both approaches is visible from the example of introducing "revision" given in Table 8.5 (compare with Table 8.3).

For illustration, we can show how the episode presented in Table 8.2 is described with terms from contemporary lexicon (Table 8.6).

Currently the Czech Lexicon consists of 10 main categories:

- stages of a lesson (the terms that allow description of a lesson from the point of view of its phases);
- organisation of a lesson (E.g. "The teacher organises the lesson.");
- teaching methods (the terms that allow description of a lesson from the point of view of teaching methods);

TABLE 8.5 Item "revision" introduced in contemporary Czech Lexicon.

Main category	Pedagogical term	Description	Examples
Teaching methods[a]	Revision and practice of knowledge and skills	The teacher revises with pupils their knowledge and skills with the aim of consolidating, generalizing, and organizing the learning contents.	Teacher asks questions requiring clarification or explanation.
			Teacher asks questions requiring one-word answer.
			Teacher asks rhetorical questions.
			Teacher conducts dialogue with several pupils.
			Teacher solves a problem, records it on the board and comments on his/her solution.
			Teacher solves a problem, records it on the board and encourages pupils to comment.

[a] Terms that allow description of a lesson from the point of view of teaching methods.

TABLE 8.6 General descriptions of entries (contemporary version of Czech Lexicon).

Speaker	Transcript of the lesson	Lexicon terms
T	Well, OK now, as some of you have mastered it already and some of you are already working on your homework to have nothing to do at home, I like that, obviously we're getting better again, so we now have it checked. Now I believe all of you are confident you have the calculations right. OK.	Assessment – non-directed
T	Let's now stop the discussion, let's listen to each other and show the right solution on the whiteboard. If you're sure you understand, you can continue solving the assigned problems. Who needs some explanations watches the whiteboard, OK? It's up to you. But, listen to each other.	Teacher's communication of information
T	So, problem a), 78, 2a. Who will show us? Klárka? Klárka, please, come here	Assigning a task
S	Because the minus two upsilon reverses.	Solution of a task provided by a pupil controlled by the teacher
T	Now, it does not reverse, you move it and use the inverse sign. Excellent!	

- pupils' individual work (E.g. "Pupils work on their own. Teacher monitors their activity.");
- processes supporting pupils' learning;
- assessment;
- homework;
- organisational forms of instruction;
- use of didactical means;
- type of tasks.

The lexicon was transformed into an open system which could be supplemented with newer and newer examples. For example, currently the category "Teaching Methods" has 10 examples and can be developed further:

- explanation in the form of a teacher's monologue;
- clarification;
- interactive explanation;
- group response;
- explanation with demonstration;
- Socrates' method;
- heuristic method;
- computer-assisted instruction;
- revision and practice of knowledge and skills;
- teacher-controlled solving of a task.

International validation

Validation of the Czech Lexicon was conducted on two levels: a review of the Czech Lexicon by the Chilean project team and a through a comparative analysis of Czech and French Lexicons. Apart from this, a lot of attention was paid to the accuracy of the used English terms and description of activities in such a way to allow comparison of lexicons from different project partner countries.

A very important and useful step towards an improvement of the quality of the Czech Lexicon was its critical review by the Chilean team. The questions and comments of Chilean colleagues were most often related to lack of clarity or ambiguity of some of the English expressions.

All Chilean comments were discussed in the Czech team and taken into consideration. Wherever it was possible, formulations in English were précised or reworded. However, in eight cases the suggested corrections would change the meaning of the term or the character of the activity. In these cases, we focused on finding a relevant illustrating example and/or non-example to clarify and precise the meaning.

Besides the validation process, the Czech Lexicon underwent two comparative studies: French – Czech (Artigue et al., 2017) and Australian – Czech (Mesiti, Novotná, Clarke, Hošpesová, & Hollingsworth, 2019a, 2019b). The methodology used for the comparison study had two parts: a formal comparison and an operational comparison of their use. The compared lexicons have the form of lists of terms or expressions with a description/ definition supplemented with some examples and non-examples. The terms are grouped into categories proper to each lexicon. As described in detail elsewhere (Artigue et al., 2017) and (Mesiti et al., 2019a, 2019b), the formal comparison of the two lexicons was organised in three main phases:

- Comparison of the structures of the two lexicons (categories and subcategories);
- Comparison of the selection of terms and the terminology used;
- Comparison of the term descriptions/definitions, of examples and non-examples.

Obstacles and rewards

While working on the lexicon, the team members were facing many difficulties that had to be solved. One of the problems encountered was the non-existence of appropriate and fitting terms, both in the area of didactics of mathematics and pedagogy. A solution that proved to be effective was to omit the term itself in an item and to use only its description. It also proved to be very difficult to label an activity universally but at the same time depicting all that was happening at that moment in the classroom, especially in case that different actors were carrying out different activities (teacher, pupils). In some cases, the existing universal terms were not subtle and precise enough to allow us to distinguish between various episodes from the analysed lesson. The result of overcoming this obstacle was inclusion of items that were not necessarily disjunctive, might overlap to a lesser or greater degree. Moreover, two-level descriptions are used. This allows us to describe the differences and similarities in the lesson more finely. A compromise had to be made between the effort to make the descriptions universal enough to allow their use for different parts of a lesson and between the need to depict the character of different parts of the lessons as accurately as possible; an example is distinguishing between different type of questions asked by the teacher and the goals the teacher wants to achieve by asking them.

The structure of entries in the Czech Lexicon is significantly affected by the fact that terms describing the course of mathematics lessons (that is, their chronological structure) formed the basis of the lexicon. The viewing of video-recorded lessons from nine different countries showed how difficult it is to use the Czech Lexicon to describe practices not typical of Czech mathematics lessons. Thus, there were many phenomena that could described and illustrated by a video excerpt but not labelled with a specific Czech term. They could be only named by their descriptions.

Entries in the Czech Lexicon focus mainly on pupils' and teachers' observable activities, mathematical content as such is less important. This means the terms can be used for lessons in any subject. We believe this is related to the discourse that is used in practice and, in certain cases, also in the research community. In this respect, Biesta (2006) speaks of a change in language used for description of education and its impact on how education is perceived: "… the language of *education* has largely been replaced by a language of *learning*." (Biesta, 2006, p. 14) According to Biesta this, on the one hand, allows for a good description of phenomena related to pupils, their learning, and cognition but, on the other hand, does not allow description of some aspects of our understanding of what is happening in education. In the Czech Republic, the language of *education* is used more by teachers who primarily speak of their activities and less about what pupils do and learn. Language of *learning* was used in some parts of Framework Educational Programme for Basic Education (2007). The fact that teachers refused to accept the reform that Framework Education Programme brought is reflected in the fact that they rejected the language in which the reform is formulated. This can also be observed in case of some subject didactic researchers (Stuchlíková & Janík, 2015). We argue that the cultivation of subject didactic discourse among the Czech education community is one of the prerequisites for the emancipation of subject didactics. By emancipation of subject didactics we mean the effort to make them stand-alone disciplines and fully respected fields of study. The discussion of the difference between the two languages and their conscious differentiation will lead to an increase in the accuracy of the thinking about education in general and the reality of the classroom.

The process of validation of the Czech Lexicon showed how difficult it is for Czech teachers to speak about lessons in general terms. This complies with the findings of Švec (2009). Czech teachers tended to pass many evaluating judgements and discuss the conception of the lesson and assess its quality rather than describe the classroom interactions in general terms. They also tended to use everyday language rather than pedagogical terminology.

This suggests that a professional lexicon is needed in the Czech teaching community. This kind of lexicon is necessary for understanding each other when communicating about what is going on in a mathematics lesson, for understanding research reports focusing on mathematics instruction if read outside of the research community and can be useful for researchers, teachers, pre-service teachers, and policy makers. Documentation of the content and structure of the Czech Lexicon is essential to deepen the Czech teaching community's understanding of the processes in a mathematics classroom.

Note

1 English translation of Framework Education program for Basic Education (with amendments as at 1. 9. 2007) could be retrieved from www.msmt.cz/file/9481_1_1/. The latest version of the document (2017) has not yet been translated into English.

References

Artigue, M., Novotná, J., Grugeon-Allys, B., Horoks, J., Hošpesová, A., Moraová, H., Pilet, J., & Žlábková, I. (2017). Comparing the professional lexicons of Czech and French mathematics teachers. In B. Kaur, W. K. Ho, T. L. Toh, & B. H. Choy (Eds.), *Proceedings of the 41st Conference of the International Group for the Psychology of Mathematics Education* (Vol. 2, pp. 113–120). Singapore: PME.

Biesta, G.J.J. (2006). *Beyond Learning. Democratic Education for a Human Future.* Boulder, CO: Paradigm Publishers.

Chevallard, Y., & Sensevy, G. (2014). Anthropological approaches in mathematics education, French perspectives. In S. Lerman (Ed.), *Encyclopedia of Mathematics Education* (pp. 38–43). New York, NY: Springer.

Clandinin, J., & Connelly, F.M. (2000). *Narrative Inquiry: Experience and Story in Qualitative Research.* San Francisco, CA: Wiley & Sons.

Clarke, D. J. (2009). Mind your language: Speaking in and about the mathematics classroom. In R.P. Hunting, T. Fitzpatrick, J.A. Milne, D.J. Itter, T.M. Mills, & T.C. Lenard (Eds.), *MAV 2009 Mathematics of Prime Importance: Proceedings of the Mathematical Association of Victoria 46th Annual Conference* (pp. 34–49). Melbourne, Australia: The Mathematical Association of Victoria.

Clarke, D. J., & Mesiti, C. (2010). The lexicon project: Accessing the pedagogical vocabulary in languages other than English. In M.M.F. Pinto, & T.F. Kawasaki (Eds.), *Proceedings of the 34th Conference of the International Group for the Psychology of Mathematics Education* (pp. 237– 238). Belo Horizonte, Brazil: PME.

Framework Educational Programme for Basic Education (with amendments as at 1. 9. 2007). Retrieved from www.msmt.cz/file/9481_1_1/

Janík, T., & Slavík, J. (2007). Fakty a fenomény v průniku didaktické teorie, výzkumu a praxe vzdělávání (Facts and phenomena in the intersection of didactic theory, research and teaching practice). *Pedagogika*, 57(3), 263–274.

Mesiti, C., & Clarke, D. (2018). The professional, pedagogical language of mathematics teachers: A cultural artefact of significant value to the mathematics community. In E. Bergqvist, M. Österholm, C. Granberg, & L. Sumpter (Eds.), *Delight in Mathematics Education* (Vol. 3, pp. 379–386). Umeå, Sweden: PME.

Mesiti, C., Novotná, J., Clarke, D., Hošpesová, A., & Hollingsworth, H. (2019a). Speaking about the mathematics classroom: A comparison of the professional lexicons of teachers in Australia and the Czech Republic. In M. Graven, H. Venkat, A. Essien, & P. Vale (Eds.), *Proceedings of the 43rd Conference of the International Group for the Psychology of Mathematics Education* (Vol. 3, pp. 89–96). Pretoria, South Africa: PME.

Mesiti, C., Novotná, J., Clarke, D., Hošpesová, A., & Hollingsworth, H. (2019b). The professional lexicons of mathematics teachers in the Czech Republic and Australia: How do they compare? In J. Novotná, & H. Moraová (Eds.), *SEMT '19 Proceedings – Opportunities in Learning and Teaching Elementary Mathematics* (pp. 475–477). Praha, Czech Republic: Univerzita Karlova.

MŠMT (2017). *Rámcové požadavky na studijní programy, jejichž absolvováním se získává odborná kvalifikace k výkonu regulovaných povolání pedagogických pracovníků* (metodický materiál k procesu posuzování vysokoškolských studijních programů, jejichž absolventi získají odbornou kvalifikaci pedagogického pracovníka) (Framework requirements on study programmes to gain qualification needed for the regulated profession of a teacher). Čj. MSMT-21271/2017-5. Retrieved from http://www.msmt.cz/vzdelavani/dalsi-vzdelavani/ramcove-pozadavky-na-studijni-programy-jejichz-absolvovanim

Novotná, J., Moraová, H., Hošpesová, A., Žlábková, I., & Bureš, J. (2016). How do we understand each other when we describe classroom activities – Lexicon. In I. Krejčí, M. Flégl, & M. Houska (Eds.), *Proceedings of ERIE 2016* (pp. 440–447). Praha, Czech Republic: CULS.

Průcha, J., Walterová, E., & Mareš, J. (2013) *Pedagogický slovník* (Pedagogical Dictionary), Praha, Czech Republic: Portál.

Rámcový vzdělávací program pro základní vzdělávání (Framework Educational Programme for Basic Education) (2017). Praha, Czech Republic: MŠMT.

Rezat, S., & Sträßer, R. (2012). From the didactical triangle to the socio-didactical tetrahedron: artefacts as fundamental constituents of the didactical situation. *ZDM*, 44(5), 641–651.

Strauss, A.L., & Corbin, J.M. (1998) *Basics of Qualitative Research: Techniques and Procedures for Developing Grounded Theory*, 2nd edition. Thousand Oaks, CA: SAGE Publications.

Stuchlíková, I., & Janík, T. (Eds.). (2015). *Oborové didaktiky: vývoj, stav, perspektivy* (Subject didactics: development, state and perspectives). Brno, Czech Republic: Masarykova Univerzita.

Švec, V. (2009). Sdílení znalostí ve školním prostředí (Sharing knowledge in school environment). *Pedagogická Orientace*, 19(2), 22–37.

Vašutová, J. (2007). *Být učitelem: Co by měl učitel vědět o své profesi* (Be teacher: What should a teacher know about their profession). Praha, Czech Republic: Univerzita Karlova.

Walterová, E. (Ed.). (2001). *Učitelé jako profesní skupina, jejich vzdělávání a podpůrný systém* (Teachers as a profession group, their education and supporting system). Praha, Czech Republic: Univerzita Karlova.

9

CZECH LEXICON

Alena Hošpesová, Hana Moraová, Jarmila Novotná, Iva Žlábková and Jiří Bureš

A Czech Lexicon

Entries in the Czech Lexicon focus mainly on pupils' and teachers' observable activities. Mathematical content as such is less important. The structure of entries in the Czech Lexicon is affected by the fact that terms describing the course of mathematics lessons were used as the basis. National validation of the Czech Lexicon showed how difficult it is for teachers to describe practices in mathematics lessons. There are many items that are described generally but not labelled by a specific Czech term.

The first lexicon table (see Table 9.1) lists all 57 terms of the lexicon in Czech with the closest English translation in brackets. The second lexicon table (see Table 9.2) includes the operational definition of each term in three columns. The first column presents the terms supplemented by their descriptions in the second column. In order to make the terms clearer, more comprehensible and reader friendly, examples and non-examples are presented in the third column. The third lexicon table (see Table 9.3) presents 11 categories into which the terms of the Czech Lexicon is grouped into categories.

The structure of the Czech Lexicon changed significantly during the process of its creation and validation (for more information, see Chapter 8). The process confirmed both the lack of terms in the domain of pedagogy and subject didactics as well as the expected difficulty for Czech teachers to speak about lessons in general terms.

The Czech Lexicon is an open system. Its use by teachers in their discussions with other teachers, teacher educators, and academics will result in its further enrichment. The Czech Lexicon is a very useful resource for teacher education as it helps to refine the language used to describe pedagogical reality. Czech teachers think in a thematic way; less about what they will teach than about the aims of their teaching. They discuss the conception of the lesson and assess its quality and they use everyday language rather than pedagogical terminology. The Czech approach corresponds with the Czech educational tradition. Teachers formulate their objectives through the teaching content their pupils are expected to master rather than by referring to the developed cognitive processes.

TABLE 9.1 Czech Lexicon – Terms.

heuristická metoda (heuristic method)	**hodnocení formativní** (formative assessment)	**hodnocení neadresné** (assessment non-directed)	**hodnocení povšechné** (assessment – general)
hodnocení sumativní (summative assessment)	**hromadné vyučování** (collective teaching)	**individualizované vyučování** (individualised teaching)	**individuální konzultace se žáky** (individual consultation with pupils)
institucionalizace (institutionalisation)	**interaktivní výklad** (interactive explanation)	**kontrola domácího úkolu** (checking homework)	**kontrola samostatné práce** (checking individual work)
kooperativní vyučování (cooperative teaching)	**linking** (linking)	**monitorování práce žáků** (monitoring of pupils´ work)	**opakování a procvičování vědomostí a dovedností** (revision and practice of knowledge and skills)
oprava didaktického testu (correction of the didactical test)	**organizační činnost žáka** (pupil's organisational activity)	**organizační otázky učitele** (teacher's organisational questions)	**organizační otázky žáka** (pupil's organisational questions)
organizační pokyn (organisational instruction)	**počítání na kalkulačce** (counting on calculator)	**pochvala žáka** (commendation)	**práce s chybou** (work with an error)
práce s interaktivní tabulí (work with smartboard)	**rozhovor učitele se žáky** (talk of the teacher with pupils)	**řešení didaktického testu** (solution of the didactical test)	**řešení úlohy řízené učitelem** (teacher-controlled solving of a task)

řešení úlohy žákem (solution of a task provided by a pupil)	**sborová odpověď** (group response)	**scaffolding** (scaffolding)	**sdělení učitele** (teacher's communication of information)
sebehodnocení (self-assessment)	**shrnutí** (summarisation)	**skupinové vyučování** (group teaching)	**Sokratovská metoda** (Socrates´ method)
udržování kázně (maintaining discipline)	**úloha důkazová** (proof problem)	**úloha problémová** (problem)	**úloha slovní** (word problem)
úloha určovací (determining problem)	**ústní zkoušení** (oral exam)	**vizualizace** (visualisation)	**vnější motivace žáka při výuce** (external motivation in a lesson)
vrstevnické hodnocení (peer-assessment)	**vyhledávání informací** (searching for information)	**výklad formou monologu** (explanation in the form of a teacher's monologue)	**výklad spojený s demonstrací** (explanation with demonstration)
vysvětlování (clarification)	**výuka podporovaná počítačem** (computer assisted instruction)	**využití modelů** (use of models)	**využití PC s dataprojektorem** (use of PC with beamer)
zadání domácího úkolu (assignment of homework)	**zadání úlohy** (assigning a task)	**zahájení vyučovací hodiny** (opening of a lesson)	**zakončení vyučovací hodiny** (concluding the lesson)
zápis informací na tabuli (written record on the board)			

TABLE 9.2 Czech Lexicon – Terms with operational definitions.

Term	Description	Examples and Non-examples
heuristická metoda (heuristic method)	Pupil discovers knowledge. Partly under their teacher's guidance, partly on their own.	Example: Problem is not solved using any school algorithm. Pupils step by step discover another procedure that can be used to solve the problem. _Non-example_: _Teacher dictates to pupil's individual steps they should make. The solving procedure may be based on some heuristic strategy._
hodnocení formativní (formative assessment)	Assessment supporting the learning process.	Example: Assessment that gives pupils feedback and directs their subsequent learning.
hodnocení neadresné (assessment non-directed)	Global assessment of the class with respect to the subject matter they have mastered.	Example: I see you have now come to an understanding of the difference between quadratic and linear equations.
hodnocení povšechné (assessment – general)	General global assessment of the whole group without taking into account individual differences in pupils' performance.	Example: You've worked really well. _Non-example_: _Assessment of individual pupils._
hodnocení sumativní (summative assessment)	Pupil is assessed for their performance, e.g. in a test.	Example: Grading.
hromadné vyučování (collective teaching)	Teacher manages all pupils´ activities, all pupils do the same activity.	Example: Teacher works with the whole class at once. _Non-example_: _Teacher communicates with one pupil._
individualizované vyučování (individualised teaching)	Teacher manages pupils´ activities, pupils learn individually (each his/her own way).	Example: Pupils solve problems on their own. _Non-example:_ _Pupil works individually without teacher intervention (a-didactical situation)._
individuální konzultace se žáky (individual consultation with pupils)	Individual consultation with a pupil after the lesson	Example: Teacher answers a pupil's question individually.

Term	Description	Examples and Non-examples
institucionalizace (institutionalisation)	Teacher reformulates pupils' inaccurate or confusing statements and thus authorises them as facts to be learned or assigns an activity in which the new item of knowledge or skill is used.	*Example:* Integration of a new piece of knowledge into knowledge structure. Teacher formulates remarks about metacognition. *Non-example:* *Pupils propose a new insight, but the teacher does not pay attention to it and continues as he/ she is prepared for the lesson.*
interaktivní výklad (interactive explanation)	Teacher communicates in a dialogical form with pupils – he/she guides pupils to the discovery of a new item of knowledge.	*Example:* Teacher communicates in a dialogical form with pupils (one pupil/group of pupils/whole class) – he/she guides pupils to the discovery of a new item of knowledge. *Non-example:* *Teacher repeats the communication unchanged.*
kontrola domácího úkolu (checking homework)	Teacher checks pupils have done the homework.	*Example:* Giving the result(s). Explanation. Teacher checks pupils' homework one by one. *Non-example:* *Teacher writes the results on the board so that any pupil can compare with his/her result.*
kontrola samostatné práce (checking individual work)	Teacher or pupils check results of pupils' individual work.	*Example:* Peer assessment. Explanatory dialogue teacher – pupils (recapitulation of the solution). Teacher presents various pupils' solutions (monologue). *Non-example:* *Teacher writes the results on the board and the pupils compare them to theirs.*
kooperativní vyučování (cooperative teaching)	Group work controlled mainly by pupils themselves. At the end, pupils reflect their work from the perspective of content and cooperation in their group.	*Example:* Pupils help one another when solving the problem. *Non-example:* *Pupils work individually.*
linking (linking)	Teacher draws attention to links with previously acquired knowledge.	*Example:* Teacher says: "We worked on this yesterday, we tried to learn this." "You learnt last year that …" *Non-example:* *Teacher does not present any links or reference and only tells the pupils what they need to know*

(*Continued*)

TABLE 9.2 (*Continued*)

Term	Description	Examples and Non-examples
monitorování práce žáků (monitoring of pupils´ work)	Teacher checks pupils' solutions. Teacher walks around the classroom.	Example: Teacher gives feedback to one solver (sotto voce). Teacher gives feedback to all pupils with respect to what he/she comes across in a pupil's/pupils' solutions (out loud). *Non-example*: *The teacher does not monitor pupils' work, e.g. sits at the desk.*
opakování a procvičování vědomostí a dovedností (revision and practice of knowledge and skills)	The teacher revises with pupils their knowledge and skills with the aim of consolidating, generalising, and organising the learning contents.	Example: Teacher asks questions. Pupils answer. Teacher solves a problem, records it on the board and encourages pupils to comment. *Non-example*: *Teacher monologically explains new knowledge to the pupils.*
oprava didaktického testu (correction of the didactical test)	The teacher explains selected errors in solutions of didactical test tasks.	
organizační činnost žáka (pupil's organisation activity)	The pupil participates in organisational activities.	Example: Pupil distributes materials, notebooks, calculators.
organizační otázky učitele (teacher's organisation questions)	Teacher poses questions with non-mathematical content.	Example: Checking that pupils have all the needed aids and equipment. *Non-example*: *Teacher asks the pupils how they had enjoyed the theatre play they had seen the day before.*
organizační otázky žáka (pupil's organisation questions)	Pupil poses questions with non-mathematical content.	Example: Pupil asks if he/she can go home or if they can go to the toilet. *Non-example*: *Pupil alerts the teacher to the point of interest he/she found in the textbook.*

Term	Description	Examples and Non-examples
organizační pokyn (organisational instruction)	Teacher gives instructions by which they organise pupils' work.	Example: "Open your notebooks.", "Start work." "Form groups." _Non-example_: _Teacher points out that a pupil is looking for a help from his/her classmate, in a book or notebook, etc._
počítání na kalkulačce (counting on calculator)	Teacher or pupils count on a calculator.	
pochvala žáka (commendation)	Expression of praise or teacher's satisfaction with a pupil's performance	Example: Teacher praises a pupil for correct solution of a problem. _Non-example_: _Teacher monitors the pupil's work. He does not say anything._
práce s chybou (work with an error)	Teacher leads the pupil to understand the cause of the error he was doing.	
práce s interaktivní tabulí (work with smartboard)	Teacher uses smartboard in their lesson, for example to demonstrate, clarify, practise.	Example: Teacher presents a problem on a smartboard.
rozhovor učitele se žáky (talk of the teacher with pupils)	Teacher asks, pupils answer.	Example: Teacher and pupils discussed the topic of the lesson based on what they did last time. _Non-example_: _Teacher's monologue._
řešení didaktického testu (solution of the didactical test)	Pupils solve the didactical test tasks.	
řešení úlohy řízené učitelem (teacher-controlled solving of a task)	Pupils solve the problem with help of their teacher.	Example: Problem is solved by pupils. The teacher gives cues about the solution. Pupil is at the board and the teacher encourages other pupils to comment on the pupil's work. _Non-example_: _Pupils solve the task individually in their workbooks._

(_Continued_)

TABLE 9.2 (*Continued*)

Term	Description	Examples and Non-examples
řešení úlohy žákem (solution of a task provided by a pupil)	Pupils solve a task on their own.	Example: Pupils practise what they have learned by solving a problem. Pupils discover something new by solving a problem. *Non-example: Pupils are watching the teacher solving the task similar to the one he/she previously explained.*
sborová odpověď (group response)	Pupils respond as a group.	Example: Teacher asks a question. He/she does not call anyone. Pupils reply collectively. *Non-example: Just one or several pupils answer, not all pupils in the class.*
scaffolding (scaffolding)	Strategies and methods that help pupils overcome barriers in learning, and techniques that allow pupils to master knowledge and skills.	Example: Teacher offers clues and strategies to pupils until they are able to cope with the activity on their own. *Non-example: Teacher tells the whole solution and/or the result.*
sdělení učitele (teacher's communication of information)	Teacher provides verbal information to pupils.	Example: Teacher says what pupils are to learn (the objective of the lesson). *Non-example: Teacher reads the assignment of the task that is to be solved immediately. He/she does not explain why they will solve the task.*
sebehodnocení (self-assessment)	Pupil assess their own performance, solution.	Example: Pupil who was about to go to present their solution says: "Oh, now I see it's incorrect." *Non-example: Pupil did not recognise the error him(her)self, he/she deduces it from the teacher's behavior.*
shrnutí (summarisation)	Teacher/pupils recapitulate and sum up verbally.	Example: Teacher recapitulates steps of the solution of the problem. Teacher formulates remarks about metacognition. *Non-example: Teacher presents another solution that had not been used.*

Term	Description	Examples and Non-examples
skupinové vyučování (group teaching)	Teacher manages pupils´ activities, pupils work in groups (or pairs).	Example: Pupils collaborate to solve the problem. *Non-example*: *Pupils work individually sitting around a table.*
Sokratovská metoda (Socrates´ method)	Presentation of teacher's precisely formulated questions to pupils. Pupils are lead to construction of their own, logically deduced knowledge.	Example: Teacher asks a sequence of questions, pupils answer. Questions and answers are consequent. *Non-example*: *Teacher's monologue.*
udržování kázně (maintaining discipline)	Teacher gives an instruction whose aim is to make pupils behave in the lesson.	Example: Teacher says: "Don't disturb."
úloha důkazová (proof problem)	Problem contains a statement, its aim is to prove its validity with the help of logical-deductive procedures	Example: Prove that the sum of two odd numbers is always an even number. *Non-example*: *Prove that the sum of two odd numbers is an even number.*
úloha problémová (problem)	Problem whose solving procedure is unknown to the pupil.	Example: Problem: Find the rule for finding the perimeter of these shapes. *Non-example*: *What is the area of the Czech Republic?*
úloha slovní (word problem)	Situation described in common language that contains some data and a question.	Example: A bricklayer builds a wall in 18 hours. Another bricklayer builds it in 20 hours. How long will it take them to build the wall if they build it together? *Non-example*: *Calculate the third power of 2.5.*
úloha určovací (determining problem)	There is an object in the task that pupils are looking for (equation, points, bodies, functions).	Example: Problem: Write the equation for the quadratic that passes through these points: (1, 15), (3,31), (−2,6). *Non-example*: *Solve the equation: $x + 5 = 8$*
ústní zkoušení (oral exam)	Teacher assigns the task to one pupil at the board. Pupil solves the role in front of the class.	
vizualizace (visualisation)	Teacher uses visualisation.	Example: Teacher makes a construction on the board.

(*Continued*)

TABLE 9.2 (*Continued*)

Term	Description	Examples and Non-examples
vnější motivace žáka při výuce (external motivation in a lesson)	External motivation of pupils by the teacher includes: setting teaching objectives, sharing teacher's attitudes and expectations towards a pupil and the class, provoking pupil's social needs, provoking pupils´ performance-related motivations, using rewards and punishments, elimination of the feeling of a bore.	Example: Teacher explains to their pupils how they can make use of knowledge from mathematics in everyday life. Game: Teacher motivates pupils to simulate real-life situations, such as family budget planning. *Non-example*: *Teacher is trying to raise interest directly through mathematical content without a link outside of mathematics.*
vrstevnické hodnocení (peer-assessment)	Assessment of a solution of a problem by classmates.	Example: Pupils swap their solutions and assess them. *Non-example*: *Each pupil checks his / her result with the result in the textbook.*
vyhledávání informací (searching for information)	Teacher or pupils search for information e.g. on internet or in literature.	
výklad formou monologu (explanation in the form of a teacher's monologue)	Teacher explains subject matter to pupils. Transmission of knowledge predominates. The aim is explanation of a new topic.	Example: Teacher explains new subject matter without giving pupils a chance to enter into their discourse. *Non-example*: *Pupils independent inquiries.*
výklad spojený s demonstrací (explanation with demonstration)	Explanation with demonstration of the object or phenomenon to pupils. Pupils observe the phenomenon in a controlled way.	Example: Teacher explains and simultaneously makes a record or draws the procedure on the board. *Non-example*: *Teacher, for example, hands out the worksheets and lets the pupils work independently.*
vysvětlování (clarification)	Teacher explains a topic familiar to their pupils in a new way.	Example: Teacher responds to pupils' failure to understand a new topic and explains it to pupils in an alternative way. *Non-example*: *Teacher repeats the communication unchanged.*
výuka podporovaná počítačem (computer assisted instruction)	In teaching/learning the teacher or pupil use computer.	Example: Use of computer for calculations. Use of some programme of dynamic geometry. Use of computer to search for information.

Term	Description	Examples and Non-examples
využití modelů (use of models)	Teacher explains a topic and shows models, e.g. of solids. Pupils can see, touch and experience.	*Example:* Teacher shows properties of solids on real 3D models.
využití PC s dataprojektorem (use of PC with beamer)	Teacher presents a visualisation created on a computer using a beamer.	*Example:* Teacher uses a PowerPoint presentation projected on a beamer.
zadání domácího úkolu (assignment of homework)	Teacher assigns work for pupils to be done at home.	*Example:* Teacher gives the page number. Teacher explains e.g. what the aim of the task is, how it will be assessed. Teacher assigns homework on worksheets. *Non-example:* *Teacher assigns homework before the end of the lesson and lets them work on before the lesson finishes.* *Teacher assigns the task.*
zadání úlohy (assigning a task)	Teacher assigns a task.	*Example:* Teacher writes the problem on the board. Teacher dictates the wording of the assignment. Teacher assigns a problem from a textbook. Teacher hands out worksheets. Teacher projects the assignment. Teacher asks pupils to read the text. *Non-example:* *Pupils pose problems.*
zahájení vyučovací hodiny (opening of a lesson)	Introductory part of a lesson.	*Example:* Greeting. Signing the class register. *Non-example:* *Teacher stands by the door and divides the pupils into the benches.* *Teacher writes grades in his/her workbook.*
zakončení vyučovací hodiny (concluding the lesson)	Final part of a lesson.	*Example:* Assigning homework. *Non-example:* *There was a bell but the teacher continues.*
zápis informací na tabuli (written record on the board)	Teacher records on the board, e.g. lesson topic, homework, instruction.	*Example:* Teacher speaks and writes on the board.

TABLE 9.3 Czech Lexicon – Terms by category.

Didaktický test (Didactical Test)	correction of the didactical test, oral exam, solution of the didactical test, work with an error
Domácí úkol (Homework)	assignment of homework, checking homework
Druhy úloh (Type of Problems)	determining problem, problem, proof problem, word problem
Fáze vyučovací hodiny (Stages of a Lesson)	concluding the lesson, opening of a lesson
Hodnocení (Assessment)	assessment - general, assessment - non-directed, commendation, self-assessment, summative assessment, formative assessment, peer assessment
Organizace vyučování (Organisation of a Lesson)	assigning a task, maintaining discipline, organisational instruction, pupil's organisational activities, pupil's organisational questions, talk of the teacher with pupils, teacher's communication of information, teacher's organisational questions, written record on the board
Organizační formy vyučování (Organisation Forms of Instruction)	collective teaching, cooperative teaching, group teaching, individual consultation with pupils, individualised teaching
Procesy podporující učení žáků (Processes Supporting Pupils' Learning)	external motivation in a lesson, institutionalisation, linking, scaffolding, summarisation
Samostatná práce žáků (Pupils' Individual Work)	checking individual work, monitoring of pupils' work, solution of a task provided by a pupil
Vyučovací metody (Teaching Methods)	clarification, computer assisted instruction, explanation in the form of a teacher's monologue, explanation with demonstration, group response, heuristic method, interactive explanation, revision and practice of knowledge and skills, Socrates' method, teacher-controlled solving of a task
Využití didaktických prostředků (Use of Didactical Means)	counting on calculator, searching for information, use of models, use of pc with beamer, visualisation, work with smartboard

Acknowledgements

This work was supported by the funds for the development of a research organisation University of South Bohemia České Budějovice and by the Mobility Fund of Charles University.

Bibliography

Presentations and Papers

The work of the Czech national team has been presented locally and internationally at the following meetings and conferences:

- Conference on applied mathematics (APLIMAT) 2017
- Congress of the European Society for Research in Mathematics Education (CERME) 2017
- Dva dny s didaktikou matematiky [Two days with didactics of mathematics] 2016
- ECEL 2018
- Efficiency and responsibility in education (ERIE) 2016
- European Conference on Educational Research (ECER) 2016
- International Congress on Mathematical Education (ICME) 2016
- International Group for the Psychology of Mathematics Education (PME) Annual Conference 2016, 2017, 2019
- International Symposium Elementary Mathematics Teaching (SEMT) 2017, 2019
- Konference České asociace pedagoigckého výzkumu (ČAPV) [Conference of Czech association for pedagogical research] 2016
- PhD Summer school, Czech Republic 2016

A selection of peer-reviewed conference publications from the Czech research team include:

Artigue, M., Novotná, J., Grugeon-Allys, B., Horoks, J., Hošpesová, A., Moraová, H., Pilet, J., & Žlábková, I. (2017). Comparing the professional lexicons of Czech and French mathematics teachers. In Kaur, B., Ho, W.K., Toh, T.L., & Choy, B.H. (Eds.), *Proceedings of the 41st Conference of the International Group for the Psychology of Mathematics Education* (Vol. 2, pp. 113–120). Singapore: PME.

Clarke, D.J., Mesiti, C., Cao, Y., & Novotna, J. (2017a). The lexicon project: examining the consequences for international comparative research of pedagogical naming systems from different cultures. In T. Dooley, & G. Gueudet (Eds.), *Proceedings of the Tenth Congress of the European Society for Research in Mathematics Education* (pp. 1610–1617). Dublin, Ireland: ERME.

Clarke, D., Mesiti, C., Novotná, J., Moraová H., & Chan, M.C.E. (2017b). Speaking in and about the mathematics classroom. In Novotná, J., & Moraová, H. (Eds.), *Proceedings of SEMT 2017* (pp. 477–479). Prague: UK-PedF.

Hošpesová, A., Novotná, J., Žlábková, I., Moraová, H. & Bureš, J. (2016). Jak si rozumíme, když popisujeme dění ve třídě – lexikon. In Žlábková, I. et al. (Eds.), *Sborník z konference ČAPV* (pp. 23–24). České Budějovice: JU.

Mesiti, C., Clarke, D.J., Roan, K., Hollingsworth, H., Cao, Y., Yu, G., Novotna, J., Zlabkova, I., & Dobie, T. (2016). Discourse about the mathematics classroom. In C. Csikos, A. Rausch, & J. Szitányi (Eds.), *Mathematics Education: How to Solve It? PME40* (pp. 357–363). Szeged, Hungary: PME.

Mesiti, C., & Novotná, J., Clarke, D.J., Hošpesová, A., & Hollingsworth, H. (2019a). Speaking about the mathematics classroom: A comparison of the professional lexicons of teachers in Australia and the Czech Republic. In M. Graven, H. Venkat, A. Essien, & P. Vale (Eds.), *Proceedings of the 43rd Conference of the International Group for the Psychology of Mathematics Education* Vol. 3 (pp. 89–96). Pretoria, South Africa: PME.

Mesiti, C., Novotná, J., Clarke, D., Hošpesová, A., & Hollingsworth, H. (2019b). The professional lexicons of mathematics teachers in the Czech Republic and Australia: How do they compare? In Novotná, J., & Moraová, H. (Eds.), *Proceedings of SEMT 2019* (pp. 475–477). Prague: UK-PedF.

Moraová, H., & Novotná, J. (2017). Differences in classroom practices in ordinary and a clil mathematics lesson. In D. Szarková, P. Letavaj, D. Richtáriková, & M. Prašílová (Eds.), *16th conference on applied mathematics APLIMAT 2017 Proceedings* (pp. 1093–1100). Bratislava: STU.

Moraová, H., & Novotná, J. (2018). Differences in classroom practices in ordinary and technology-supported mathematics lessons. In Ntalianis, K., Andreatos, A., & Sgouropoulou, C. (Eds.), *Proceedings ECEL 2018* (pp. 393–399). Reading: ACPIL.

Novotná, J. & Moraová, H. (2016). Jak vidíme vyučovací hodinu? Jedna vyučovací hodina, různé pohledy. In Vondrová, N. (Ed.), *Dva dny s didaktikou matematiky* (pp. 119–123). Praha: UK-PedF.

Novotná, J. & Moraová, H. (2019). Czech professional lexicon of mathematics teachers at the primary level: What is specific for it? In Novotná, J., & Moraová, H. (Eds.), *Proceedings of SEMT 2019* (pp. 477–479). Prague: UK-PedF.

Novotná, J., Moraová, H., Hošpesová, A., Žlábková, I. & Bureš, J. (2016). How do we understand each other when we describe classroom activities - Lexicon. In M. Flégl, M. Houška & I. Krejčí Eds.), *Proceedings of the 13th International Conference Efficiency and Responsibility in Education* (pp. 440–447). Prague: CUL.

10

THE FINNISH MATHEMATICS TEACHERS' LEXICON

A focus on organisation and relationships

Markku S. Hannula, Fritjof Sahlström and Jani Kiviharju

Introduction

Teachers' professional language reflects their pedagogical thinking. The richer and more nuanced teachers' professional language is, the more possibility there is for elaborate reflections and discussions about teaching and learning (Mesiti et al., 2016). In this chapter we examine the pedagogical language of Finnish mathematics teachers. An overview of the Finnish education system is provided first to provide a context for this examination.

Classroom culture in Finland appears to be based on two strong and conflicting, but also interweaving discourses: a tradition of formal and social pedagogy, and a top-down implemented individualist didactic of the basic school (Simola, 1998). Schooling in Finland has historically been compulsory, with the aim of educating the masses as future citizens. Finnish teacher training was dominated by a strong Herbart-Zillerian tradition until the late 1940s. Pedagogical individualism entered into Finnish educational discourse later, as a top-down education reform designed to complete a social education mission. The principle of individualised teaching was not part of the Finnish pedagogical vocabulary until the 1960s.

Interestingly, there is limited research about what happens in Finnish classrooms. From the 1980s, empirical research based on videotaped lessons concluded that the model of verbal interaction in classrooms seems to have remained the same for a long period of time; the teacher talks more than two thirds of the time and the pupils give short responses (Leiwo, Kuusinen & Kuusisto, 1981; Leiwo, Kuusinen, Nykänen & Pöyhönen, 1987). One characterisation of the Finnish comprehensive school classroom might be considered crushing: A "wasteland not only of intelligence but also of emotions" (Leiwo et al, 1987, p. 169).

Nearly a decade later, a British evaluation team reported on their observations of Finnish classrooms. The team visited, observed and interviewed principals, teachers and students in 50 schools that were selected because they were pilot schools or otherwise interested in curriculum reform. They concluded that "in both the lower and upper comprehensive schools, we did not see much evidence of, for example, student-centered learning or independent learning" (Norris, Asplund, MacDonald, Schostak, & Zamorski, 1996, p. 29).

A few studies on Finnish teachers from the 2000s have presented a clear picture of a profession that is committed to its traditional work in the classroom, and resistant to and strongly critical of innovations. According to a survey, eight out of ten Finnish teachers see their work as rewarding, like it, and are strongly involved in it (Santavirta, Aittola, Niskanen, Pasanen, Tuominen, & Solovieva, 2001; see also Virta & Kurikka, 2001). What appears to stress them the most are the required meetings, planning and reporting, not the basic classroom work. Syrjäläinen (2002) interviewed teachers about their experiences of and attitudes toward recent school reforms and innovations. She summarised their critical thinking as follows: reforms mean too heavy a work load; teachers have no say in the innovations; the development work is too often chaotic; the sphere of teachers' responsibilities has been extended too far; only lip service is paid to professional responsibility and competence; and there is too much unrealistic and even dangerous development work (Syrjäläinen, 2002, pp. 90–100).

Traditional teaching appears to be prevalent in Finland. According to the comparative TALIS 2013 study (Taajamo, Puhakka, & Välijärvi, 2014), compared to teachers in the other 34 participating countries, Finnish lower secondary teachers: give fewer different tasks to different students (37% and 44%, respectively); prefer less group work (34% and 47%, respectively); refer less often to everyday problems in teaching (64% and 68%, respectively) and give less literal feedback to students (25% and 54%, respectively). In the most recent TALIS 2018 (OECD, 2019), Finnish teachers show little change between 2013 and 2018 in many of the studied domains, such as in the use of teaching practices pertaining to clarity of instruction (OECD, 2019, p. 57) and in working hours (OECD, 2019, p. 71). Finnish teachers also continue to report a strong sense of their profession being appreciated in society.

Simola and her colleagues (Simola, Kauko, Varjo, Kalalahti & Sahlström, 2017) state that the dynamics of Finnish classroom cultures seems to combine two discourses: a strong Finnish paternalistic pedagogical tradition and pupil-centred progressivism. The progressivism has mainly been a top-down process emanating from the national curriculum and teacher education for comprehensive school, whereas the paternal pedagogy is at the core of a traditional approach to schooling. This classroom pedagogy is paternalistic in the sense that teachers see themselves as adults keeping a professional distance from pupils and parents; it is progressive in its heavy commitment to the "no child left behind" ideology that is strongly supported in state educational discourse, efficient special-education and remedial teaching systems, school healthcare and other welfare services and free school meals for all pupils (Simola et al., 2017, p. 111).

In PISA studies (e.g. OECD, 2013), Finnish students have been above the OECD average in their performance, with a gender difference clearly in favor of girls. The Finnish education system has small between-school differences, but quite large within-school differences. Students appear to have relatively low anxiety and low enjoyment, and report a realistic self-concept. Correlation between perseverance and mathematics performance is strong.

The formal qualification for a mathematics teacher in Finland is a Master level degree that includes specific credits of mathematics and specific credits of pedagogical studies. Some experienced teachers have a previously recognised Bachelor level degree. A national evaluation found that almost all teachers are formally qualified and that most of them teach two or more subjects (Hannula & Oksanen, 2013). The typical teaching subject combinations reported were mathematics and physics, and/or chemistry. Gender balance for mathematics teachers was fairly equal (43% male).

The Finnish language is one of the Finno-Ugric languages, related to Hungarian, Estonian and some smaller languages. Finnish is unlike most European languages as it is not an Indo-European language. On the analytic-synthetic dimension, Finnish language is quite synthetic, although not at the most extreme end of the spectrum (Miestamo, 2006). As a synthetic language, Finnish composes (synthesises) multiple concepts into each word, unlike analytic languages (such as English) that break up (analyse) concepts into separate words. For example, Finnish language has plenty of inflectional morphemes (e.g. **in** *my* mind – miele**ssä***ni*), compound words (e.g. lesson – oppitunti, lit. knowledge hour) and derived words (e.g. oppi – knowledge → oppia – to learn, opettaa – to teach, opiskella – to study, oppilas: a student, etc.).

Although the Finnish language is grammatically and structurally different from the Indo-European languages, many of the words have been adopted from other languages. These influences are visible in the Finnish Lexicon. Important sources of influence have been Swedish, German and Russian, and more recently English. The oldest educational terminology arrived through Sweden and Russia, when Christianity arrived in Finland from east and west.

When Finnish mathematics teachers were asked to describe themselves as a teachers using metaphors, the chosen metaphors revealed that many identified primarily as experts in mathematics teaching (51%), while some saw themselves as experts in pedagogy (14%) and only a few focused on their role as mathematics experts (6%) (Oksanen, Portaankorva-Koivisto & Hannula, 2014).

The following sections outline the process for generating the first version of the Finnish Lexicon and its national validation, and describe some of its characteristic features.

Method

The initial development of the Finnish Lexicon

In Finland, the initial development of the lexicon was made by a team consisting of the first author and three experienced mathematics teachers (including the third author), who alternated between viewing and annotating video events, and engaging in discussions to reach consensus on the relevance of each term. When watching and discussing the Finnish stimulus video jointly, we generated 47 terms. We then each watched two stimulus videos from other countries and individually identified terms to describe what happened during these lessons. When meeting again, we also brainstormed possible additional relevant terms. Altogether we generated 290 entries, with some terms identified more than once. In the next phase we discussed the list, removing multiple entries, and deciding which of the closely related terms to keep and which would be sufficiently relevant for the Lexicon. Through this process we ended up with 89 terms that the team agreed to be relevant. After an international Lexicon project meeting, the Finnish lexicon team considered another 23 terms inspired by the discussions in the meeting, leading to a refined list of 105 terms. For each of these terms we wrote descriptions, and specified examples and non-examples, and then conducted a national review.

In Finland, this process of naming events led to a realisation that many of the important events that teachers name in mathematics lessons are not activities, as suggested by the original protocol. For example, the term *kertaus* (revision) is not used primarily as a name for an event, but rather as a qualifier for several different things, such as a revision lesson or a revision task. Other terms that did not refer to activities were *oivaltaminen* (realisation), *tuntisuunnitelma* (lesson plan), and *keventäminen* (use of humour to lighten the atmosphere).

We had a few terms that we found difficult to translate into English as the translation didn't quite capture the same meaning. For example, the translation of *opetuskeskustelu* (questioning) shifts from the original dialogic discussion promoting learning into the teacher testing whether the students have the right knowledge. Another difficult term to translate was *työrauha* (good working climate), where we decided against translating it as "discipline" or "classroom management." Moreover, the Finnish word *ohjaus* (guidance) means "to steer." As the Finnish term relates metaphorically to movement rather than constructing, we decided not to use the English translation "scaffolding."

Procedure for national review of the Finnish Lexicon

For the national review of the Finnish Lexicon, we conducted an electronic survey with Finnish mathematics teachers in November–December 2016. The aims of the national review were to determine: the familiarity of the lexicon terms among Finnish mathematics teachers, how frequently they use these terms, and how well they recognise them from the descriptions and examples provided. Moreover, we asked teachers to suggest new lexical terms to be included as well as improvements for the names and descriptions we had generated.

The Finnish national review survey consisted of six sections: (1) Demographics; (2) Questions about the terms (How familiar is the term? How often do you use the term? How often do your colleagues use the term? How often does the phenomenon referred to by the term happen?); (3) Suggesting lexical terms matched to term descriptions; (4) Familiarity of terms, when the full descriptions were given and suggestions for improvements requested; (5) Suggestions for additional terms (6) Thank you and contact information. We used a 5-point scale for sections 2 and 4. Four parallel versions of the survey were developed, rotating all lexical terms through sections 2 to 4. In each version each of the sections included 26 terms. Because we were worried about the length of the survey, we encouraged participants to skip the open response items and respond to the multiple-choice items if they were in a hurry.

National review data

The survey was distributed through the national mathematics teachers' union's weekly newsletter to 4400 recipients. The survey was also sent to about 200 recipients through the mailing list of the Finnish Mathematics and Science Education Researchers' Association, as well as to about 20 teachers the first author knew personally. The four different versions of the survey were randomised by asking the respondent to select one of four possible links based on the month of their birthday.

A total of 72 responses were received from mathematics teachers, all meeting the formal qualifications. As typical for Finnish teachers, most taught more than one subject. The secondary subjects taught were typically physics and/or chemistry (53), or computer science (11). Nine respondents taught only mathematics. Most respondents (45) taught at lower secondary level, 26 at upper secondary level, two at elementary level, six at vocational education level and three at tertiary education level. Fifteen respondents taught at more than one level.

The four different versions of the survey received 11 to 25 responses each, suggesting that the randomisation was not always followed. However, the respondents for the different versions represented variation in geography and age. Many respondents skipped the open response items, particularly towards the end of the survey. The number of suggested term names ranged from 4 to 20, and 140 improvements to the term descriptions were suggested. Based on the national review, we refined the Finnish lexicon. There were suggestions for adding a total of 49 new terms to the lexicon. After a review by the research team, we selected 40 terms that we intend to develop for a later national review.

Analysis

We analysed and evaluated the teachers' familiarity and use of terms according to a fixed criterion of at least two thirds of the respondents considering the term familiar or very familiar. When that criterion was not met, we considered the other survey results to decide whether or not to include the term in the Finnish Lexicon.

In addition to identifying whether the lexical terms were acceptable for the respondents, we also attempted to identify the most important terms. For the term to be important, we considered four different criteria: high familiarity, frequent usage, the typicality of the event in the class, and easy production of the term based on the description. Moreover, we considered that some terms might be important for teacher language even if not all of the four criteria are met. For example, a term describing a rarely occurring but influential event may be important. Based on these ideas, we defined a term to be important when it met at least two of the following four criteria:

- Rather or very familiar to over 90% of respondents
- Most (>50%) respondents reported that they or their colleagues used the term frequently (2 highest options)
- The respective event occurs frequently (2 highest options) in most respondents' classes (>50%)
- Most respondents (> 50%) are able to produce the correct term or its synonym based on the description provided.

Results

The results section includes an overall summary of the responses and the analysis identifying the most familiar terms.

The refined Finnish Lexicon includes 99 terms that are organised into six categories: Kasvatus (Upbringing: 15 terms), Organisointi (Organising: 12 terms), Arviointi (Assessment: 19 terms), Pedagogiset ratkaisut (Pedagogical tools and approaches: 31 terms), Matemaattiset termit (Mathematics Specific Terms: 13 terms), and Vuorovaikutus (Interaction: 9 terms). Only 25 terms were expressed by a single word, 26 were expressed as compound words and the remaining 48 terms were expressed as short phrases.

It may be a characteristic of Finnish language that it was not always easy to name events. First, we often had an option to choose between a noun (kehu, a praise) and a verb (kehua, to praise) but sometimes there was an option between different derived versions of the same basic word. For example, instead of *oivaltaminen* (realisation) we could have used the word "oivallus" which, depending on the context, could mean either a novel idea, or the event of realising an idea. For 12 terms we decided to provide alternative, synonymous names (e.g. *johdanto/orientointi/pohjustaminen*; introduction/orientation).

Familiarity and usage

The national review showed that teacher respondents were familiar with the terms, but not all terminology was in frequent use (Table 10.1). As the familiarity responses did not depend on whether we provided teachers with the term only or a longer description, we combined the two familiarity responses for further analysis. Similarly, the frequency of use was rather similar whether we asked how often the respondent or their colleagues use the term and we decided to combine these data in our future analysis.

Most of the terms in the national review were very familiar to almost all respondents, with 72 terms reaching over 4.5 on the 5-point scale. There were three terms that were very familiar to all respondents: *koe* (tests), *kertaus* (revision), and *demonstraatio* (demonstration).

Nine terms did not meet the 67% familiarity threshold. These terms were *oppilaan pilkkaaminen* (mocking a student), *käänteinen opetus* (flipped classroom), *luokan tai oppilaan antama kannustus* (peer encouragement), *vastauksen vahvistaminen* (confirming a response), *ratkaisun hylkääminen* (rejecting a solution), *työtavan pohjustus* (orienting for a working mode), *opettajan suosikki* (teacher's pet), *avustaja (oppilas opettajan apuna)* (helper; student helping the teacher), and *oppilastyö* (project work). Yet, even these terms were familiar to most teachers, with at least 52% of the respondents finding them familiar or very familiar.

While passive recognition of terms was generally good, some of the terms were not in the active vocabulary of the respondents. When asked about how often the teachers themselves, or their colleagues use these terms, we identified 21 terms that were used seldom or very seldom (i.e. average score smaller than 2.5). These included the nine words below the threshold for familiarity. Among the least frequently used ten words were also *vastauksen toistaminen* (teacher repeating the student response), *ryhmätyön purku* (group work debriefing), and *brainstorming/aivoriihi* (brainstorming). The ten most frequently used terms (in the order of frequency in use), were *koe* (tests), *eriyttäminen* (differentiating), *itsenäinen työskentely* (individual work), *sanallinen tehtävä* (word problem), *kertaus* (revision), *työrauha* (good working climate), *ongelmatehtävä tai pulma* (a problem or a puzzle), *soveltava tehtävä/sovellustehtävä* (application task), *itsearviointi* (self-evaluation), and *oivaltaminen* (realisation).

TABLE 10.1 The mean values and standard deviations for different survey item types.

Survey item type		$\bar{x}$	SD
Term only	How familiar?	4.5	0.89
	How often you use?	3.1	1.23
	How often your colleagues use?	3.0	1.14
	How often this thing happens?	3.7	1.14
Full description	How familiar?	4.5	0.85

It is important to note that all words that were familiar to the teachers were not frequently used. For example, the word *demonstraatio* (demonstration) was familiar to all respondents, but its use was just slightly above average at 3.31. The familiarity of the word was probably due to so many respondents teaching physics or chemistry (in addition to mathematics), where demonstrations are a key teaching method. Moreover, the seldom used terms *ryhmätyön purku* (group work debriefing) (familiarity score 4.00), and *brainstorming/aivoriihi* (brainstorming) (familiarity score 3.93), were well known by the respondents.

Important terms

While most reviewed terms were familiar to the respondents, we also used the more selective criteria for identifying important terms in the lexicon (Table 10.2).

The important terms included many that relate to the good relationship between teacher and students, such as *keventäminen* (use of humour to lighten the atmosphere), *luokkahenki* (classroom atmosphere), *työrauhan ylläpito* (cultivating good working climate), and *läsnäolo/välittäminen* (presence/caring).

Many important terms relate to how the lesson can be organised. For example, the following terms more or less define a typical Finnish mathematics lesson: *kotitehtävien tarkistaminen* (checking homework), *johdanto/orientointi/pohjustaminen* (introduction/orientation), *malliratkaisu* (worked-out example), *ohjeistus* (providing instructions), *materiaalin jakaminen* (distribution of materials), *itsenäinen työskentely* (individual work), *ohjaus* (guidance), *eriyttäminen* (differentiating), *koonti/yhteenveto* (making a summary), and *kotitehtävien antaminen* (homework assignment). Of course, there is some variation, as the terms *oppilaiden yhteistyö* (student collaboration), *ryhmätyö* (group work), *teknologian hyödyntäminen* (use of technology), *kertaus* (revision), and *vihkotyöskentely* (notebook work) indicate.

With respect to teacher–student interaction during guidance, we see here some interesting specificity of terminology: *oppilaan kysymys* (student's question), *perustelujen vaatiminen* (request for justification), and *oivaltaminen* (realisation).

Few words specific to mathematics met the criteria of important terminology. There were three terms for specific types of mathematical tasks as well as the term *oikea terminologia* (correct terminology). Furthermore, the terms *matematiikan teoria* (mathematical theory) and *asioiden rinnastaminen* (connect and contrast) teachers recognised quite well, but used very little.

Refining the Finnish Lexicon

Although the majority of the terms were well recognised, some terms or descriptions had to be reconsidered, because the names generated by the respondents did not always match the name we had chosen. For example, most suggested the names "class spirit" or "group spirit" for our description of "classroom climate" (see *luokkahenki* (classroom atmosphere)). As another example, 91% of the respondents recognised the term "exact mathematical language" as familiar, yet none of the respondents were able to produce the same exact term based on the description (see *oikea terminologia* (correct terminology)).

In the validation, we identified 22 terms that received ambiguous evaluation, i.e. the familiarity, usage and term generation provided conflicting results. For example, the three

TABLE 10.2 The most important terms in the Finnish Lexicon.

Category (in Finnish and English) important terms/all terms	Term (in Finnish)	Term (in English)
Kasvatus (Upbringing) 8/15	*kannustaminen/ tsemppaaminen*	encouragement and pep
	keventäminen	use of humour to lighten the atmosphere
	kiusaaminen	bullying
	koulun järjestyssäännöt	school rules
	luokan ilmapiiri	classroom climate
	läsnäolo/välittäminen	caring
	työrauha	good working climate
	työrauhan ylläpitäminen	cultivating good working climate
Organisointi (Organising) 8/12	*aikataulutus*	scheduling
	istumajärjestys	sitting arrangement
	läsnäolijoiden tarkastus	roll call
	ohjeistus	providing instructions
	tunnin aloitus	beginning the lesson
	tuntisuunnitelma	lesson plan
	välineet	equipment
Arviointi (Assessment) 10/19	*arvioinnista kertominen ja keskustelu*	explaining and discussing assessment
	itsearviointi	self-evaluation
	koe	tests
	kokeen palautus	returning assessed tests
	kotitehtävien antaminen	homework assignment
	kotitehtävien tarkistus	checking homework
	palautteen antaminen	providing feedback
	perustelujen vaatiminen	request for justification
	positiivisen palautteen antaminen / kehuminen	providing positive feedback / praising
	tavoitteiden asettaminen	setting assessment goals

Category (in Finnish and English) important terms/all terms	Term (in Finnish)	Term (in English)
Pedagogiset ratkaisut (Pedagogical Tools and Approaches) 11/31	eriyttäminen	differentiating
	itsenäinen työskentely	individual work
	johdanto/orientointi/ pohjustaminen	introduction/orientation
	kertaus	revision
	koonti/yhteenveto	making a summary
	malliratkaisu	worked-out example
	oivaltaminen	realisation
	ryhmätyö	group work
	teknologian hyödyntäminen	use of technology
	verkkomateriaalin käyttö	use of online material
	vihkotyöskentely	notebook work
Matemaattiset termit (Mathematics Specific Terms) 4/13	oikea terminologia	correct terminology
	päässälasku	mental calculation
	sanallinen tehtävä	word problem
	soveltava tehtävä/ sovellustehtävä	application task
Vuorovaikutus (Interaction) 6/9	kannustaminen eli tsemppaaminen	encouragement and pep
	ohjaus	guidance
	oppilaan kysymys	student's question
	oppilaan vastaus	student response
	oppilaiden yhteistyö	student collaboration
	viittaaminen	student raises hand

terms that describe an undesirable event in the classroom: "mocking a student" (see *oppilaan nolaaminen* (embarrassing a student)), "teacher's pet" (see *suosiminen* (favoritism)) and *lunttaaminen* (cheating in test) were not used by the teacher and these were not seen to happen in the class. However, teachers were very familiar with these terms and could even name these events correctly. We believe that it is important that the lexicon includes also terminology for undesired events, and hence we decided to keep these terms in the lexicon. However, we used the validation information to refine terminology.

Two other terms that received contradictory evaluations were specific pedagogical practices that seemed to be unevenly distributed among teachers: *käänteinen opetus* (flipped classroom) and *henkilökohtainen palautekeskustelu* (personal feedback discussion). These terms seem to relate to an emerging practice that some teachers already use, some teachers are aware of but do not yet use, and some teachers are not yet aware of.

For some terms, we had been unsuccessful in giving a name familiar to the teachers. For example, the respondents recognised the term "Lesson structure" but they preferred using the term *tuntisuunnitelma* (lesson plan), instead.

Conclusion

The national review of the Finnish mathematics teachers' lexicon indicated that most terms were familiar to the teachers. We have identified and validated 93 terms and included an additional six "nearly validated" terms for a Finnish lexicon for mathematics teachers. Based on teachers' responses and with special attention to the terms suggested for verbal descriptions, we made several refinements to the initial lexicon. For example, seven of the terms that were validated we renamed for the published lexicon:

- *opettajan suosikki* (teacher's pet) → *suosiminen* (favoritism)
- *oppilaan pilkkaaminen* (mocking a student) → *oppilaan nolaaminen* (embarrassing a student)
- *opettaja luennoi* (teacher lectures) → *luennointi* (lecturing)
- *oppilastyö* (student work) → *projektityö* (project)
- *eksakti matemaattinen kieli* (exact mathematical language) → *oikea terminologia* (correct terminology)
- *rutiinitehtävä* (routine task) → *perustehtävä* (fundamental task)
- *ongelmatehtävä tai pulma* (problem) → *ongelmanratkaisu* (problem-solving)

Furthermore, we combined two validated terms "hoputus" (hurrying) and "tuntitehtävien tarkastaminen" (checking classwork) with another similar term.

We realise that the number of teacher respondents to the national review survey was not high, (especially respondents who completed the least popular version of the survey (11)). Therefore, it is important to get additional validation data to make a more informative judgement regarding unfamiliar terms. In addition, the respondents suggested additional terms for the Finnish lexicon. Out of these we identified 41 terms that we intend to validate at a later stage. The suggested terms include, for example, "open task," "concept map," "learning to learn," "surface learning," "peer assessment," and "responsibility." Hence, the Finnish Lexicon is not yet in its final form, although we anticipate that it covers quite well the terminology teachers typically use.

Examining the Finnish Lexicon, it suggests that Finnish mathematics teachers conceptualise their teaching primarily through their relationship and interaction with their students, rather than through the teaching of mathematical content. One might argue that the extent of terminology related to a topic is not necessarily an indication of the perceived importance of that topic. However, if there is significant and continued attention and discussion on a topic, would that not inevitably lead to a more detailed vocabulary to foster such discussions?

When comparing these results with the earlier metaphor study conducted by Oksanen et al. (2014), we can see that the results of both studies suggest a primary focus on teachers' expertise in organising and orchestrating mathematics teaching, while some attention is given to general pedagogy ("Kasvatus"), and rather little attention is placed on mathematical content knowledge. Taken together, these studies indicate that the main focus of Finnish mathematics teachers – at least as expressed in their language – is on the act of teaching. They do pay some attention to student learning, but quite little to the mathematical content. Furthermore, in Oksanen and colleagues' study (2014) the metaphors were dominantly about learning as movement and teaching as guidance. This is nicely aligned with the Finnish term for scaffolding *ohjaus* (guidance) included in the Finnish Lexicon.

Hänninen, Iltanen and Öz (2018) have already used the draft lexicon to review the use of the terms from the Finnish mathematics teachers' lexicon in Finnish National curricula from 2004 and 2014. They found multiple differences in the use of different terms between these curricula. Most notably in the new curriculum, the terms *kannustaminen* (encouragement and pep) and *eriyttäminen* (differentiating) had replaced *päättely* (deduction) and *ongelmanratkaisu* (problem-solving) in the list of five most frequently used lexical terms in the curriculum as a whole. *Ohjaus* (guidance) in a general sense was mentioned significantly more frequently in the new curriculum (1686 mentions) than in the old curriculum (131 mentions). They also found some of the terms only in the new curriculum, for example: *läsnäolo* (presence), *oppilaan kysymys* (student's question), *tutkimustehtävä* (investigation), and *kasvatuskeskustelu* (discussion about student behaviour). Their conclusion was that the changes demonstrate a change towards a more individual approach concerning students.

While the Finnish Lexicon provides a progressive view regarding teacher–student relations, it also reflects a rather conservative view regarding teaching methods in Finnish mathematics classrooms. Yet, new teaching and assessment methods are emerging in Finnish classrooms, reflecting the new National Curriculum (Finnish National Board of Education, 2014). A broader view of assessment focusing more on guidance and other ways of formative assessment are highlighted in the new curriculum, as well as larger cross-subject projects and phenomenon-based studying. Some terms that did not meet the familiarity criteria of the validation but were well recognised by some teachers may well be on the verge of becoming mainstream educational vocabulary, or they are stabilising their place in Finnish educational vocabulary as terms of phenomena that are already present and expanding in Finnish schools. The Finnish Lexicon may provide teachers the needed vocabulary to help the transition to new methods and support discussions about them among colleagues.

Acknowledgments

Two teachers, Rita Järvinen and Maarit Rossi, were essential in creating the first version of the Finnish Lexicon. We also wish to acknowledge the important work of our research assistants Verneri Valasmo, Enrique Garcia Moreno-Esteva and Emmanuel Bofah, as well as Josephine Moate who provided helpful language reviewing throughout the project.

References

Finnish National Board of Education. (2014). *National core curriculum for basic education*. Helsinki, Finland: National Board of Education.

Hänninen, S., Iltanen K., & Öz, M. (2018). Matematiikan opettajien käyttämä kieli opetussuunnitel-mateksteissä. Suomalaisen matematiikan opettajilta kerätyn termistön vertailua vuosien 2004 ja 2014 perusopetuksen opetussuunnitelmissa. [Mathematics teachers' language in curriculum documents: Comparison of Finnish mathematics teachers' terminology in 2004 and 2014 comprehensive education curricula.] Unpublished seminar work.

Hannula, M.S., & Oksanen, S. (2013). Opettajamuuttujien yhteys osaamisen muutokseen. [The relation between teacher variables and change in learning outcomes.] In J. Metsämuuronen (Ed.), *Perusopetuksen matematiikan oppimistulosten pitkittäisarviointi vuosina 2005–2012* [The longitudinal assessment of mathematics learning outcomes in basic education years 2005–2012] (pp. 255–296). (Koulutuksen seurantaraportit; No. 2013:4). Helsinki, Finland: Opetushallitus.

Leiwo, M., Kuusinen, J., & Kuusisto, A. (1981). *Opettajan ja oppilaan kielellinen vuorovaikutus: 1, Opetusdiskurssin kuvaus.* [The linguistic interaction between teacher and student: 1, Description of the teaching discourse.] Jyväskylä, Finland: Jyväskylän yliopisto.

Leiwo, M., Kuusinen, J., Nykänen, P., & Pöyhönen, M.R. (1987) *Kielellinen vuorovaikutus opetuksessa ja oppimisessa II. Peruskoulun luokkakeskustelun määrällisiä ja laadullisia piirteitä.* [Linguistic Interaction in Teaching and Learning II. Classroom Discourse and its Quantitative and Qualitative Characteristics]. Publication Series, Research Report 3. Jyväskylä: Institute for Educational Research, University of Jyväskylä.

Mesiti, C., Clarke, D.J., Roan, K., Hollingsworth, H., Yiming, C., Guowen, Y., Novotna, J., Zlbkova, I., & Dobie, T. (2016). Discourse about the mathematics classroom. In C. Csikos, A. Rausch, & J. Szitányi (Eds.), *Mathematics Education: How to Solve It? PME40* (pp. 357–363). Szeged, Hungary: PME.

Miestamo, M. (2006). Suomi maailman kielten joukossa eli mikä suomen rakenteessa onkaan erityistä. [Finnish among the World languages, or what is that is specific in the structure of Finnish.] In M. Harmanen & M. Siiroinen (Eds.), Kielioppi koulussa. [Grammar in schools.] Helsinki, Finland: Äidinkielen opettajain liitto.

Norris, N., Asplund, R., MacDonald, B., Schostak, J., & Zamorski, B. (1996). *An independent evaluation of comprehensive curriculum reform in Finland.* Helsinki, Finland: National Board of Education.

OECD (2013). *PISA 2012 Results: Excellence through equity: Giving every student the chance to succeed* (Volume II). Paris, France: OECD.

OECD (2019). *TALIS 2018 Results (Volume I): Teachers and School Leaders as Lifelong Learners.* Paris, France: OECD.

Oksanen, S., Portaankorva-Koivisto, P., & Hannula, M.S. (2014). Teacher metaphors - differences between Finnish in-service and pre-service mathematics teachers. In P. Liljedahl, S. Oesterle, C. Nicol, & D. Allan (Eds.), *Proceedings of the Joint Meeting of PME 38 and PME-NA 36* (Vol. 4, pp. 361–368). Vancouver, Canada: PME.

Santavirta, N., Aittola, E., Niskanen, P., Pasanen, I., Tuominen, K., & Solovieva, S. (2001). *Nyt riittää. Raportti peruskoulun ja lukion opettajien työympäristöstä, työtyytyväisyydestä ja työssä jaksamisesta* [Enough! A Report on the Work Environment, Satisfaction and Stress of Teachers at Comprehensive and Upper Secondary School]. Helsinki, Finland: University of Helsinki.

Simola, H. (1998). Decontextualizing teachers' knowledge: Finnish didactics and teacher education curricula during the 1980s and 1990s. *Scandinavian Journal of Educational Research,* 42(4), 325–338.

Simola, H., Kauko, J., Varjo, J., Kalalahti, M., & Sahlström. F. (2017). *Dynamics in education politics – Understanding and explaining the Finnish case.* London: Routledge.

Syrjäläinen, E. (2002). *Eikö opettaja saisi jo opettaa? Koulun kehittämisen paradoksi ja opettajan työuupumus* [Can't teachers just teach? The paradox of developing school and the exhaustion of teachers]. Tampere, Finland: Tampereen yliopisto.

Taajamo, M., Puhakka, E., & Välijärvi, J. (2014). *Opetuksen ja oppimisen kansainvälinen tutkimus TALIS 2013. yläkoulun ensituloksia* [International study on teaching and learning TALIS 2013]. Helsinki, Finland: Opetus- ja kulttuuriministeriö.

Virta, A., & Kurikka, T. (2001). Peruskoulu opettajien kokemana. [Comprehensive school as expereiced by the teachers.] In T.E. Olkinuora & E. Mattila (Eds.), *Miten menee peruskoulussa? Kasvatuksen ja oppimisen edellytysten tarkastelua Turun kouluissa* [How is it going in comprehensive school? Inspecting the conditions for teaching and learning in Turku schools] (pp. 55–86). Turku, Finland: Turun yliopisto.

11

FINNISH LEXICON

Markku S. Hannula, Fritjof Sahlström, Jani Kiviharju, Maarit Rossi and Rita Järvinen

A Finnish Lexicon

The lexicon presented in this chapter consists of 99 terms considered familiar by Finnish teachers in the mathematics education community. All terms are presented in the original language (Finnish); a literal translation into English; and a closest English translation. Included is a general description of the classroom phenomena with a classroom illustration. These examples and non-examples are provided in English. Together, these elements help give a more complete operational definition of the term.

The first lexicon table (see Table 11.1) lists the terms of the lexicons in Finnish, the literal translation is given in quotation marks and the closest English translation is listed below these. The literal translation is only provided if it differs from the closest English translation. The second lexicon table (see Table 11.2) includes the operational definition of each term arranged into three columns. The first column lists the detail of the term itself, the second column gives its description while examples and non-examples from the classroom are found in column three (Table 11.3).

The 99 terms of the Finnish Lexicon are organised into six categories: Arviointi (Assessment) (19 terms), Kasvatus (Upbringing) (15 terms), Matemaattiset Termit (Mathematics Specific Terms) (13 terms), Organisointi (Organising) (12 terms), Pedagogiset Ratkaisut (Pedagogical Tools and Approaches) (31 terms), and Vuorovaikutus (Interaction) (9 terms).

The Finnish Lexicon has been arrived at following negotiations between the researchers in the national team, mathematics teachers, and members of the mathematics education community. As discussed in *Chapter 10*, the Finnish Lexicon reflects terminology that is more focused on teacher–student interaction and lesson organisation in comparison with terms that name specific aspects of mathematics teaching. Our hope for this lexicon is that it contributes to conversations about teaching and can be used for the study of teacher practice.

TABLE 11.1 Finnish Lexicon – Terms.

aiempaan viittaaminen "referring to prior content" (linking)	**aikataulutus** "timetabling" (scheduling)	**ajanhallinta** "time control" (lesson time management)	**arkimatematiikka** (everyday math)
arvioinnista kertominen ja keskustelu "telling and discussing assessment" (explaining and discussing assessment)	**arviointikeskustelu** (assessment discussion)	**asioiden rinnastaminen** "positioning things side by side" (connect and contrast)	**avustaja (oppilas opettajan apuna)**[1] "assistant (student as teacher's help)" (student as teacher's assistant)
brainstorming/ aivoriihi/ideariihi "idea workshop" (brainstorming)	**demonstraatio** (demonstration)	**eriyttäminen** (differentiating)	**etenemisen seuranta** (monitoring progress)
havainnollistaminen "visualisation" (visualisation/ concretisation)	**henkilökohtainen palautekeskustelu** (personal feedback discussion)	**istumajärjestys** (sitting arrangement)	**itsearviointi** (self-evaluation)
itsenäinen tehtävän tarkistaminen "autonomous control of task" (independently checking solutions)	**itsenäinen työskentely** "autonomous working" (individual work)	**johdanto/orientointi/ pohjustaminen** "leading in/ orientation" (introduction/ orientation)	**kannustaminen eli tsemppaaminen** "spurring/ (promoting) fighting spirit" (encouragement and pep)
kasvatuskeskustelu "upringing ('kasvatus') discussion" (discussion about student behaviour)	**kehotus/aktivoiminen/ hoputtaminen** "suggesting/activating" (prompting to work)	**kertaus** "repetition" (revision)	**keventäminen** "lightening up (light as opposite to heavy)" (use of humour to lighten the atmosphere)
kiusaaminen "antagonising" (bullying)	**kodin ja koulun yhteistyö**[2] "collaboration of home and school" (collaboration between the teacher and guardians)	**koe** (tests)	**kokeen palautus** "returning test" (returning assessed tests)
koonti/yhteenveto "collecting/pulling together" (making a summary)	**kotitehtävien antaminen** "giving home tasks" (homework assignment)	**kotitehtävien tarkistus** "control of home tasks" (checking homework)	**koulun järjestyssäännöt** "school's order rules" (school disciplinary rules)

kuri (discipline)	**käänteinen opetus/flipped classroom**[1,2] "flipped teaching" (flipped classroom)	**luennointi**[2] (lecturing)	**lunttaaminen** (cheating (in a test))
luokan haltuun ottaminen "taking class into control" (getting the students' attention)	**luokan säännöt ja käytännöt** "classroom rules and practices" (classroom rules and norms)	**luokan tai oppilaan antama kannustus**[1] "spurring by the class or another student" (peer encouragement)	**luokkahenki** "class spirit" (classroom atmosphere)
läsnäolijoiden tarkistus "controlling those being present" (roll call)	**läsnäolo/välittäminen** (presence/caring)	**malliratkaisu** "model solution" (worked-out example)	**matematiikan teoria** "theory of mathematics" (mathematical theory)
materiaalin jakaminen "sharing material" (distribution of materials)	**ohjaus** "steering" (guidance)	**ohjeistus** "instructioning" (providing instructions)	**oikea terminologia**[2] "right terminology" (correct terminology)
oivaltaminen "epiphany" (realisation)	**ongelmanratkaisu**[2] (problem-solving)	**opettaja kiertää luokassa** "teacher goes around in class" (teacher monitors and guides classwork)	**opetuskeskustelu** "teaching discussion" (questioning)
opetusvideo tai –animaatio "teaching video or animation" (video or animation)	**oppikirjan lukeminen** "reading the study book" (reading the textbook)	**oppilaan kysymys** (student's question)	**oppilaan nolaaminen**[1,2] (embarrassing a student)
oppilaan ratkaisun esittäminen "presentation of student solution" (presentation of student work)	**(oppilaan) ratkaisun läpikäynti** "going through (student) solution" (assessment of (student) solution)	**oppilaan vastaus** (student response)	**oppilaiden yhteistyö** (student collaboration)
oppimispeli/ oppimisleikki "learning game, learning play" (play or game)	**oppitunnin häiriöt** "lesson disturbances" (disturbances)	**palautteen antaminen** (giving feedback)	**perustehtävä**[2] "basic task" (fundamental task)
perustelujen vaatiminen "requesting justification" (request for justification)	**pistokoe, pistari** "stabtest" (quiz)	**positiivisen palautteen antaminen/kehuminen** (giving positive feedback /praising)	**projektityö**[1,2] "project work" (student project)

(Continued)

TABLE 11.1 (*Continued*)

pänttääminen/ulkoa opettelu "cramming/learning (from) out(side)" (learning by heart)	päässälasku "counting in head" (mental calculation)	päättely "making ends/concluding" (deduction)	ryhmien muodostaminen "forming groups" (grouping)
ryhmätyö (group work)	**ryhmätyön purku** "group work deconstruction" (group work debriefing)	**sanallinen tehtävä** "worded problem" (word problem)	**soveltava tehtävä/sovellustehtävä** "applied task" (application task)
suosiminen[1,2] "favouring" (favoritism)	**tapakasvatus** "manners education" (promotion of good manners)	**tavoitteiden asettaminen** "setting goals" (setting assessment goals)	**teknologian hyödyntäminen** "benefitting from technology" (use of technology)
tunnin aloitus "lesson's beginning" (beginning the lesson)	**tunnin lopetus** "lesson's ending" (ending the lesson)	**tunnin rakenne (punainen lanka)** "structure of the lesson (a red thread)" (lesson structure (common thread))	**tuntisuunnitelma** "hour plan" (lesson plan)
tuntitehtävä "hour task" (classwork)	**tutkimustehtävä** "research task" (investigation)	**työrauha** "work peace" (good working climate)	**työrauhan ylläpito** "work peace maintenance" (cultivating good working climate)
vastauksen toistaminen "repeating response" (teacher repeating the student's response)	**vastauksen täsmentäminen** "making response precise" (clarifying and making a student's response precise)	**verkkomateriaalin käyttö** "using internet material" (use of online material)	**vihje** (hint)
vihkotyöskentely "notebook working" (notebook work)	**viittaaminen** "signaling/pointing" (student raises hand)	**virheen korjaaminen/käsittely** "repairing an error/handling an error" (correction or treatment of an error)	**väittely** "argumenting" (debate)
välineet (equipment)	**yhteistoiminnallinen oppiminen** (cooperative learning)	**yleistäminen** (generalising)	

[1] Term was not validated as familiar or very familiar.
[2] Term was revised after validation.

TABLE 11.2 Finnish Lexicon – Terms with operational definitions.

Term	Description	Examples and Non-examples
aiempaan viittaaminen "referring to prior content" (linking)	The teacher refers to prior content and shows its relationship to current content.	*Example:* The teacher points out similarities and differences between a solution under discussion and one presented earlier in the same lesson. *Non-example:* *Revision of an earlier studied content.*
aikataulutus "timetabling" (scheduling)	The teacher determines time for a task and monitors students' progress with it. If necessary, may extend time.	*Example:* The teacher gives a task for students and tells that they have ten minutes to work on it. *Non-example:* *The teacher schedules the lesson but does not inform students how much time they have for each task.*
ajanhallinta "time control" (lesson time management)	The teacher's competence in planning and controlling the lesson events so, that there is ample time for all activities.	*Example:* The teacher manages to complete all planned activities. *Non-example:* *The teacher begins and ends the lesson in time.*
arkimatematiikka (everyday math)	A mathematical exercise that is directly linked to a common situation outside school.	*Example:* Students calculate the net salary of a summer job taking into account the tax rate. *Non-example:* *Graphing functions in a coordinate system.*
arvioinnista kertominen ja keskustelu "telling and discussing assessment" (explaining and discussing assessment)	The teacher talks about criteria and practices for evaluation.	*Example:* The teacher discusses with students about the weight of a collaborative task in determining final grade. *Non-example:* *The teacher discusses with students about the correct solutions in test.*
arviointikeskustelu (assessment discussion)	The teacher discusses with a student in order to get an understanding of the student's competences.	*Example:* The teacher invites the student to present and discuss his or her portfolio for the semester to determine his or her grade. *Non-example:* *The teacher tells the student the reasons for the grade he or she gave the student.*

(Continued)

TABLE 11.2 (*Continued*)

Term	Description	Examples and Non-examples
asioiden rinnastaminen "positioning things side by side" (connect and contrast)	To compare two things in order to identify differences and similarities.	Example: When learning about the area of a circle, the teacher points the similarities and differences with circumference of a circle. *Non-example: Building an equation.*
avustaja (oppilas opettajan apuna)[1] "assistant (student as teacher's help)" (student as teacher's assistant)	The teacher assigns a student to assist the teacher.	Example: A student works as a scribe on board as the teacher dictates. *Non-example: A volunteering student is giving out the material in class.*
brainstorming/aivoriihi/ ideariihi "idea workshop" (brainstorming)	Shared generation of ideas and recording them without immediate evaluation.	Example: The teacher asks students to name games that have a random element as an introduction to probability. *Non-example: The teacher gives mental computations to students to solve.*
demonstraatio (demonstration)	The teacher shows something to the class and students are expected to notice something through their observation.	Example: The teacher uses rice and hollow solids to show that the volume of a cube is three times the volume of a pyramid. *Non-example: The teacher shows a solution method.*
eriyttäminen (differentiating)	The teacher gives different guidance, questions or tasks according to student competence and disposition.	Example: Allowing students to choose between easier of more challenging tasks. The teacher gives different guidance to different students, even for the same task.
etenemisen seuranta (monitoring progress)	The teacher checks how students are making progress.	Example: The teacher walks around the class to see how far students have advanced. *Non-example: The teacher assesses a test.*
havainnollistaminen "visualisation" (visualisation/ concretisation)	Mathematical concepts or operations are represented in ways that allow the learner to observe and manipulate the representation.	Example: Student studies geometric shapes using Geogebra. *Non-example: Student draws the graph of a given function.*

Term	Description	Examples and Non-examples
henkilökohtainen palautekeskustelu (personal feedback discussion)	The teacher discusses privately with a student about his/her progress in studies.	Example: Students are invited to individual feedback discussion regarding their progress. *Non-example: The teacher discusses a student's behaviour.*
istumajärjestys (sitting arrangement)	The designation of students to their seats in the class.	Example: The teacher asks a disturbing student to come sit to the first row. The teacher asks students to form groups of four and to arrange the seats accordingly.
itsearviointi (self-evaluation)	The teacher requests students to evaluate their own learning.	Example: After students have worked on individual tasks, the teacher asks them to raise their hands if they felt that the tasks were easy. *Non-example: The teacher invites the students to ask clarifying questions.*
itsenäinen tehtävän tarkistaminen "autonomous control of task" (independently checking solutions)	Student uses the internet, textbook or a specific results book to see the correct answer and compares it with his or her own result.	Example: The teacher positions a results book on a desk where students can go and check if they have a correct result. *Non-example: Students tell their solutions and the teacher confirms if they are correct.*
itsenäinen työskentely "autonomous working" (individual work)	Students work independently and at their own pace. Students can commonly collaborate.	Example: The teacher gives students a number of tasks to work on and advance in individual pace. *Non-example: Doing homework.*
johdanto/orientointi/ pohjustaminen "leading in/orientation" (introduction/ orientation)	Before actually going into the topic, the teacher orients students to the area.	Example: Showing news that use graphs to introduce statistics. *Non-example: The teacher presents the course schedule.*
kannustaminen eli tsemppaaminen "spurring/(promoting) fighting spirit" (encouragement and pep)	The teacher instills confidence in his/her student(s).	Example: The teacher tells that the new topic may seem confusing at first, but that he or she fully trusts that students will eventually get it.

(Continued)

TABLE 11.2 (*Continued*)

Term	Description	Examples and Non-examples
kasvatuskeskustelu "upringing ('kasvatus') discussion" (discussion about student behaviour)	The teacher discusses with a student about their behaviour intending change.	Example: The teacher asks a student to come outside the class, where they have a discussion about the student's behaviour. *Non-example: The teacher sends a message to a student's parents.*
kehotus/aktivoiminen/ hoputtaminen "suggesting/activating" (prompting to work)	The teacher prompts a student or students to begin, continue, or speed up work.	Example: The teacher notifies students that the lesson is about to end soon and encourages students to complete any unfinished task hastily.
kertaus "repetition" (revision)	Revision of previously taught content.	Example: At the beginning of new content area, the teacher revises the relevant previously taught content. *Non-example: The teacher repeats something that was just said.*
keventäminen "lightening up (light as opposite to heavy)" (use of humour to lighten the atmosphere)	Teaching includes elements that are intended to amuse students.	Example: The teacher repeatedly uses their own pet in examples for tasks.
kiusaaminen "antagonising" (bullying)	A student is repeatedly and systematically target of another student's actions that aim to harm him or her.	Example: Verbal abusing in class. *Non-example: Two well-matched rivalling students quarrel or fight with each other.*
kodin ja koulun yhteistyö[2] "collaboration of home and school" (collaboration between the teacher and guardians)	The teacher's communication with a student's legal guardians, their joint discussions, and collaboratively agreed principles for the student's upbringing.	Example: The teacher makes a reference to things agreed at the parents' meeting. *Non-example: The teacher gives homework.*
koe (tests)	A task or a series of tasks that each student gives solutions to individually.	Example: An intermediate test about halfway through a larger content area.

Term	Description	Examples and Non-examples
kokeen palautus "returning test" (returning assessed tests)	The teacher returns tests to students, where tasks are corrected, errors marked and grade given.	Example: The teacher presents solutions to all exam tasks and discusses different solutions with students, then returns their exam papers. *Non-example: The teacher posts exam results on the electronic platform.*
koonti/yhteenveto "collecting/pulling together" (making a summary)	Teacher sums up taught content.	Example: After a multiple step solution, the teacher summarises the stages.
kotitehtävien antaminen "giving home tasks" (homework assignment)	The teacher tells what tasks to do at home and possibly gives some instruction.	Example: The teacher assigns specific tasks to do as homework. *Non-example: Assigning lesson tasks.*
kotitehtävien tarkistus "control of home tasks" (checking homework)	Presenting solutions for homework and discussing them.	Example: Volunteering students write their solutions to homework on the board and the teacher checks them.
koulun järjestyssäännöt "school's order rules" (school disciplinary rules)	The school level official rules determining the limits of appropriate behaviour and consequences for breaking the rules.	Example: Coming late to lessons three times leads to detention.
kuri (discipline)	Controlling of student actions based on school rules, threat of punishment, and/or the teacher's authority.	Example: The teacher uses stern voice to tell a student to stop disturbing. *Non-example: The teacher praises a student for their good behaviour.*
käänteinen opetus/flipped classroom[1,2] "flipped teaching" (flipped classroom)	A model of blended learning, which uses learning material on the Internet for home studying prior to class.	Example: The teacher has provided a teaching video for students to see through as homework. In class, students start to solve tasks straight on, the teacher monitors their progress and helps students.

(*Continued*)

TABLE 11.2 (*Continued*)

Term	Description	Examples and Non-examples
luennointi[2] (lecturing)	A teacher-centred teaching method, where the teacher is presenting a broader content to the whole class.	Example: The teacher presents a new topic, asking some questions to activate students. *Non-example: Teacher-led teaching discussion, where essential thoughts are elicited from students.*
lunttaaminen (cheating (in a test))	A student cheats in a test.	Example: A student is peaking at another student's solution in a test. *Non-example: Students do their homework together.*
luokan haltuun ottaminen "taking class into control" (getting the students' attention)	The teacher summons students" attention to an activity. Typically when class moves from a student-centred to a teacher-centred working mode.	Example: During a collaborative activity the teacher realises that the groups need additional instruction and they calls the class" attention to the board where they give additional instruction. *Non-example: The teacher takes the attention of a single student.*
luokan säännöt ja käytännöt "classroom rules and practices" (classroom rules and norms)	Norms and rules determined by the teacher or negotiated by the teacher and students that are not in the official documents.	Example: Mobile phones are to be set silent and kept in bags for the duration of the lesson. *Non-example: Laws and school disciplinary rules.*
luokan tai oppilaan antama kannustus[1] "spurring by the class or another student" (peer encouragement)	A student or students cheer or applaud another student's good performance.	Example: After a student's presentation peers applaud and cheer. *Non-example: A student says to another student, that their solution is correct.*
luokkahenki "class spirit" (classroom atmosphere)	The teacher's and students' social cohesion and level of feeling comfortable in class.	Example: There is continuous need for the teacher to control students' behaviour in class (poor atmosphere). *Non-example: The students' academic performance in the class is good.*
läsnäolijoiden tarkistus "controlling those being present" (roll call)	The teacher checks who are present and records it.	Example: While students work on a task, the teacher records students who are present.

Term	Description	Examples and Non-examples
läsnäolo/välittäminen (presence/caring)	The teacher takes contact with students in a way that builds confidential and caring relationship between teacher and students.	Example: The teacher asks the student to stay after the lesson and then asks if everything is OK. *Non-example: The teacher interrupts bullying and penalises the bully.*
malliratkaisu "model solution" (worked-out example)	Teacher-led production of a solution that is intended to serve as a model for students.	Example: The teacher writes a new type of task on a board and solves it while elaborating each step. *Non-example: The teacher presents a solution for a task that students were not able to solve as homework.*
matematiikan teoria "theory of mathematics" (mathematical theory)	Things to be taught that relate to mathematical concepts and their relationships.	Example: Definition of power and derivation of the formulae to calculate with them. *Non-example: Calculating the ratio of volume and area.*
materiaalin jakaminen "sharing material" (distribution of materials)	Giving material to students in class.	Example: The teacher hands out a new notebook for the students.
ohjaus "steering" (guidance)	The teacher guides and aids students' learning process, commonly with questions.	Example: During collaborative project the teacher gives hints for the solution. *Non-example: The teacher instructs which tasks students should do.*
ohjeistus "instructioning" (providing instructions)	The teacher instructs students.	Example: The teacher tells what manipulatives to use for the activity. *Non-example: The teacher instructs a solutions method.*
oikea terminologia[2] "right terminology" (correct terminology)	The teacher uses precise language and terminology and trains students for precise expression.	Example: The teacher rejects "shifting terms" as an answer and requests a more accurate expression.

(Continued)

TABLE 11.2 (*Continued*)

Term	Description	Examples and Non-examples
oivaltaminen "epiphany" (realisation)	A student learns/ understands new content.	Example: While studying a new topic, a student suddenly says, "Now I got it!" *Non-example:* *A student solves a task through imitation of a model solution.*
ongelmanratkaisu[2] (problem-solving)	A process utilising mathematical knowledge to find a solution to a problem that at first doesn't have a solution method.	Example: Student groups ponder a word problem that has no given method solution method, and which requires consideration and insight. *Non-example:* *Solving a multistep problem using a method that was taught recently.*
opettaja kiertää luokassa "teacher goes around in class" (teacher monitors and guides classwork)	Teacher monitoring and support during student-centred work.	Example: During group work the teacher walks around in the class and monitors students' progress.
opetuskeskustelu "teaching discussion" (questioning)	A teaching method, where the teacher asks questions and continuation questions from the class, a group of students, or from an individual student.	Example: After students have drawn graphs of a second-degree polynomial functions with GeoGebra, the teacher leads a discussion about the relationships between the function's graph and a corresponding equation.
opetusvideo tai –animaatio "teaching video or animation" (video or animation)	The teacher shows to the whole class or students watch individually a video or an animation.	Example: Students watch an animated proof for a Pythagorean theorem using their smartphones. *Non-example:* *Students generate a video or an animation.*
oppikirjan lukeminen "reading the study book" (reading the textbook)	The teacher or a student reads from the textbook.	Example: Students are asked to read what their textbook tells about Pythagoras. *Non-example:* *A student uses a textbook example as a model when solving a task.*
oppilaan kysymys (student's question)	Student utterance, which the teacher is expected to answer.	Example: A student interrupts the teacher's presentation to ask for clarification. *Non-example:* *A student asks the teacher to confirm if a solution is correct.*

Term	Description	Examples and Non-examples
oppilaan nolaaminen[1,2] (embarrassing a student)	The teacher or student's peers disrespect a student by saying hurtful things or setting them up to be laughed at.	Example: When the teacher points out an error in a student's solution, a peer mocks the student. *Non-example:* *A student makes a funny comment about their own mistake and the class laughs.*
oppilaan ratkaisun esittäminen "presentation of student solution" (presentation of student work)	The teacher or a student presents student's own work in the class.	Example: The teacher presents a solution of a student using a document camera.
(oppilaan) ratkaisun läpikäynti "going through (student) solution" (assessment of (student) solution)	The teacher (possibly with other students) examines a student's output.	Example: The teacher edits a student's solution to a task to improve clarity by adding intermediate stages.
oppilaan vastaus (student response)	A student responds to a question posed in the class.	Example: A student gives an answer to a question posed by the teacher. *Non-example:* *A student gives advice to a peer.*
oppilaiden yhteistyö (student collaboration)	A student receives help from a peer or helps a peer.	Example: During individual work a student helps a peer.
oppimispeli/oppimisleikki "learning game, learning play" (play or game)	Play or game that is intended to promote learning of mathematics.	Example: A board game that can be played in small groups to learn mathematics. *Non-example:* *A team building game with no mathematical content.*
oppitunnin häiriöt "lesson disturbances" (disturbances)	Different incidents that break the normal flow of the lesson.	Example: A student leaves in the middle of the lesson (for example, to a dentist). *Non-example:* *A pair of students discusses a task lively and noisily.*
palautteen antaminen (giving feedback)	The teacher tells to a student or students what were the strengths and weaknesses of their work or behaviour and how they could improve.	Example: The teacher tells that an idea presented by a student was creative.

(*Continued*)

TABLE 11.2 (*Continued*)

Term	Description	Examples and Non-examples
perustehtävä[2] "basic task" (fundamental task)	A mathematical exercise with a straightforward method of solution that all students should learn to solve.	Example: Solving an equation with a taught method.
perustelujen vaatiminen "requesting justification" (request for justification)	After a student's response the teacher asks the student to tell the reasons for the answer, and possibly to develop or elaborate.	Example: After a student's response, the teacher asks: "How do you know it?" *Non-example: The teacher requires writing intermediate stages for a solution.*
pistokoe, pistari "stabtest" (quiz)	A short test given to students without advance warning.	Example: A few simple tasks testing if the students got the new topic.
positiivisen palautteen antaminen/kehuminen (giving positive feedback /praising)	The teacher thanks or praises based on what has happened.	Example: The teacher tells a student that he or she has made progress over the semester.
projektityö[1,2] "project work" (student project)	A student or students work for a longer time on a project and produce a tangible output.	Example: Students make an animation of the values of circumference and area of a circle as a function of the radius.
pänttääminen/ulkoa opettelu "cramming/learning (from) out(side)" (learning by heart)	Determined studying, where one tries to learn something by heart through repetition.	Example: A student prepares for the exam through repeating certain formulae until they know it by heart.
päässälasku "counting in head" (mental calculation)	A student produces a result directly without help of tools or intermediate stages.	Example: A student calculates without the use of tools how much an item costing 80 euros costs with a 20% discount.
päättely "concluding" (deduction)	Thinking that advances from known facts to new conclusions through logically sound steps.	Example: A word problem is solved logically step by step.
ryhmien muodostaminen "forming groups" (grouping)	The teacher determines that there will be work in groups and instructs the students how to form the groups.	Example: The teacher asks students to form groups of four for an activity.

Term	Description	Examples and Non-examples
ryhmätyö (group work)	Students work in groups with a shared product as the outcome.	Example: Students make a survey in the class and produce a graphical representation of the results.
ryhmätyön purku "group work deconstruction" (group work debriefing)	The teacher goes through core content of group tasks with whole class.	Example: Each group presents their solution and the teacher facilitates a whole class discussion on it. *Non-example:* *The product of the group work is displayed.*
sanallinen tehtävä "worded problem" (word problem)	A mathematical exercise that is presented as text rather than in mathematical notation.	Example: A task from the textbook that is presented in an everyday context and vocabulary.
soveltava tehtävä/ **sovellustehtävä** "applied task" (application task)	Task where acquired knowledge is used to solve a problem in a realistic context.	Example: Students calculate expenses of an imaginary mobile subscription with given data usage. *Non-example:* *Students solve word problems from the textbook.*
suosiminen[1,2] "favouring" (favoritism)	A certain student receives more teacher's attention or is otherwise treated better by the teacher than other students.	Example: The teacher gives one student substantially more attention than to other students.
tapakasvatus "manners education" (promotion of good manners)	The teacher aims to improve students' manners through models of good habits and expects students to exhibit them.	Example: The teacher greets students in a friendly manner and expects them to greet in response.
tavoitteiden asettaminen "setting goals" (setting assessment goals)	The teacher tells what students are expected to know after the topic has been studied. Alternatively, students may set their own goals.	Example: The teacher explains that by the end of the semester, they will be able to multiply and divide fractions. Students may set goals, such as solving basic tasks, solving basic and moderate tasks, or solving difficult tasks. *Non-example:* *The teacher explains the assessment criteria.*

(*Continued*)

TABLE 11.2 (*Continued*)

Term	Description	Examples and Non-examples
teknologian hyödyntäminen "benefitting from technology" (use of technology)	The teacher or a student uses calculator, computer, or another similar tool.	Example: A student draws graphs of functions with GeoGebra.
tunnin aloitus "lesson's beginning" (beginning the lesson)	An event that determines the beginning of the lesson.	Example: The teacher says, "Good morning!" to the class.
tunnin lopetus "lesson's ending" (ending the lesson)	An event that determines the end of the lesson.	Example: The teacher has assigned individual work until the lesson ends. The bell rings to mark the end of the lesson.
tunnin rakenne (punainen lanka) "structure of the lesson (a red thread)" (lesson structure (common thread)	The relation between the different parts of the lesson, that is intended to form a balanced whole.	Example: The lesson consists of a) an introductory activity motivating to the topic, b) a worked out example, c) individual work, and d) application tasks aiming at procedural fluency.
tuntisuunnitelma "hour plan" (lesson plan)	The teacher's plan for the activities for the lesson.	Example: The teacher's rough plan for the lesson events that is not written down anywhere.
tuntitehtävä "hour task" (classwork)	Task that is given to students during a lesson and that students are expected to solve during class.	Example: The teacher writes a task on board for the students to solve in class. *Non-example:* *The teacher uses questioning to solve a task on board together with the class.*
tutkimustehtävä "research task" (investigation)	Students investigate a mathematical phenomenon as instructed by the teacher.	Example: Group task on measuring perimeter and diameter for different circles and then calculating their relationship.
työrauha "work peace" (good working climate)	Conditions in the class allow focused studying.	Example: Students work in pairs, focused on their work and discussing with their pair in low voice. *Non-example:* *Students are quiet but disengaged.*

Term	Description	Examples and Non-examples
työrauhan ylläpito "work peace maintenance" (cultivating good working climate)	The teacher guides students to behave in a way that respects other students' work or intervenes in the case of disturbing behaviour.	Example: The teacher moves to stand next to a student who disturbs the lesson. *Non-example:* *The teacher penalises the student for not doing their homework.*
vastauksen toistaminen "repeating response" (teacher repeating the student's response)	The teacher repeating verbatim or almost verbatim a student's response.	Example: A student responds unclearly or quietly and the teacher repeats the response so that the class can hear it. *Non-example:* *The teacher completes a student's response.*
vastauksen täsmentäminen "making response precise" (clarifying and making a student's response precise)	The teacher correcting and specifying the response of a student.	Example: The teacher accepts a student's response and further specifies (e.g. "Yes, two negatives make a positive, WHEN MULTIPLIED with each other). *Non-example:* *A student rounds off the result to incorrect accuracy and the teacher corrects this mistake.*
verkkomateriaalin käyttö "using internet material" (use of online material)	The teacher guides or instructs students to use material in the web.	Example: The teacher asks students to open a GeoGebra-applet and gives the address. *Non-example:* *The teacher presents material he or she has found on the web.*
vihje (hint)	Additional information given during the solution process, which is intended to aid in solving the problem.	Example: The teacher indicates where a student has made an error but does not specify what the error was.
vihkotyöskentely "notebook working" (notebook work)	A student writes notes in their notebook from the board or according to oral delivery.	Example: The teacher asks students to copy from the blackboard according to specific instructions. *Non-example:* *Highlighting text in textbook.*

(*Continued*)

TABLE 11.2 (*Continued*)

Term	Description	Examples and Non-examples
viittaaminen "signaling/pointing" (student raises hand)	A student raises his or her hand to get teacher's attention.	Example: A student raises his or her hand to indicate a willingness to answer teacher's question.
virheen korjaaminen/ käsittely "repairing an error/ handling an error" (correction or treatment of an error)	Recognising and repairing the teacher's or a student's mistake, and subsequent discussion.	Example: The teacher notices that there is an error in a student's solution and asks students to discuss that solution in groups. *Non-example: The teacher assesses exams and marks the mistakes.*
väittely "argumenting" (debate)	Teaching method, where there are (usually) two conflicting options and the students discuss about these using arguments.	Example: Students disagree about the solution to a task and the teacher lets students discuss through which is the correct solution.
välineet (equipment)	Physical items related to learning (organising them, etc.).	Example: Handing out rulers and compasses.
yhteistoiminnallinen oppiminen (cooperative learning)	Learning dependent upon students working together and their shared construction of knowledge.	Example: A group work to create a concept map. *Non-example: Group work without positive interdependence.*
yleistäminen (generalising)	The teacher expands and elaborates a specific example into a more general issue or concept.	Example: After presenting a graphical method for solving a linear equation, the teacher mentions that also other equations can be solved using a graphical method.

[1] Term was not validated as familiar or very familiar.
[2] Term was revised after validation.

TABLE 11.3 Finnish Lexicon – Terms by category.

Arviointi (Assessment)	assessment discussion, assessment of (student) solution, checking homework, clarifying and making a student's response precise, explaining and discussing assessment, homework assignment, monitoring progress, peer encouragement, personal feedback discussion, providing feedback, providing positive feedback/praising, quiz, self-evaluation, setting assessment goals, tests, returning assessed tests, request for justification, teacher repeating the student's response
Kasvatus (Upbringing)	bullying, caring, classroom climate, classroom rules and norms, collaboration between the teacher and guardians, discipline, discussion about student behaviour, disturbances, embarrassing a student, encouragement and pep, favoritism, good working climate, maintaining good working climate, promotion of good manners, school rules, use of humor to lighten the atmosphere
Matemaattiset Termit (Mathematics Specific Terms)	application task, connect and contrast, correct terminology, deduction, everyday math, fundamental task, generalising, linking, mathematical theory, mental calculation, problem-solving, visualisation / concretisation, word problem
Organisointi (Organising)	(student as teacher's) assistant, beginning the lesson, demonstration, distribution of materials, ending the lesson, equipment, grouping, lesson plan, lesson time management, providing instructions, roll call, scheduling, sitting arrangement
Pedagogiset Ratkaisut (Pedagogical Tools and Approaches)	brainstorming, classwork, cooperative learning, debate, differentiating, flipped classroom, group work, group work debriefing, hint, independently checking solutions, individual work, introduction/orientation, investigation, learning by heart, lecturing, lesson structure (common thread), making a summary, notebook work, play or game, presentation of student work, questioning, reading the textbook, revision, worked-out example, realisation, revision, student project, teacher monitors and guides classwork, use of online material, use of technology, video or animation, worked-out example
Vuorovaikutus (Interaction)	correction or treatment of an error, encouragement and pep, getting the students' attention, guidance, prompting to work, student collaboration, student's question, student raises hand, student response

Acknowledgements

This work was funded by research funding made available by the University of Helsinki and by Åbo Akademi University.

Bibliography

The work of the Finnish national team has been presented locally and internationally at the following meetings and conferences:

- Ainedidaktiikan symposium 2017 [The Symposium for Subject Education 2017] 2017.
- The Eighth Nordic Conference on Mathematics Education, NORMA 17 2017.
- European Association for Learning and Instruction (EARLI) Biennial Conference 2017, 2019.

12

IDENTIFYING THE PROFESSIONAL LEXICON OF MIDDLE-SCHOOL MATHEMATICS TEACHERS

The French case

Michèle Artigue, Brigitte Grugeon-Allys, Julie Horoks and Julia Pilet

Introduction

As non-English native speakers, we are sensitive to the limitations resulting from the hegemony of the English language in international communication in mathematics education. We thus immediately adhered to the goal of The International Classroom Lexicon Project of addressing this issue, while progressing towards a shared professional language where terms have a precise and agreed meaning. This is not the case in most countries and contributes to maintain the teacher profession in a semi-professional status (Adler, 2016; Chevallard & Cirade, 2010).

However, the story told in this chapter is particular, shaped by specific contextual influences. In this chapter, we first present some characteristics of the French educational context having influenced our vision of the project and our practical work. We then introduce the methodology used to build the French Lexicon before presenting its main characteristics, and the local analyses carried out. We end with a reflection on what has been achieved and perspectives for future work.

Contextual elements and their influence

In the French educational system, compulsory education traditionally began at six. From September 2019, it has been revised down to three to include the years of kindergarten already attended by nearly all children. It thus covers kindergarten (three grades), primary education (five grades), middle school (four grades), and the first year of high school. With few exceptions, all students follow the same curriculum until grade 10 when differentiation between vocational and general education begins. The curriculum is a national curriculum. Primary teachers are generalists, but middle-school mathematics teachers only teach mathematics. The standard preparation for these takes place in university institutes where future teachers enter with a three-year university degree (Licence), and both prepare a national competition, organised by the Ministry of Education, and a Master degree of teaching. If successful to both, they become civil servants and are appointed to schools. For a long time, the preparation of secondary teachers has focused on the disciplinary area they would teach. Since 1990, however, it has become more professional, integrating better mathematical, didactical and pedagogical knowledge and competences. However, the priority given to the knowledge of the subject matter still characterises the system. Teacher professional development takes different forms, for the most part on a voluntary basis. Traditionally in mathematics, the IREMs (Institutes of Research on Mathematics Teaching) are essential providers of professional development.

An important aspect of the French educational system regarding mathematics is the IREM network.[1] Operating as independent university units, aligned closely with mathematics departments, IREMs have an original mode of functioning. Their activities indeed take place in thematic groups gathering a diversity of expertise: university mathematicians, teachers, teacher educators, researchers in mathematics education (called in France didacticians) and historians of mathematics, all with equal status in the groups. These are not permanent groups and participants only work part-time in IREMs, maintaining thus a strong link with their field of practice. Groups develop reflection and action-research, prepare teacher training sessions based on their reflection and research, and produce resources for teaching and teacher education. The IREM structure has strongly influenced the development of French didactic research. As explained in (Artigue et al., 2019), IREMs have nurtured institutional and scientific relationships between didacticians and mathematicians, and supported the strong sensitivity of the French didactic community to epistemological and historical issues. They have allowed research to develop in contact with teachers and classrooms. Journals published by the network contribute to the dissemination of research constructs and results within accessible forms of discourse. It must be added that the didactics of mathematics emerged in France, in the early seventies, as a genuine research field distinct from pedagogy.

These elements have influenced our vision and implementation of the project. More than general pedagogical terms, we have identified didactic terms "reasonably" shared by experienced and reflective teachers, such as those involved in IREM groups. The two middle-school teachers who are members of our Lexicon research team, are members of the IREM of Poitiers and the video was recorded in one of their grade eight classes. The four researchers of our team are members of a didactic research laboratory, the LDAR (Laboratoire de Didactique André Revuz), and all have close connection with the IREM of Paris. Three of them actually work in teacher education and collaborate with middle-school teachers in research projects.

Building the French Lexicon

Building the lexicon was a long-term process involving successive cycles. In this section, we share how we created the first version, then, briefly describe its successive refinements.

Building the first version

This building combined a priori reflection and the use of lesson videos. We first collectively tried to identify candidates for the lexicon in the vocabulary available to describe mathematics classroom sessions, guided by the following principles:

- Select terms used by experienced teachers and whose meaning seems reasonably consensual;

- Select terms capturing forms of mathematics teachers' expertise;
- Include terms from didactic research having reasonably permeated teachers' discourse.

Then we looked at the French lesson video, trying to capture the quality of the classroom mathematical activity and main aspects of the role of the teacher. Using the coding template provided by the Australian team, we introduced a three-level structure showing the main phases of the lesson planned by the teacher in his preparation, sub-phases when pertinent, and episodes deserving coding. After agreeing on this coding scheme and a first list of terms, we distributed the work creating a pair researcher-teacher for the coding of each video. Finally, we collectively discussed the outcome of this work. The process resulted in the first version presented at the international meeting of the project, in November 2015. This lexicon was made of 100 terms organised in nine categories, illustrated as much as possible, by short episodes from different videos.

The review and revision process

The review and revision process started in January 2016 and was completed in November 2017. It was conceived as a move from local to national review, guided by the collective decisions taken at the partner meetings, but it presented also specificities linked to contextual opportunities and constraints. It involved a multiplicity of actors. The different steps are presented in Table 12.1.

TABLE 12.1 The review and revision process.

Time	Phases	Lexicon Versions
January–February 2016	First revision based on clarity and non-circularity check, and discussions at the first Melbourne meeting (Nov. 2015)	Version 1 → Version 2
March–April 2016	Local review of version 2 through questionnaire 1	
May–June 2016	Enlarged review of version 2	Version 2 → Version 3
September–November 2016	National review of version 3 through on-line questionnaire 2	
January–April 2017	Revision of version 3 following the analysis of the on-line questionnaire, and discussions at the second Melbourne meeting (Dec. 2016)	Version 3 → Version 4
April–June 2017	Cross-checking by the Australian team of version 4 and language editing; subsequent revision.	Version 4 → Version 5
September–November 2017	Final adjustments following the Beijing meeting of July 2017, and a teacher workshop in October 2017	Version 5 → Version 6

The evolution of the lexicon

As shown in Table 12.2, the first review processes led to a reduction in the number of categories; some distinctions, for instance between the pedagogical and didactical management of the classroom, were identified as an artificial distinction. From version three, the structure stabilised to include six categories: General terms, Nature of tasks (activities), Lesson phases, Forms of pedagogical organisation, Mathematical activities, Didactical and pedagogical management of the classroom, with three sub-categories: organisation, interaction, exploitation and evaluation. The number of terms, in and of itself, did not substantially change. However, this rather stable number corresponds to a twin process of both expansion and reduction. Some major evolutions included the introduction of nine dual terms built on significant oppositions, such as *outil/objet* (tool/object) or *exercice/problème* (exercise/problem) in version 3, and the removing of terms that did not reach the level of 2/3 of familiarity a priori agreed for inclusion in the national review, mainly research terms such as the terms *devolution* and *ostension*, from the theory of didactic situations. However, for research terms, considering the goal of contributing to the development of professional discourse, we only removed terms with less than 50% of familiarity.

Another substantial change regarded definitions. At the first Melbourne meeting, indeed, initial comparison revealed that our definitions were longer than those in the other lexicons, and we were asked to reduce them; indeed we did so for version two. However, in order to maintain the quality of information, we added comments to some definitions (see next section). Alongside the review process, definitions and comments were revised and new comments were added. From version four to version five also, following language editing by Australian colleagues, changes were introduced in the English version, some of them leading in return to simplifications in the French formulations. At the first Melbourne meeting, it was also decided to move from the illustration of terms by excerpts of videos to descriptive examples. We thus began to include such examples/non-examples in version two, and their number regularly increased alongside the revision process. In the final version, only one term is missing example and 13 terms are missing non-examples. Among these, five are dual terms and the discriminating examples proposed play the role of non-examples. Changes were also made to increase the accuracy of examples/non-examples, especially for non-examples. To conclude the section, we illustrate this evolution with the term *institutionalisation* in Table 12.3.

TABLE 12.2 Quantitative evolution of the French Lexicon.

	Version 1	*Version 2*	*Version 3*	*Version 4*	*Version 5*	*Version 6*
Categories	9	8	6	6	6	6
Terms	100	111	115	119	120	116
Comments	0	42	79	76	78	80
Examples	0	135	168	185	188	189
Non-examples	0	42	56	102	107	103

TABLE 12.3 Evolution of the term *institutionalisation*.

Institutionalisation	
Version 1	*Version 6*
Definition: Process guided by the teacher through which the pieces of knowledge put in action in students' activities are decontextualised and linked to the institutional knowledge at stake. Institutionalisation can be local or global. It usually implies a written trace, and the students can be more or less associated with its production.	**Definition:** Process under the teacher's responsibility through which the pieces of knowledge put in action in students' activity are decontextualised and linked to the scholarly knowledge at stake. **Comment:** Institutionalisation can be local or global. The students can be more or less associated with the production of this institutional discourse, and generally have to record it in their notebook. **Example:** After a situation involving affine functions, the teacher institutionalises the fact that the graph of a linear function is a straight line. This straight line passes through the origin if and only if the function is linear.
Illustrative video excerpts: FR: 00:10:51–00:12:00 FR: 00:49:59–00:55:00	**Non-example:** The teacher asks students to copy in their notebook the solution of an exercise written on the board.

The current French Lexicon

The current version of the French Lexicon contains 116 terms distributed into six categories. The list of terms with associated definitions, examples, and non-examples is provided in the next chapter. However, due to space restrictions, commentaries are not included, the number of examples has been reduced and some non-examples omitted. In what follows, we outline some characteristics of this lexicon and illustrate these with selected examples, privileging terms specific to this lexicon.

Mathematical orientation

One important characteristic of the French Lexicon is its mathematical orientation. It is visible through the existence of a category for mathematical activities (19 terms), the number of terms qualifying the nature of mathematical tasks (17 terms), examples most often with explicit mathematical content, and also through the didactic nature of this lexicon, which looks designed to give account of the specificities of mathematics classrooms. This characteristic is in line with the mathematical orientation of the professional discourse of French mathematics teachers, which is confirmed by the high level of familiarity of the terms denoting mathematical tasks and activities (see next section). The richness of this terminology contrasts with the limited number of terms attached to the affective dimension of teaching and learning, and classroom atmosphere.

Comments

Another characteristic is the existence of comments for 80 terms. Analysing these comments, we have distributed them into seven categories according to the type of information they provide. In Table 12.4, we give one example for each category.

TABLE 12.4 Examples of comments.

Category of comment	Example of comment
Clarifying origin and meaning	**Situation didactique/adidactique (Didactic/ adidactic situation):** The concepts of didactic / adidactic situation from the theory of didactic situations are used to model teaching situations. By extension, they are used to model phases or episodes of teaching situations.
Clarifying function	**Problème ouvert (Open problem):** Open problems are used to practice transversal competences and to develop a scientific approach, i.e. to explore, to experiment, to conjecture, to test, and to prove.
Clarifying forms and modalities	**Progression (Progression):** A progression can be personal to a teacher or shared by teachers teaching the same grade. It can be linear or spiral.
Clarifying teacher' and students' roles	**Changement de cadre (Change of setting):** Changes of setting are generally initiated by teachers but they may also be initiated by students, for instance to come back to a more familiar setting.
Connecting terms	**Mise en commun (Kneading-up):** Kneading-up can lead to a synthesis and an institutionalisation
Giving synonyms	**Activité introductrice (Introductory activity):** Synonym: Introductory situation
Referring to curricular conditions or expectations	**Problématisation (Problematisation):** The importance of problematisation and the importance of involving students in this process is more and more emphasised in curricular documents.

In the final steps of the revision process, some comments were transformed into examples.

Use of dual terms

Another characteristic is the presence of nine dual terms, when opposing two terms helps make sense of each of them. These are mainly in the category General terms, as illustrated in the example below:

> **Mobilisable/available knowledge**
> Definition:
> Mobilisable knowledge: can be implemented when its use is requested by the task or by the teacher. Available knowledge: can be implemented without indication, when relevant.
> Example:
> To compute lengths in a geometrical figure, students must use Thales' theorem. If the text mentions "using Thales' theorem", the theorem must be mobilisable knowledge; if not, it must be available knowledge.

Influence of didactic research

One characteristic of the French context is the existence of a strong didactic research community. Its professional discourse has been progressively disseminated through teacher education and curricular resources. Most of them come from didactic theories, such as *contrat didactique* (didactic contract), and *institutionnalisation* (institutionalisation) from the theory of didactic situations (TDS) (Brousseau, 1997), *outil/objet* (tool/object) from Douady's theoretical approach (Douady, 1984), *registre de représentation* (register of representation) from Duval's semiotic approach (Duval, 1995), *aide constructive/procédurale* (constructive/procedural help) from the double approach, ergonomic and didactic, of teachers' practices by Robert and Rogalski (2002) or *moments de l'étude* (study moments) from the anthropological theory of the didactic (ATD) (Chevallard, 2019). The current lexicon contains 18 such terms, a small number however when compared to the didactic glossary initiated by Balacheff.[2] Moreover, often, the research technical definitions have been simplified.

Some research terms also come from innovative practices that emerged in an IREM group and appear to have been adopted widely. This is the case for the terms *narration de recherche* (narrative of research) that emerged in the IREM of Montpellier (Sauter, 1998), *problème ouvert* (open problem) in the IREM of Lyon (Arsac, Germain & Mante, 1991) and *débat scientifique* (scientific debate) in the IREM of Grenoble (Legrand, 1993). Local analyses (see next section) show that they have a higher level of familiarity than theoretical terms. All research terms did not reach a sufficient level of familiarity to be part of the final version, which includes 30 research terms, 20 of them coming from didactic research. In some cases, we added a close word more commonly used, for instance, *aider* (helping) was added to *étayer* (scaffolding), *motiver* (motivating) to *enrôler* (engaging); in other cases, we just mentioned the existence of a connected research term in a comment, as for instance for the terms *commentaire métacognitif* (metacognitive comment), *orchestration instrumentale* (instrumental orchestration), *dévolution* (devolution) and *ostension* (ostension). Doing so, we hope to contribute to make the lexicon a bridging tool between research and practice.

Local analyses

In this section, we present a synthesis of the analysis of the online questionnaire. According to what had been collectively decided (see Chapter 1), the questionnaire began with a demographic data section, then distinguished three types of entries:

- A term and description entry associated with questions regarding familiarity, frequency of use and frequency of corresponding practice (Likert-scale response with 5 levels);
- A description entry and question asking to propose a term for this description, when possible;
- A term entry and question asking to propose a description for the term if known and the same Likert-scale questions as for the first type of entry.

To make the task-load reasonable, the 115 terms were distributed among four questionnaires and the three types of entries, respecting a balanced distribution of the 6 categories in each of them. About 30 terms were thus proposed in each questionnaire, about

10 for each type of entry. To compensate this limitation, we added a question in which the 115 terms were listed and participants were asked if they felt very familiar, familiar or non-familiar with them. Respondents were also given the opportunity of suggesting improvements for descriptions, examples, and non-examples, to suggest new terms, to make general comments on the questionnaire. The questionnaire was disseminated using various channels: IREM network, APMEP, and IFé (French Institute for Education). We received 174 responses, with some variation in numbers for the different modalities.[3] The lexicon version used for it was version three.

The respondents' profiles

As shown in Figure 12.1, the demographic data collected show that answers come from a diversity of academies (18 over 26), with an over-representation of the three academies of the Ile-de-France region and Académie de Poitiers, where the team is located. There is

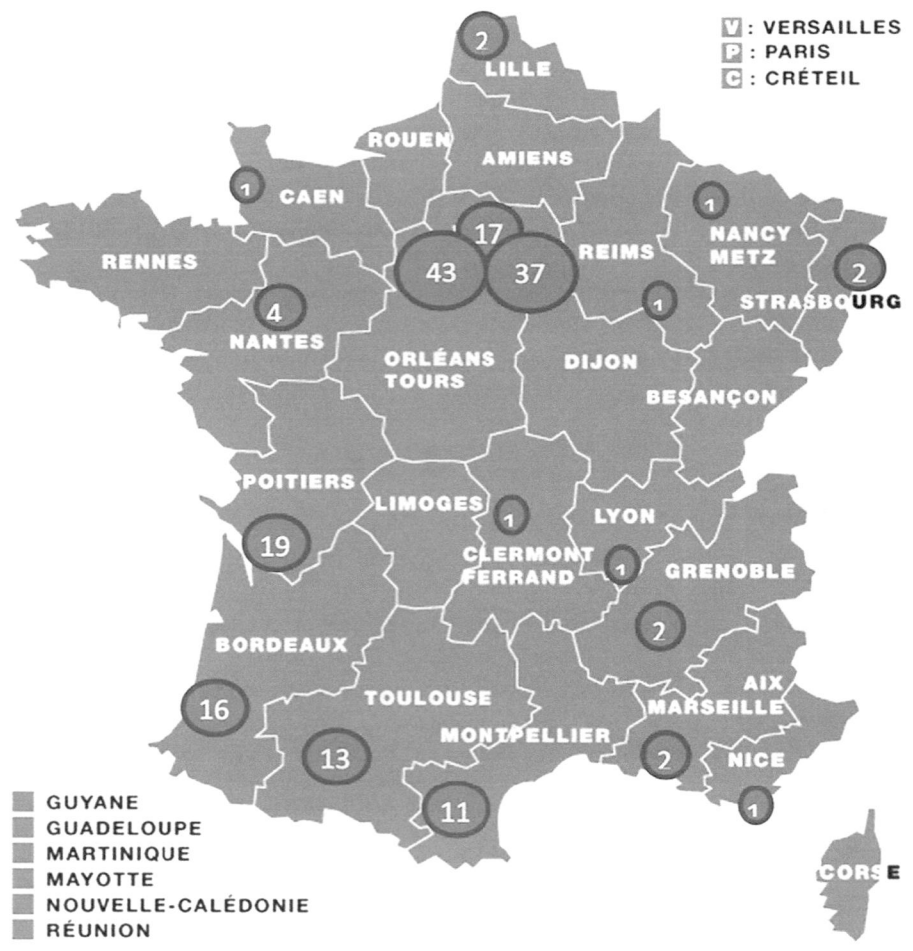

FIGURE 12.1 Geographical distribution of respondents.

some gender imbalance as 104 are women (60%) while the ratio of female mathematics teachers in middle school is about 54%. The mean for age distribution is 43.6 years, which is in line with national statistics. The distribution of age shows that most respondents are experienced as expected, with, respectively, 12.7%, 20.7%, 34.5%, and 32.2%, of respondents in the categories: less than three years, between four and ten years, between 11 and 20 years, and over 20 years of experience. The great majority are teachers: 134, while 26 are teacher educators, and only ten researchers; four ticked the box "other." Moreover, 161 declare to have an experience of teaching mathematics in secondary school and 127 in middle school.

Only 47 participants have participated in working groups of one of the existing professional networks, mainly in IREM groups (38/47), and this participation is not equally distributed: less than 5% participation for respondents at the beginning of their career, and more than 37% for those over 20 years of experience. The online questionnaire has thus reached a diverse population, beyond teachers involved in action-research and reflective groups or involved in teacher education activities.

Familiarity with lexicon terms

In the last part of the questionnaire, respondents were asked to rate their level of familiarity with all lexicon terms, according to three levels: Not familiar, Familiar, Very familiar. We assigned a familiarity score to each term in the questionnaire, from 0 to 3, according to the following criteria:

- Score 0 (Not familiar): more than half of the responses were *Not familiar*;
- Score 1 (Somewhat familiar): *Not familiar* responses between one-third and one-half;
- Score 2 (Familiar): more than two-thirds of *Familiar* and *Very familiar* responses and less than two-thirds of *Very familiar* responses;
- Score 3 (Very familiar): more than two thirds of *Very familiar* responses.

For each category, a global score of familiarity was then established. The distribution of scores among terms and categories are given in Tables 12.5 and 12.6.

TABLE 12.5 Distribution of familiarity categories among terms.

Familiarity categories	Not familiar	Somewhat familiar	Familiar	Very familiar
Number of terms	10	11	18	76

TABLE 12.6 Distribution of scores among categories.

Categories	General terms	Nature of tasks	Lesson phases	Pedagogical organisation	Mathematical activities	Pedagogical and didactical classroom management
Scores	1.64	2.35	2.54	2.36	2.76	2.42

Table 12.5 shows that 21 terms are in the lowest categories. These are all research terms. All those with score 0 were removed from the final version of the lexicon with one exception: *situation d'action* (situation of action), whose rate of familiarity was close to 50%, to give a similar treatment to the three connected terms of TDS: situation of action, situation of formulation, situation of validation. The category Mathematical activities has the highest average score, while the category of General terms has the lowest score. This is not surprising considering the mathematical orientation of the teacher culture and the fact that the General terms category includes many research terms. We investigated whether participation in a research or action group influenced levels of familiarity, especially for terms with scores 0–2. Such influence is visible but not systematic at all.

Relationship between familiarity, practice and use

To study the relationship between familiarity, practice and use, we used parts 1 and 3 of the four questionnaires. In these parts, for ten terms, participants were asked to rate from 1 to 5 their level of familiarity with the term, how much they used it in conversations with colleagues, and how much it corresponded to frequent practices. We associated average rates for familiarity, use and practice to these rates for the 80 terms involved. Table 12.7 shows the differences between the average rates of familiarity and practice.

As expected, for a majority of terms (64/80), familiarity is higher than the frequency of corresponding practices. However, the difference is generally rather small, less than 0.5 in absolute value. Surprisingly, for 16 terms it is negative, 13 among these being research terms with scores 0 or 1. While not familiar to teachers, these terms seem corresponding to practices they recognise when reading definitions and examples. We thus decided to make those removed present in the lexicon, when possible, through comments or examples associated with connected terms. For 14 terms, conversely, the difference between familiarity and practice is strongly positive. These terms, in fact, correspond to tasks and forms of pedagogical organisation, mathematical activities or forms of assessment, well known from teachers, and promoted in curricular documents, but remaining rather marginal in actual practices, such as *tâche complexe* (complex task), *différencier* (differentiating), *expérimenter* (experimenting) and *évaluation diagnostique* (diagnostic assessment). We also find two different terms: *exercice de révision* (revision exercise), and *cours magistral* (lecturing). We interpret their presence as the effect of the disqualification of these traditional practices in the institutional discourse.

Regarding the relationship between familiarity and use, all differences are positive, and distributed in a rather balanced way among the categories used in Table 12.7. For 29 terms, the difference is less than 0.5, for 30 it lies between 0.5 and 1, and for 21 it lies between 1 and 1.6, the extreme case being the term *cours magistral* just mentioned, which tends to confirm our interpretation. Globally the differences are thus bigger than those observed between familiarity and practice. Among the 21 terms of the last category, we find terms corresponding to

TABLE 12.7 Difference between familiarity and practice.

Difference F/P	$d < -1$	$-1 \leq d < -0.5$	$0.5 \leq d < 0$	$0 \leq d \leq 0.5$	$0.5 < d \leq 1$	$d > 1$
Number of terms	6	3	7	38	12	14

innovative practices or mathematical activities as was the case for familiarity/practice, the five terms associated with specific forms of assessment, and two types of help (*aide méthodologique* (methodological help), *aide instrumentale* (instrumental help)). Surprisingly, we find also terms such as *exercice de mise en route* (starting exercise) or *gérer le temps* (managing time). These differences call for further research to better understand the responses given.

Associating terms with definitions

For this analysis, we could only use part 2 of the questionnaire and the data thus regarding 40 terms. Asking to propose a term for a given definition was the most difficult task. The majority of proposals were compatible with the definitions, but often not exactly the terms used in the lexicon. Synonyms were used, and also other phrasing especially for terms not reduced to a single word. We also observed discrepancies due to the fact that a definition was understood as referring to another category, which induced a different formulation. Some proposals also labelled modalities of management of a task or a lesson phase instead of the nature of the task or the phase itself. The dual terms were those creating most difficulties with a few exceptions such as *outil/objet* (tool/object), which is not surprising as teachers were not informed of the existence of such terms.

Creating definitions for terms

The task asking to propose a definition for given terms was less problematic. The definitions proposed by participants were, globally, reasonably close to those of the lexicon, and even richer in some cases, pointing out relationship or distinctions with other notions, or better incorporating the students' perspective. We give one example below; the translation of the proposed definition is ours.

Managing time.

Lexicon definition: Dealing with time constraints, personal or institutional, during a lesson or more globally.

Proposed definition:

For the teacher: to organise the rhythm of a lesson, a sequence.

For students: to plan the time to spend on each exercise of an assessment, on the different phases of an activity (experiment, demonstration, writing) in order to reach the objective within a set time.

However, some confusion was observed between terms with close meaning, such as *expliquer* (explain) and *expliciter* (making explicit), *formuler* (formulating) and *reformuler* (reformulating), or close terms belonging to different categories.

Additional terms suggested

Participants were also proposed to suggest new terms in part 4 of the questionnaire where terms were grouped in four categories: General terms and Nature of tasks, Lesson phases and Forms of pedagogical organisation, Mathematical activities, Pedagogical and didactical classroom management. 153 terms were proposed with the following distribution among these four categories: 23, 28, 80 and 22 terms. Once again, the category Mathematical activities was the most attractive with proposals such as *chercher* (researching), *communiquer* (communicating), corresponding to mathematical competences listed in curricular documents, or specific mathematics activities such as *construire* (constructing), *étudier ou trouver des exemples* (studying or finding examples). In the first category, some respondents proposed to add new research terms, and especially terms coming from ATD. Some had been envisaged in the initial phase of the elaboration of the lexicon, but discarded because they could not be supposed reasonably shared by experienced teachers. In the last category, interestingly some respondents mentioned the lack of terms referring to students with special needs and associated classroom management.

In the four categories, many proposals mentioned terms already present in the lexicon but in another category, or some rephrasing of existing terms, or modalities of action at times already present as examples. No more than a few respondents simultaneously proposed the same terms. For that reason, only a few terms were added.

Reflections and perspectives

In the sections above, we have first explained our motivation for our participation in The International Classroom Lexicon Project, and described characteristics of the French education context having influenced its realisation. We have then presented how the current version of the French Lexicon has been progressively built, with the combination of collective decisions taken in the regular meetings of all Lexicon teams and local decisions of the French team. We have outlined the main characteristics of the French Lexicon, and summarised the results of the local analyses carried out, exploiting data collected through the online questionnaire.

There is no doubt that the French Lexicon is a unique object, highly influenced by the French educational context and culture. Its mathematical orientation, outlined above, reflects the importance attached to the disciplines in this culture. Its didactic orientation reflects the existence of a strong didactic tradition in mathematics distinct from pedagogy. The presence of technical terms from didactic research shows the connection of this tradition to school practice thanks to specific institutions such as IREMs or IFé. Conversely, the limited number of terms directly related to the affective dimension of teaching and learning processes reflects the limitations of French didactic research in this area. To these characteristics adds the existence of commentaries, specific to this lexicon. They express our vision that the meaning of most lexicon terms cannot be conveyed in one short descriptive sentence. What the lexicon terms encapsulate, in one or a few words, and make them important elements of the teachers' professional discourse is much more, and commentaries help to make this more visible.

Reflecting on the whole enterprise, there is no doubt that building the French National Lexicon has been a challenging task. Not only the selection of terms and their precise formulation among the huge diversity of terms used by teachers to exchange about their classroom practices, but also the elaboration of definitions trying to capture a reasonably shared understanding, the selection of examples and non-examples, the decisions to be taken regarding the number and diversity of feedback received during the review process. The current lexicon is not the unique one that could have resulted from this long process, and it is influenced by the team that produced it. However, today, this lexicon patiently built and refined is a tangible object on which we can rely in order to progress towards establishing the kind of professional lexicon that is needed if we want the profession of mathematics teacher to be more than a semi-profession. Of course, focused on the terms used to describe and make sense of a mathematics lesson, this lexicon cannot pretend to cover the global needs of such a professional lexicon that should take in charge all dimensions of teachers' professional practices, in and outside the classroom,[4] but this is a first step in this direction.

This first step being achieved, another challenging and exciting step, already started, can fully develop: the comparison with other lexicons, the reflection on how these might contribute to enrich our professional discourse. This is in line with the vision we have of the French Lexicon as a dynamic object.

For these comparisons, English versions are necessary. Creating the English version of the French Lexicon, we could rely on existing English versions for many terms due to their use by French authors in English publications. The main difficulties were observed with terms in the category Nature of tasks, and they were solved by using periphrases (for instance *problème d'approfondissement* (problem for deepening knowledge)) or a word by word translation as for instance for *narration de recherche* (narrative of research). We also had to choose how to express actions. In French, actions are expressed using the corresponding verb in infinitive form. In the English version, we systematically used the possibility offered by the gerundial infinitive for actions, interactions and mathematical activities, which certainly introduces slightly different connotations. However, our international meetings have made clear that some obvious translations are in fact more problematic than the difficult ones, because according to our contexts and cultures, the same terms may have different meanings. The terms *problème ouvert* and *débat scientifique* are good examples. Their translation is a priori straightforward: open problem and scientific debate. However, these terms have a particular meaning in the French context, not necessarily shared by those using these terms in Anglophone contexts. Hence the importance of definitions, examples and non-examples in the lexicons.

We can compare lexicons, but lexicons are in some sense only bricks for the teachers' professional discourse. When teachers describe a mathematics classroom session, they tell a story intertwining mathematical and pedagogical-didactical discourses. Such stories are essential to capture the reality of mathematics classroom practices and also to approach the quality of these practices. For that reason, coming back to the lexicon videos and the initial coding of these, which were a spur for the building of the lexicon, we have decided to associate narratives to them. As explained in (Artigue et al., 2017), these narratives are a team collective production. They embed lexicon terms into a fully developed discourse; they restore the mathematical-didactical-pedagogical imbrication specific of mathematics classrooms. They seem a promising tool both for comparing our respective lexicons, their potential and limitation, and for using the lexicons in professional development activities.

Acknowledgements

This project has been funded by a Discovery Grant from the Research Council of the Australian Government (ARC-DP140101361). We deeply thank the two teachers of the French Lexicon team: Florence Debertonne-Dassule and Thierry Chevalarias, our foreign colleagues in the international Lexicon teams, all those who have accepted to take part in the review process, and also the IREMs of Poitiers and Paris, and our research laboratory, the LDAR, for their support.

Notes

1 http://www.univ-irem.fr
2 https://dico-ddm.blogspot.fr
3 Participants were asked to select a modality according to the trimester of their birthday date. These were not equally distributed among the four semesters (42, 62, 42, 27).
4 Regarding this point, one interesting incentive inspired by The International Classroom Lexicon Project is the one recently launched by researchers from eight different countries and languages, with the objective of identifying the words used by teachers in their respective countries to speak about their documentational activity and the resources they use (Trouche, 2019).

References

Adler, J. (2016). Mathematics discourse in instruction (MDI): A discursive resource as boundary object across practices. In G. Kaiser (Ed.), *Proceedings of the 13th International Congress on Mathematical Education* (pp. 125–144). New York, NY: Springer.

Artigue, M., Novotná, J., Grugeon-Allys, B., Horoks, J., Hospesová, A., Moraová, H., Pilet, J., & Žlábková, I. (2017). Comparing the professional lexicons of Czech and French mathematics teachers. In B. Kaur, W.K. Ho, T. L. Toh & B.H. Choy (Eds.), *Proceedings of PME 41* (Vol. 2, pp. 113–120). Singapore: PME.

Artigue, M., Bosch, M., Chaachoua, H., Chellougui, F., Chesnais, A., Durand-Guerrier, V., Knipping, C., Maschietto, M., Romo-Vázquez, A., & Trouche, L. (2019). The French didactic tradition. In W. Blum, M. Artigue, M.A. Mariotti, R. Strässer, & M. Van den Heuvel-Panhuizen (Eds.), *European traditions in didactics of mathematics* (pp. 11–56). New-York, NY: Springer Open.

Arsac, G., Germain, M., & Mante, M. (1991). *Problème ouvert et situation problème*. Lyon, France: IREM de Lyon.

Brousseau, G. (1997). *Theory of didactical situations in mathematics*. Dordrecht, The Netherlands: Kluwer Academic Publishers.

Chevallard, Y. (2019). Introducing the anthropological theory of the didactic: An attempt at a principled approach. *Hiroshima Journal of Mathematics Education*, 12, 1–44.

Chevallard, Y., & Cirade, G. (2010). Les ressources manquantes comme problème professionnel. In G. Gueudet & L. Touche (Eds.), *Ressources vives. Le travail documentaire des enseignants* (pp. 41–55). Rennes, France: Presses Universitaires de Rennes.

Douady, R. (1984). Jeux de cadres et dialectique outil-objet. *Recherches en Didactique des Mathématiques*, 7(2), 5–31.

Duval, R. (1995). *Sémiosis et pensée humaine*. Bern, Switzerland: Peter Lang.

Legrand, M. (1993). Débat scientifique en cours de mathématiques et spécificité de l'analyse. *Repères-IREM*, 10, 123–159.

Robert, A., & Rogalski, J. (2002). Le système complexe et cohérent des pratiques des enseignants de mathématiques: une double approche. *Revue Canadienne de l'Enseignement des Mathématiques, des Sciences et des Technologies*, 2(4), 505–528.

Sauter, M. (1998). Narrations de recherche: Une nouvelle pratique pédagogique. *Repères-IREM*, 30, 9–21.

Trouche, L. (2019). Evidencing missing resources of the documentational approach to didactics. Towards ten programs of research/development for enriching this approach. In L. Trouche, G. Gueudet, & B. Pepin (Eds.), *The "resource" approach to mathematics education* (pp. 447–490). New York, NY: Springer.

13

FRENCH LEXICON

Michèle Artigue, Thierry Chevalarias, Florence Debertonne-Dassule, Brigitte Grugeon-Allys, Julie Horoks and Julia Pilet

A French Lexicon

The lexicon presented in this chapter consists of 116 pedagogical-didactical terms that we can consider reasonably shared by experienced middle-school mathematics teachers in France, and constituting basic elements of their professional discourse together with mathematical terms. This lexicon has been developed through a long process of successive evaluations and revisions, detailed in *Chapter 12*. Not only terms have been selected, but for each term an operational definition has been proposed, often complemented by a comment, and also examples and non-examples, helping make sense of the term and prevent possible misunderstanding. The terms of this lexicon denote different entities from lesson phases, types of tasks, and of mathematical activities to the forms that interactions take in the classroom. For that reason, and quite early, we found useful to organise them into categories. They are currently distributed into six categories. The original version of the lexicon, in French, has also been progressively translated into English (British spelling), with approximate versions for the terms without equivalent in English, and the translation process has also been submitted to evaluation and revision.

The three tables in this chapter, respectively, give access to:

- The terms listed in the alphabetic order of their English version, with the French original terms in parentheses (Table 13.1)
- The terms listed by alphabetic order in French, with their English version, definition, associated examples, and non-examples (Table 13.2).
- The distribution of the terms into categories and sub-categories (Table 13.3).

In Table 13.2, the code R indicates that the term comes from academic research, the codes T, respectively, S, denote more a teacher, respectively, a student practice, and T, S a shared practice. Note also that in this table, due to length restriction, the comments are not included, and examples and non-examples have been limited to those considered most useful by the research team. Information about comments can be found in *Chapter 12*.

TABLE 13.1 French Lexicon – Terms.

auto-évaluation (self-assessment)	**activité introductrice** (introductory activity)	**activité préparatoire** (preparatory activity)	**aide conceptuelle** (conceptual help)
aide constructive/ procédurale (constructive/ procedural help)	**aide instrumentale** (instrumental help)	**aide matérielle** (help with materials)	**aide méthodologique** (methodological help)
aide technique (skills help)	**aider, étayer** (helping, scaffolding)	**argumenter** (arguing)	**attirer l'attention sur, insister sur un point mathématique** (drawing the attention on, stressing a mathematical point)
bilan, synthèse (summary, synthesis)	**cadre** (setting)	**calculer** (calculating)	**changer de cadre** (changing the setting)
changer de registre de représentation (changing the register of representation)	**choisir les variables didactiques** (selecting didactic variables)	**circuler dans la classe** (circulating the classroom)	**classe inversée** (flipped classroom)
clôture de séance (lesson closing)	**coder (un dessin, un schéma, un graphique)** (noting mathematical symbols on a drawing, schema, graphics)	**commenter** (commenting)	**comparer, hiérarchiser des productions** (comparing, ranking productions)
conjecturer (conjecturing)	**connaissance mobilisable/disponible** (mobilisable/available knowledge)	**contrat didactique** (didactic contract)	**correction** (correction)
cours magistral (lecture)	**débat scientifique** (scientific debate)	**décontextualiser** (decontextualising)	**définir** (defining)
démontrer (proving)	**différencier** (differentiating)	**donner du travail à faire hors classe** (assigning homework)	**donner les consignes** (giving instructions)
encourager (encouraging)	**enrôler, motiver** (engaging)	**enseignement dialogué, cours dialogué** (dialogic teaching)	**entrée dans la classe et début de séance** (starting the class)

envoyer un (des) élève(s) présenter son (leur) travail à la classe (sending student(s) present work to the class)	estimer (estimating)	évaluation diagnostique (diagnostic assessment)	évaluation entre pairs (peer-assessment)
évaluation formative (formative assessment)	évaluation sommative (summative assessment)	évaluer (assessing)	exercice d'application (application exercise)
exercice de mise en route (warm-up exercise)	exercice de révision (revision exercise)	exercice en contexte (in-context exercise)	exercice technique (technical exercise)
exercice/problème (exercise/ problem)	expérimenter (experimenting)	expliciter (making explicit)	expliquer (explaining)
exposé d'élèves (students' presentation)	exposition des connaissances, cours (lecturing)	faire des connexions (connecting)	formuler (formuler)
généraliser (generalising)	gérer des erreurs (dealing with errors)	gérer l'organisation, le matériel (dealing with the material organisation of the class)	gérer la discipline (disciplining)
gérer la prise de notes (supporting the taking of notes)	gérer le ou les tableaux (managing the board(s))	gérer le temps (managing time)	gérer un incident didactique/non didactique (dealing with an incident didactical/ non-didactical)
institutionnalisation (institutionalisation)	introduction d'une séance (lesson introduction)	justifier (justifying)	lancement d'une activité (launching an activity)
milieu (milieu)	mise en commun (kneading-up)	modéliser (mathématiquement) (mathematical modelling)	moments de l'étude (study moments)

(Continued)

TABLE 13.1 (*Continued*)

montrer (demonstrating)	narration de recherche (narrative of research)	observer (repérer) le travail des élèves (observing students' work)	outil/objet (tool/object)
phase de rappel (phase of recall)	phase de résolution, recherche (solving, research phase)	poser des questions d'ordre général (asking general questions)	poser des questions d'ordre mathématique (asking mathematical questions)
présenter, rédiger des solutions (presenting, writing out a solution)	problématiser (problematising)	problème d'approfondissement (problem for deepening knowledge)	problème de réinvestissement (problem for reinvesting knowledge)
problème ouvert (open problem)	progression (progression)	raisonner (reasoning)	rappeler (recalling)
rédiger (writing out)	reformuler (reformulating)	registre (register)	relancer une question (re-asking a question)
répondre à une question (answering a question)	représenter (representing)	séquence / séance (sequence/lesson)	simuler (simulating)
situation d'action (situation of action)	situation de formulation (situation of formulation)	situation de validation (situation of validation)	situation didactique/ adidactique (didactic/adidactic situation)
situation-problème (problem situation)	susciter des idées, des stratégies (stimulating ideas or strategies)	tâche complexe (complex task)	tâche/activité (task/activity)
transition entre phases (transition between phases)	travail collectif (whole class work)	travail en demi-classe (half-class work)	travail en groupe autonome/ accompagné (autonomous/guided group work)
travail individuel autonome/ accompagné (autonomous/ guided individual work)	valider/invalider des productions (validating/invalidating productions)	valoriser (valuing)	vérifier (verifying)

TABLE 13.2 French Lexicon – Terms with operational definitions.

Term	Description	Examples and Non-examples
auto-évaluation (self-assessment)	Assessment form in which students assess their own production, and possibly grade it.	Example: After their work, students can look at the question answers and give themselves a score. *Non-example:* Peer-assessment.
activité introductrice (introductory activity)	Activity aimed at motivating the introduction of a new notion, making it appear as the appropriate tool to solve a problem or class of problems.	Example: Activity showing the inadequacy of arithmetic processes to solve linear equations of the type ax + b = cx + d and the need to introduce an algebraic approach. *Non-example: Activity introducing a new concept without motivating it as a tool to solve a problem or a class of problems.*
activité préparatoire (preparatory activity)	Activity ahead of the introduction of a new concept or theme, refreshing associated prerequisites.	Example: Solving linear equations by arithmetic processes before the introduction of algebraic solving. *Non-example: Activity already involving the new concept.*
aide conceptuelle T (conceptual help)	Help regarding the understanding of mathematical concepts.	Example: Teacher shows the difference between Thales' theorem and its converse, to help students make the right choice in solving a series of exercises. *Non-example: Teacher suggests to code a geometrical figure to visualise the hypotheses of a problem.*
aide constructive/ procédurale R, T (constructive/ procedural help)	Constructive help is support offered to improve student learning, while procedural help is support directed towards solving the task.	Example: (procedural help) Teacher indicates the theorem to apply. (constructive help) Teacher asks students to think about the conditions that allow to use a theorem in an exercise.
aide instrumentale R, T (instrumental help)	Help regarding the use of tools (geometry instruments, technological tools ...)	Example: Teacher explains the syntax of the calculator commands to formally solve an equation. *Non-example: Teacher proposes a calculator to students having forgotten theirs (see material help).*

(Continued)

TABLE 13.2 (*Continued*)

Term	Description	Examples and Non-examples
aide matérielle T (help with materials)	Help providing access to material resources or tools.	Example: Teacher proposes a 3D model to help students visualise a 3D geometry situation. *Non-example: Teacher explains how to use a square to draw parallel lines (see instrumental help).*
aide méthodologique T (methodological help)	Help regarding strategies which can be used for solving a task.	Example: Suggesting students to use a diagram. *Non-example: Suggesting a zoom to better visualise a graphical representation.*
aide technique T (skills help)	Help regarding the implementation of a technique.	Example: Showing the class how to simplify a given fraction. *Non-example: Suggesting students to study a particular case.*
aider, étayer R, T (helping, scaffolding)	Restricting the complexity of tasks, allowing students to solve tasks they cannot solve alone.	Example: Guiding students by marking out the steps for solving a problem. *Non-example: Giving extra-time to students to solve a task.*
argumenter T, S (arguing)	Developing a rational discourse to convince of the truth or falsity of a statement or result, or of the relevance of an action.	Example: Justifying the choice of a calculation procedure, invoking the peculiarities of the numbers involved. *Non-example: Using an authority argument.*
attirer l'attention sur, insister sur un point mathématique T (drawing the attention on, stressing a mathematical point)	Highlighting a mathematical point particularly important to notice and remember.	Example: Drawing students' attention to the fact that, when solving an equation, the position of the unknown, to the left or right of the equal sign, does not matter. *Non-example: Asking students to note what is written on the board.*
bilan, synthèse (summary, synthesis)	Phase to identify the important points to remember from the mathematical activity carried out.	Example: Teacher finishes a sequence about the Pythagorean theorem and its applications by an overall summary. *Non-example: Teacher ends a phase of individual solving, distributing a written solution.*

Term	Description	Examples and Non-examples
cadre **R** (setting)	Set of objects of a mathematical field, the relationships between them, their diverse formulations and the associated mental images	Examples: Numerical, algebraic, functional, geometric settings. *Non-example:* *Graphical representation (see register).*
calculer **T, S** (calculating)	Using mathematical operations to get a result.	Examples: Mentally calculating the sum of two numbers. Calculating the probability of obtaining a given sum by throwing two dices. *Non-example:* *Solving a system of linear equations graphically.*
changer de cadre **R, T, S** (changing the setting)	Expressing an exercise or problem in another setting to make accessible new interpretations and new solving tools.	Example: Selecting an algebraic setting to solve a problem set up in the geometric setting. *Non-example:* *Switching from point to fractional writing for a decimal number (see changing register).*
changer de registre de représentation **R, T, S** (changing the register of representation)	Associating to mathematical objects expressed in a given register representations in another register.	Example: Moving from the algebraic representation of a function to a graphical or tabular one. *Non-example:* *Drawing the graph of the sum of two functions from their respective graphs (treatment internal to the graphic register).*
choisir les variables didactiques **R, T** (selecting didactic variables)	Selecting the parameters of a situation, the values of which influence the validity, efficiency or cost of the solving strategies.	Examples: Selecting the size of numbers to make inefficient a calculation procedure. In geometry, selecting the tools available for the construction of a figure. *Non-example:* *Choosing to work individually or by pair.*
circuler dans la classe **T** (circulating the classroom)	Circulating the classroom during autonomous or guided work of the students.	Example: Circulating the classroom to make sure that all students are working. *Non-example:* *The teacher stays at her table, marking homework while the students solve an exercise.*

(*Continued*)

TABLE 13.2 (*Continued*)

Term	Description	Examples and Non-examples
classe inversée (flipped classroom)	Form of teaching which reverses the nature of learning activities at home and in class.	Example: Asking students to watch a video presenting Pythagoras' theorem at home and to answer online an associated quiz. In class then, proposing students to solve exercises in groups, providing them the necessary support. *Non-example: Students use in class a site of exercises on line to work on the Pythagoras' theorem, after an introduction to the topic.*
clôture de séance (lesson closing)	Phase during which the teacher ends the work, possibly giving information about the next session and homework to the students.	Example: The teacher asks students to read, before the next session, a document she has put on the school platform and to prepare questions. *Non-example: Time being over, the session ends abruptly and students go out.*
coder (un dessin, un schéma, un graphique) T, S (noting mathematical symbols on a drawing, schema, graphics)	Adding signs to a drawing to specify its properties.	Example: Adding codes to a geometric drawing to indicate equal lengths and angles. *Non-example: Building a diagram to represent a situation.*
commenter T, S (commenting)	Developing a discourse regarding the tasks proposed, solving strategies, attitudes and so on.	Example: The teacher comments that when solving an application problem, it is important to verify that the mathematical solutions found are plausible in the context of the problem. *Non-example: Validating a student's solution and asking those who used another method to copy it in their notebook.*
comparer, hiérarchiser des productions T, S (comparing, ranking productions)	Differentiating solving strategies, techniques, representations, results, and possibly classifying them.	Examples: Asking students to compare their solving methods, as regards efficacy and cost. *Non-example: Presenting different productions, simply one after the other.*

Term	Description	Examples and Non-examples
conjecturer **T, S** (conjecturing)	Formulating mathematical assertions assumed to be true but for which one does not yet have a mathematical proof.	<u>Example:</u> After exploration with dynamic geometry software, students conjecture that the triangle of maximum area that can be inscribed in a circle is equilateral. <u>Non-example:</u> *Deciding whether a given mathematical statement is true or false.*
connaissance mobilisable/ disponible **R** (mobilisable/ available knowledge)	Mobilisable knowledge: can be implemented when its use is requested by the task or the teacher. Available knowledge: can be implemented, without indication, when relevant.	<u>Example:</u> To compute lengths in a geometrical figure, students need to use Thales' theorem. If the text mentions "using Thales' theorem," the theorem must be mobilisable knowledge, if not, it must be available knowledge.
contrat didactique **R** (didactic contract)	The mutual expectations (only partly explicit) regarding mathematical knowledge that link teacher and students in class, and determine their respective responsibilities.	<u>Example:</u> Grade 8 students are expected to explicit the successive transformations when solving linear equations. <u>Non-example:</u> *The rules of behaviour to be respected in the classroom (not specific of mathematical knowledge)*
correction (correction)	Phase aimed at presenting expected solutions and possibly rectifying errors.	<u>Example:</u> After distribution of a model answer, students working in pairs exchange their productions and correct the production of their partner. <u>Non-example:</u> *Kneading-up students' proposals.*
cours magistral (lecture)	Form of teaching where teachers present the mathematical content to the students.	<u>Example:</u> Lecture on the solving of systems of linear equations in two variables, the teacher presenting the different cases and interpreting them graphically. <u>Non-example:</u> *Proposing to students different systems to solve, then discussing similarities and differences with them before summarising the different possible cases.*

(Continued)

TABLE 13.2 (*Continued*)

Term	Description	Examples and Non-examples
débat scientifique **R** (scientific debate)	Form of teaching starting from problematic assertions that students have to collectively discuss, and possibly correct, using forms of mathematics rationality, within a process orchestrated by the teacher.	Example: A collective debate on the validity of conjectures about radicals, leading students to differentiate between examples, non-examples and counter-examples. *Non-example:* *Proposing an open problem and organising a discussion about possible strategies.*
décontextualiser **R, T** (decontextualising)	Detaching mathematical knowledge from the specifics of the particular situation in which it was developed.	Example: Detaching the following property from a context of enlargment of figure: in a situation of proportionality, the image of the sum is the sum of images. *Non-example:* *Complexifying a proportionality task by changing the data without changing the context.*
définir **T, S** (defining)	Naming and specifying a mathematical object/concept by the statement of one of its characteristic properties.	Example: Defining an even number as an integer n for which there exists an integer k such that $n = 2\,k$. *Non-example:* *Introducing a mathematical object by showing examples.*
démontrer **T, S** (proving)	Validating an assertion or a result by using the form of justification (logic and deductive) specific to mathematics.	Examples: Proving a numerical property by algebraic reasoning. Proving an existence theorem by providing an example. *Non-example:* *Validating a property on an infinite set by checking it for a finite number of cases.*
différencier **T** (differentiating)	Taking into account the heterogeneity of students' learning when teaching.	Example: After a diagnostic test, proposing different tasks on the same topic adapted to students' knowledge.
donner du travail à faire hors classe **T** (assigning homework)	Giving extra work to complement - or prepare for - the classroom work.	Example: Closing a session, the teacher gives exercises to solve for the next session. *Non-example:* *Giving extra work to the students having completed the proposed tasks.*

Term	Description	Examples and Non-examples
donner les consignes **T** (giving instructions)	Giving directives to students regarding the tasks to perform.	Example: Introducing a geometrical task, the teacher asks the class to draw the corresponding figure before solving the task.
encourager **T** (encouraging)	Helping students gain confidence in their abilities, especially by underlying their progress, even small.	Example: A student shows the teacher a very partial solution. The teacher congratulates him and adds: "I'm sure you can continue quite well." *Non-example: In the same context, the teacher simply says: "ok, go on."*
enrôler, motiver **R, T** (engaging)	Arousing the interest of students and their engagement in the mathematical work.	Example: The teacher, seeing a student "dreaming," asks him to read aloud the statement of the next task. *Non-example: Isolating a student disrupting the class.*
enseignement dialogué, cours dialogué **R** (dialogic teaching)	Interactive and collective teaching device, the teacher making knowledge progress by asking students questions and relying on their answers.	Example: In a synthesis phase, asking students to describe their solutions and reworking these descriptions with them to get a correct mathematical expression. *Non-example: When lecturing, asking a question simply calling for an agreement, then continuing without asking more to the students.*
entrée dans la classe et début de séance (starting the class)	Phase preceding the actual mathematical activity, during which the teacher welcomes students and manages various issues.	Example: Students enter into the class and sit down. The teacher checks student attendance, then asks them to take out their notebook and show homework. *Non-example: Warm-up exercise at the beginning of a lesson.*
envoyer un (des) élève(s) présenter son (leur) travail à la classe **T** (sending student(s) present work to the class)	Asking one or several students to present their work, generally at the board.	Example: Screens of several students are projected on the board, and these are asked to explain their solutions. *Non-example: The teacher writes a student's solution on the board under his dictation.*

(Continued)

TABLE 13.2 (*Continued*)

Term	Description	Examples and Non-examples
estimer **T, S** (estimating)	Giving an order of magnitude or an approximate value of a numerical result, without trying to determine it exactly.	Example: Estimating the result of a numerical calculation by replacing the numbers involved with simpler numbers. *Non-example:* *After performing a calculation, giving an approximation of the result with a given accuracy.*
évaluation diagnostique **T** (diagnostic assessment)	Assessment form taking place before a teaching process, to identify students' knowledge and adapt the teaching strategy if necessary.	Example: Before introducing linear systems of equations, the teacher gives a diagnostic test about linear equations.
évaluation entre pairs **S** (peer-assessment)	Assessment form in which students assess each other's work.	Example: Group productions are displayed on the board, and students must determine their respective strong and weak points.
évaluation formative **R, T, S** (formative assessment)	Assessment form during the teaching process allowing students to situate their work with respect to expectations, and the teacher to regulate her teaching.	Example: During a lesson, after a quiz, the teacher decides to propose new exercises on the same topic. *Non-example:* *The teacher corrects a problem without taking into account the students' solutions.*
évaluation sommative **R, T** (summative assessment)	Assessment form, after the teaching process, aiming at assessing students' learning and compare it with expectations.	Example: The teacher gives a common test, built with her colleagues, at the end of a teaching sequence. *Non-example:* *A diagnostic test.*
évaluer **T** (assessing)	Gathering information on students' mathematical activity and/or knowledge.	Examples: Diagnostic, formative, summative assessment.
exercice d'application (application exercise)	Exercise, the solving of which mobilises knowledge previously taught with minimal adaptation.	Example (application of the Pythagorean theorem): In a right triangle, calculate the length of a side knowing those of the two other sides. *Non-example:* *Problem of comparison of areas, the solving of which requires Pythagoras' and Thales' theorems, proposed without any hint.*

Term	Description	Examples and Non-examples
exercice de mise en route (warm-up exercise)	First mathematical activity of a lesson aiming to set students to work.	Example: The lesson starts with a short collective activity of mental calculation, before moving to the main theme: geometry. *Non-example: The lesson starts with the management of organisational issues, then the teacher starts teaching a new topic.*
exercice de révision (revision exercise)	An exercise for practicing skills already taught, recently or not, making them and associated knowledge mobilisable or even available.	Example: After completing a sequence on Pythagorean theorem, the teacher gives students ten exercises on this theme to prepare the written test, a week later. *Non-example: Recall made during the solving of an exercise.*
exercice en contexte (in-context exercise)	Exercise set in an extra-mathematical context, and needing some mathematisation.	Example: Exercise asking to associate first degree equations to daily life situations and to solve them. *Non-example: Exercise asking to associate graphical representations and linear equations.*
exercice technique (technical exercise)	Exercise proposed to work on one or a few given techniques, to master these.	Example: Exercises of development or factorisation of algebraic expressions. *Non-example: Exercise in context.*
exercice/problème (exercise/problem)	Exercise: task whose solving only requires to apply well identified techniques, possibly with some adaptation. Problem: task whose solving requires the search for a solving strategy.	Examples: Exercises: technical exercise, application exercise. Problems: open problem, problem for deepening knowledge.
expérimenter T, S (experimenting)	Conducting an organised action to explore a problem, to test an idea, an approach, a conjecture.	Example: Building a geometric figure with a dynamic geometry software and moving free points to test a conjecture on alignment of points. *Non-example: Trials randomly done to become familiar with a situation.*

(*Continued*)

TABLE 13.2 (*Continued*)

Term	Description	Examples and Non-examples
expliciter **T, S** (making explicit)	Making explicit something implicit until then.	Example: Expliciting the multiplied sign implicit in products like (x + 1) (x + 2). *Non-example:* *Introducing a new notation.*
expliquer **T, S** (explaining)	Making something more intelligible.	Example: Explaining why a quadratic expression can take the same value for two different values of x, using a graphical representation. *Non-example:* *Repeating a hint just given.*
exposé d'élèves (students' presentation)	Pedagogical form where a student or group of students present work prepared on a given theme outside the classroom to their classmates.	Example: A group of students gives a presentation on number systems, prepared from texts provided by the teacher and personal search on the Internet. *Non-example:* *Students present their production at the end of a phase of group work during a lesson.*
exposition des connaissances, cours (lecturing)	Phase during which the teacher presents the knowledge content in a more or less structured way.	Example: After a research phase, the teacher points out the new knowledge engaged in it, and presents it in a structured way at the board; students take notes. *Non-example:* *After an introductory activity, teacher and students jointly build an institutionalisation.*
faire des connexions **T, S** (connecting)	Establishing links between mathematical domains, tasks, techniques, representations, statements...	Example: Connecting Thales' theorem to homothety in geometry, proportionality situations in the numerical domain, and linear functions in algebra. *Non-example:* *Generalising to new numbers the search for divisibility criteria.*
formuler **T, S** (formuler)	Expressing an approach, a reasoning, a property... with a natural or symbolic language.	Examples: Asking students to express the exercise solution in their own words. Symbolising the rule used in a calculation.
généraliser **T, S** (generalising)	Extending a mathematical property, method of resolution, concept.	Example: Generalising the relationship on the sum of angles of a triangle to the sum of the interior angles of a convex polygon. *Non-example:* *Introducing a new technique for solving systems of linear equations.*

Term	Description	Examples and Non-examples
gérer des erreurs **T** (dealing with errors)	Taking care of students' errors.	Example: Invalidating a wrong answer by giving a counter-example. *Non-example:* *Pointing out an error without further comment or action.*
gérer l'organisation, **le matériel** **T** (dealing with the material organisation of the class)	Organising the classroom work, the positioning of the students and the use of material resources.	Example: Using an interactive white board to project and discuss students' productions.
gérer la discipline **T** disciplining)	Ensuring respect for work instructions and the social rules in the classroom.	Example: Telling a student that his attitude in class is not correct, and if necessary applying some sanction.
gérer la prise de **notes** **T** (supporting the taking of notes)	Facilitating the taking of notes by students.	Example: Writing on the board the summary of the lesson with different colors to highlight important aspects and facilitate note taking. *Non-example:* *Taking notes is free and entirely under the responsibility of the students.*
gérer le ou les **tableaux** **T** (managing the board(s))	Organising the spatial and chronological use of the board(s), digital or not, and the content written or projected on them.	Example: The teacher uses one part of the board to project documents, another one as working space, and the last one to write what has to be copied in the notebook.
gérer le temps **T** (managing time)	Dealing with time constraints, personal or institutional, during a lesson or more globally.	Example:The teacher announces the precise time the students are given to complete the task. *Non-example:* *Having not enough time to sum up, the teacher ends the activity abruptly.*
gérer un incident **didactique/non** **didactique** **T** (dealing with an incident didactical/ non-didactical)	Reacting to an incident occurring in the classroom.	Examples: Non-didactic incident: A student asks to go out. The teacher asks a classmate to accompany her. Didactic incident: A student rejects another student's correct answer. The teacher organises a collective discussion about this answer.

(*Continued*)

TABLE 13.2 (*Continued*)

Term	Description	Examples and Non-examples
institutionnalisation **R** (institutionalisation)	Process under teacher's responsibility aiming at decontextualising the knowledge engaged in students' activity and linking it to institutional knowledge.	Example: After a situation involving affine functions, the teacher institutionalises that the graph of an affine function is a straight line, passing through the origin only for linear functions. *Non-example:* *Asking students to copy the solution of an exercise in their notebook.*
introduction d'une séance (lesson introduction)	Phase introducing the goals of the lesson, its mathematical content and the planned organisation.	Example: The teacher says that the lesson is about applications of Pythagorean theorem, that he will distribute a list of exercises to be solved individually, and will end by a synthesis. *Non-example:* *Presenting an isolated task during a lesson.*
justifier **T, S** (justifying)	Producing reasons or proof to support an opinion, an action or a decision.	Example: Justifying the choice of the unknown in an equation layout. *Non-example:* *Description of a solution.*
lancement d'une activité (launching an activity)	Phase for introducing and motivating an activity, to engage students in associated work.	Example: The teacher gives a written task to students, and asks them to read it silently. Then, she asks one student to read it aloud and other students to reformulate the work to be done. *Non-example:* *The teacher distributes a worksheet, just saying that students have to solve it individually.*
milieu **R** (milieu)	The elements of a learning situation with which the students interact, and provide them objective feedback.	Example: In a situation of construction with dynamic geometry, the software is a component of the milieu offering objective feedback: if the construction is based on perception, it does not resist to the dragging of its free points. *Non-example:* *The distribution of students in the classroom.*

Term	Description	Examples and Non-examples
mise en commun (kneading-up)	Phase following a research or solving phase, and aiming to share, compare and possibly validate students' proposals or solutions.	Example: After 10 minutes of individual research on a problem, the teacher organises a kneading-up to review the strategies proposed by students before launching a phase of group work. *Non-example:* *After a solving phase, the teacher just presents the expected solution.*
modéliser (mathématiquement) T, S (mathematical modelling)	Expressing mathematically an extra-mathematical situation to answer questions regarding it.	Example: Modelling the evolution of a population by a sequence, assuming doubling time approximately constant, to make predictions. *Non-example:* *Expressing algebraically a geometrical problem to solve it (see change of setting).*
moments de l'étude R (study moments)	Different didactic moments associated with the teaching of a mathematical theme in the Anthropological Theory of the Didactic (ATD).	Example: First meeting with a type of task
montrer T, S (demonstrating)	Making visible, illustrating.	Examples: Introducing parallelograms by showing examples and non-examples. Showing how to solve a particular type of exercise with an example. *Non-example:* *Evoking different types of polygons already met.*
narration de recherche R (narrative of research)	Activity where students are invited to work on a problem and then to write a report on their research describing all the tracks followed, including those unsuccessful.	Example: Students, working in groups, search the number of diagonals of a pentagon, a hexagon and then a polygon with 100 sides. Then, each group writes a narrative of its research, collected by the teacher and discussed in the next lesson. *Non-example:* *Problem-solving activity in which the students must write only their solution.*

(Continued)

TABLE 13.2 (*Continued*)

Term	Description	Examples and Non-examples
observer (repérer) le travail des élèves **T** (observing students' work)	Collecting information on students' activity to identify their strategies and regulate didactic action.	Example: During a phase of autonomous group work, the teacher circulates around the classroom taking notes of students' strategies and errors on her digital tablet, to organise a synthesis. *Non-example:* *Circulating in the classroom to give the students back some homework.*
outil/objet **R** (tool/object)	A mathematical concept is seen as tool when one focuses on its use to solve problems, as object when considered as a cultural object, element of scholarly knowledge.	Example: (the concept of equation) Tool use: solving problems with equations. Object use: working on changes preserving the equivalence of equations.
phase de rappel (phase of recall)	Phase for reminding students already taught knowledge and making it mobilisable, or for linking with a previous lesson.	Examples: Teacher organises a recall on the solving of systems of linear equations, before introducing a new solving method. Observing students' difficulties, the teacher stops a research phase and recalls properties taught the previous year. *Non-example:* *Interacting with a student, the teacher reminds her a useful property previously learned.*
phase de résolution, recherche (solving, research phase)	Phase during which students search and develop approaches for solving a task.	Example: Students work in groups on the solving of a problem after a phase of individual appropriation. *Non-example:* *The teacher gives hints very quickly before students have had sufficient time to move forward in their research.*
poser des questions d'ordre général **T, S** (asking general questions)	Asking questions not directly linked to mathematical activity.	Examples: Teacher: "Are you ready?" Student: "What notebook to use?" *Non-example:* *Teacher to a student: "Can you justify your answer?"*

Term	Description	Examples and Non-examples
poser des questions d'ordre mathématique **T, S** (asking mathematical questions)	Asking questions regarding the mathematical activity at play.	Examples: Teacher to student: "Can you explain your answer?" Student to teacher: "Do I have to prove this property?" *Non-example:* *A student asks if he can get out.*
présenter, rédiger des solutions **T, S** (presenting, writing out a solution)	Formulating, verbally or in written form, the result of mathematical activity according to shared classroom rules.	Example: Writing out the solution of a problem, detailing its different steps. *Non-example:* *Taking personal notes.*
problématiser **T, S** (problematising)	Associating one or several mathematical questions to a given theme or context.	Example: Formulating questions from data regarding smartphone subscriptions, with or without purchase. *Non-example:* *Giving students a specific problem to solve.*
problème d'approfondissement (problem for deepening knowledge)	Problem allowing students to understand more deeply a piece of knowledge.	Example: After studying some examples of quadratic functions, proposing a problem about the number of intersection points with the axes of the graphs of a family of quadratic functions depending on one parameter. *Non-example:* *In the same context, studying a new numerical example.*
problème de réinvestissement (problem for reinvesting knowledge)	Problem aimed at using notions already studied in new contexts or at solving complex problems where these notions are combined with other concepts.	Example: Use of algebra in a new setting, for instance in a geometry problem. *Non-example:* *Application exercise to consolidate a method just taught.*
problème ouvert **R** (open problem)	Problem with the three following characteristics: a short statement which doesn't induce a method or result, different possible approaches, a conceptual domain familiar to students.	Example: Find the last two digits of 2^{222} *Non-example:* *A problem structured in several parts.*

(*Continued*)

TABLE 13.2 (*Continued*)

Term	Description	Examples and Non-examples
progression (progression)	Teaching organisation covering one or more topics, specifying the distribution of content and competencies.	Example: Teachers working at the same level agree on a common progression, alternating the teaching of different mathematical domains. *Non-example:* *Organisation of assessment modalities in the classroom.*
raisonner **T, S** (reasoning)	Using rational forms of thinking to carry out a (mathematical) activity.	Example: Using the fact that an algebraic expression is a sum of squares to show that it is always positive. *Non-example:* *Carrying out a given program of calculation.*
rappeler **T, S** (recalling)	Recalling old knowledge or past episodes useful for the progression of the current mathematical activity.	Example: Making students recall the methods used to solve similar exercises in a previous lesson. *Non-example:* *Using knowledge seen during the previous lesson without explicit linking.*
rédiger **T, S** (writing out)	Expressing in a written text a solving process, the solution of a problem or exercise, respecting the codes for mathematical expression.	Example: Collectively writing the solution of an exercise. *Non-example:* *Students' drafts.*
reformuler **T, S** (reformulating)	Expressing instructions, properties, statements with some variation in the wording.	Example: Reformulating a student's solution in a more accurate mathematical language. *Non-example:* *Repeating a student's formulation without modifying it.*
registre **R** (register)	A representation system with rules allowing the three operations of construction, treatment and conversion of representations.	Examples: Registers of algebraic expressions, graphical representations, natural language. *Non-example:* *Gestures accompanying mathematical activity (not a register due to the absence of such rules).*

Term	Description	Examples and Non-examples
relancer une question **T** (re-asking a question)	Asking a question again, changing more or less its phrasing.	Example: Teacher asks again a question, specifying it to guide students' attention to a particular detail. Non-example: Repeating a question just asked.
répondre à une question **T, S** (answering a question)	Providing information in response to a question.	Example: Teacher asks: "What is the variable?." A student answers. "The length AM." Non-example: Teacher replies to a student's question: "What do you think about it?"
représenter **T, S** (representing)	Associating symbols, geometrical drawings, diagrams, graphs… to mathematical objects, relationships or processes.	Example: Representing statistical data with bar or circular charts, box plots ... Non-example: Associating another problem with a problem, by analogy.
séquence / séance (sequence/lesson)	A sequence is a set of coordinated lessons on a single theme.	Example: Sequence of lessons on statistics, starting the data collection with a questionnaire. Non-example (of sequence): Succession of lessons alternating numbers and geometry over a school week.
simuler **T, S** (simulating)	Elaborating, a representation, often virtual, of a device or process to facilitate its study.	Example: Simulating a poll with a spreadsheet to highlight sampling fluctuations. Non-example: Taking a picture of a complex geometric object to visualise it.
situation d'action **R** (situation of action)	Situation where the students' interaction with the milieu allows the emergence of models of action (strategies), and the test of their validity.	Example: The situation of Brousseau's puzzle (see didactic/ adidactic situation). Models of action appear, first the additive model (adding 3 cm to all measures) and it is disqualified because the pieces do not fit. Non-example: A situation of construction where the students depend on the teacher to know if their construction is right.

(*Continued*)

TABLE 13.2 (*Continued*)

Term	Description	Examples and Non-examples
situation de formulation **R** (situation of formulation)	A situation where students have to formulate the mathematical knowledge involved to be successful, and where the interaction with the milieu allows them to test the effectiveness of their formulation.	Example: Communication situation for the reproduction of geometric figures. Issuers write a message (without drawing) describing the geometric figure and receivers construct it. The pair "receiver / issuer" wins if the two figures can be superposed. *Non-example:* *Writing out the solution of an exercise after its collective solving.*
situation de validation **R** (situation of validation)	A situation for which the stake is to validate assertions. Students' interactions with the milieu must allow them to determine by themselves if the assertions are valid or not.	Example: The situation of validation in the race to 20 (https://sites.math.washington. edu/~warfield/Inv%20to%20Did66%20 7–22-06.pdf) *Non-example:* *Asking students to prove that two lines are parallel, using the reciprocal of Thales' theorem.*
situation didactique/ adidactique **R** (didactic/adidactic situation)	The concept of didactic situation models a situation organised to enable students learn some given (mathematical) knowledge through their interaction with an appropriate milieu. Adidactic qualifies a situation in which students momentarily behave as if ignoring the teacher's didactic intention.	Example: The situation of puzzle enlargement by Brousseau: students working in groups must enlarge a puzzle. Each student builds a piece of the puzzle. The group succeeds if the pieces fit together. This situation is adidactic if the students work without wondering what operation the teacher expects of them.
situation-problème **R** (problem situation)	Situation aimed at making the students find by themselves how to solve a given problem, and develop new knowledge.	Example: The situation of bordered squares with the necessity of symbolism of algebraic type resulting from increase in the size of squares. *Non-example:* *Application exercise.*

Term	Description	Examples and Non-examples
susciter des idées, des stratégies T (stimulating ideas or strategies)	Making emerge ideas or strategies to approach a problem or working out a question.	Example: After reading the statement of a problem and identifying important elements, the teacher asks students their ideas about how to start the search. *Non-example:* *Proposing students to study a particular case.*
tâche complexe (complex task)	Task for which the solving obliges students to take initiative, and often to use external resources (documents, the Internet, etc.).	Example: Determining the most interesting phone price from promotional advertising and tariff information. *Non-example:* *Exercise of application on affine functions.*
tâche/activité R (task/activity)	Task: what to do under certain conditions. Activity: what individuals do to carry out a task.	
transition entre phases (transition between phases)	Phase, often short, in which the teacher manages the entry in a new phase of the lesson.	Example: Teacher ends a warming-up phase and tells students that she will now introduce a new geometrical concept. *Non-example:* *A new phase starts but the teacher does not explicitly mention it.*
travail collectif (whole class work)	Work with the whole class orchestrated by the teacher.	Example: A collective review of quadrangle properties with the help of dynamic geometry software.
travail en demi-classe (half-class work)	Pedagogical organisation with class split in two halves.	Example: Half the class works on the computers.
travail en groupe autonome/ accompagné (autonomous/ guided group work)	Students work in groups for a while, without teacher intervention/with teacher guidance.	Example (autonomous work): Students work in small groups on an open problem, and produce a poster presenting their work. The teacher circulates around the room, without intervening. *Non-example:* *Students arranged in groups work on personalised sheets of exercises, at their own pace.*

(Continued)

TABLE 13.2 (*Continued*)

Term	Description	Examples and Non-examples
travail individuel autonome/ accompagné (autonomous/ guided individual work)	The students work alone for a while, without teacher intervention/with teacher guidance.	Example (guided work): Students work individually on an exercise. The teacher distributes prepared hints to students when needed. Non-example *(autonomous work)*: *The teacher gives students an exercise and asks them to solve it individually but, after 30 seconds, asks who has a strategy to offer.*
valider/invalider des productions T, S (validating/ invalidating productions)	Deciding whether an answer, product, reasoning is correct and/or pertinent, or not.	Example: The teacher asks students to decide on the validity of the answer given by one student. *Non-example:* *The teacher just gives the expected procedure.*
valoriser T (valuing)	Valuing the quality of the students' productions, even unfinished, or their attitudes, using explicit criteria.	Example: The teacher congratulates a student for the accuracy of his writing. *Non-example:* *A student answers a question and the teacher simply says: "ok."*
vérifier T, S (verifying)	Checking mathematical work (computation, reasoning…) to ensure its validity.	Example: Checking the solving of an equation by substituting the solutions found in the equation.

As mentioned above, at a very early stage, the members of our research team found it useful to structure the French Lexicon (see *Chapter 12* for details). Lexicon terms are currently distributed into six categories: General terms (11 terms); Nature of tasks, activities (17 terms); Lesson phases (12 terms); Forms of pedagogical organisation (9 terms); Mathematical activities (19 terms); and Pedagogical and didactical management of the classroom (48 terms). For this last category, the biggest one, three sub-categories were created: organisation; interactions; exploiting and assessing, with, respectively, 12, 23, and 13 terms. Table 13.3 shows this organisation.

The French Lexicon is the outcome of a long process, which has involved about two hundred people beyond the six teachers and researchers, who were members of the Lexicon team: teachers, teachers educators, researchers in mathematics education, and a few mathematics inspectors. This enterprise represents the first attempt made at documenting, in a systematic way, the pedagogical-didactical terminology reasonably shared by middle-school mathematics teachers in France. As shown in *Chapter 12*, the French Lexicon reflects important characteristics of our mathematics education culture, and also the current strengths and limitations of the teachers' professional discourse. We hope that it will become a tool in the hands of teachers and all those involved in their preparation and professional development, generating insightful reflections, and that, together with the comparison with the other lexicons, these will contribute to its progressive refinement and enrichment.

TABLE 13.3 French Lexicon – Terms by category.

Termes généraux (general terms)	didactic contract; exercise/problem; milieu; mobilisable/available knowledge; progression; register; sequence/lesson; setting; study moments; task/activity; tool/object
Nature des tâches, activités (nature of tasks, activities)	application exercise; complex task; didactic/adidactic situation; in-context exercise; introductory activity; narrative of research; open problem; preparatory activity; problem for deepening knowledge; problem for reinvesting knowledge; problem situation; revision exercise; situation of action; situation of formulation; situation of validation; technical exercise; warm-up exercise
Phases de séance (lesson phases)	correction; institutionalisation; kneading-up; launching an activity; lecturing; lesson closing; lesson introduction; phase of recall; solving, research phase; starting the class; summary, synthesis; transition between phases
Formes d'organisation pédagogique (forms of pedagogical organisation)	autonomous/guided group work; autonomous/guided individual work; dialogic teaching; flipped classroom; half-class work; lecturing; scientific debate; students' presentation; whole class work
Activités mathématiques (mathematical activities)	arguing; calculating; changing the register of representation; changing the setting; noting mathematical symbols on a drawing, schema, graphics; conjecturing; connecting; defining; estimating; experimenting; generalising; mathematical modelling; problematising; proving; reasoning; representing; simulating; verifying; writing out
Gestion pédagogique et didactique de la classe (pedagogical and didactical management of the classroom):	
1. organisation (organisation)	circulating the classroom; dealing with the material organisation of the class; differentiating; disciplining; giving homework; giving instructions; managing the board(s); managing time; observing students' work; selecting didactic variables; sending student(s) present work to the class; supporting note-taking
2. interactions (interactions)	answering a question; asking general questions; asking mathematical questions; demonstrating; drawing the attention on, stressing a mathematical point; conceptual help; constructive/procedural help; encouraging; engaging; explaining; formulating; helping, scaffolding; help with materials; instrumental help; justifying; reasking a question; making explicit; methodological help; recalling; reformulating; skills help; stimulating ideas or strategies; valuing
3. exploitation et évaluation (exploiting and assessing)	assessing; commenting; comparing, ranking productions; dealing with an incident did actical/non-didactical; dealing with errors; decontextualising; diagnostic assessment; formative assessment; peer-assessm ent; presenting, writing-out solutions; self-assessment; summative assessment; validating/invalidating productions

Acknowledgements

This project has been funded by a Discovery Grant from the Research Council of the Australian Government (ARC-DP140101361). We deeply thank our foreign colleagues in the international Lexicon team, all those who have accepted to take part in the review process, and also the IREMs of Poitiers and Paris, and our research laboratory, the LDAR, for their support.

Bibliography

Presentations and publications

The work of the French national team has been presented locally and internationally at the following meetings and conferences:

- APMEP (French association of mathematics teachers of public education) Annual Meeting 2015, 2017
- International Group for the Psychology of Mathematics Education (PME) Annual Conference, 2017
- American Educational Research Association (AERA) Annual Meeting, 2018
- EMF (Francophone Mathematical Space) 2018 Conference
- International Seminar of Mathematics Education, Universidad Antonio Nariño, Colombia, 2018
- LDAR Seminar, Université Paris-Diderot, 2018
- Res(s)ources 2018 Conference
- IREM Seminar, Université de la Réunion, 2019

Publications from the French research team include:

Artigue, M. (2016). L'enseignement des mathématiques au carrefour des cultures. *Bulletin de l'APMEP*, 518, 153–170.

Artigue, M. (2019). Reflection on a theoretical approach from a networking perspective: the case of the documentational approach to didactics. In L. Trouche, G. Gueudet, & B. Pepin (Eds.), *The 'resource' approach to mathematics education* (pp. 89–118). New York, NY: Springer.

Artigue, M., Chevalarias, T., Debertonne-Dassule, F., Grugeon-Allys, B., Horoks, J., & Pilet, J. (2019). Approcher la diversité culturelle dans l'enseignement des mathématiques à travers le filtre du langage professionnel des enseignants. In M. Abboud (Ed.), *Actes du Colloque EMF 2018* (pp. 860–868). Paris, France: IREM de Paris.

Artigue, M., Novotná, J., Grugeon-Allys, B., Horoks, J., Hospesová, A., Moraová, H., Pilet, J., & Žlábková, I. (2017). Comparing the professional lexicons of Czech and French mathematics teachers. In B. Kaur, W.K. Ho, T.L. Toh & B.H. Choy (Eds.), *Proceedings of PME 41* (Vol. 2, pp. 113–120). Singapore: PME.

14

DOCUMENTING AND DEVELOPING THE CURRENT GERMAN LEXICON

Jenny Christine Cramer, Christine Knipping, David A. Reid and Birte J. Specht

Introduction

The German mathematics education research community is characterised by a long history of collaboration within the German-speaking community, including but not limited to *Stoffdidaktik*. This German Didaktik-tradition is related to similar traditions in many European countries (see Blum, Artigue, Mariotti, Sträßer & Van den Heuvel-Panhuizen, 2019) and its roots can be traced back to the Didactica Magna (The Great Didactic) of the Czech pedagogue Comenius (Hudson & Meyer, 2011, Kaiser, 2017). Stoffdidaktik refers to an approach to didactics strongly focused on mathematics as a subject (Hefendehl-Hebeker, vom Hofe, Büchter, Humenberger, Schulz & Wartha, 2019). It can be traced back to W. Lietzmann and his seminal work "teaching methods for mathematics teaching" (orig. "Methodik des mathematischen Unterrichts," Lietzmann, 1923), working in the tradition of Felix Klein.

> ...his purpose was to provide practicing teachers with a detailed insight into the subject matter, to propose appropriate formulations and illustrations such as possible learning paths, and to indicate obvious difficulties.
>
> (Hefendehl-Hebeker et al., 2019, p. 27)

Arnold Kirsch's lecture at ICME-3 in Karlsruhe advanced this approach (Kirsch, 1977, 2000) and has continued to evolve since (e.g, Kirsch 1987). Another strong theme in German mathematics education research is applications and modelling, represented, for example, by the lecture given by Werner Blum at ICME-12 in Seoul (Blum, 2015, see also Vorhölter, Greefrath, Borromeo Ferri, Leiß & Schukajlow, 2019). Design science is a further theme that is strong in the German didactic tradition. It aims to bridge between theory and practice and is exemplified by Erich Wittmann's lecture at ICME-9 in Tokyo (Wittmann, 2004, see also Nührenbörger, Rösken-Winter, Link, Prediger & Steinweg, 2019). Further useful summaries of didactics in the German-speaking community have appeared in recent years (e.g., Jahnke & Hefendehl-Hebeker, 2019, Sträßer, 2019).

In addition to the long tradition of didactics within the German-speaking community, this also reflects that German researchers are also very active in international conferences, especially ICME (International Congress on Mathematical Education) and CERME (Congress of the European Society for Research in Mathematics Education), which brings them into contact with other approaches and makes us aware of the use among researchers of different words, that reflect different conceptual frameworks. This awareness fuelled our interest in the Lexicon project, as it seemed likely that among teachers there might be even wider differences in the conceptual frameworks used in teaching, that could be revealed by differences in the vocabulary used to describe mathematics lessons.

It was important for us to clarify the kinds of terminology we were interested in. German tends to be a very concrete language, with technical terms often built out of concatenations of everyday terms. This makes meanings more transparent, but also risks that some aspects of technical terms might be overlooked. For example, the German word for quadrilateral is *Viereck*, literally "four-corner." Learners are unlikely to forget that a Viereck has four corners, unlike English speaking learners who might not strongly associate "quad" with "four." But such differences in terminology, while interesting, were not our focus. Rather we wished to look for words used or understood by teachers to describe what goes on in mathematics lessons, and that might reveal the sort of conceptual differences that might underlie subtle differences in teaching practice.

Country-specific features

The educational system in Germany shares some general features, but is also very diverse. This reflects the jurisdiction of the 16 Federal States (*Länder*) over education. School forms, curricula, and teacher education are all determined at the state level, and state governments take responsibility for funding education (OECD, 2019). The federal government has limited influence, and the main Germany wide body that has an influence on education is the Standing Conference of the Ministers of Education and Cultural Affairs of the Länder in the Federal Republic of Germany (KMK: *Kultusminister Konferenz*). It is a consortium of ministers responsible for education and schooling, and institutes of higher education and research and cultural affairs. In this capacity, the consortium formulates the joint interests and objectives of all 16 Federal States. Two significant documents produced by the KMK are the *Bildungsstandards* (Education Standards) and the *Standards für die Lehrerbildung* (Teacher Education Standards) (see KMK 2008, KMK & IQB 2010). The *Bildungsstandards* (first enacted in 2003, published by KMK in 2004, see also Klieme et al., 2003) outline curriculum recommendations and have been adapted (but not adopted) by the Federal States as they have revised their own curricula. The *Standards für die Lehrerbildung* outline recommendations for teacher education including both subject matter background and didactical competences to be developed in pre-service teacher education. Detailed information about education systems in Germany is available from the KMK at https://www.kmk.org/kmk/information-in-english.html.

The general structure of German schools traditionally consisted of:

- Primary schools (*Grundschulen*), beginning at age 6, compulsory, and taught by generalist teachers. In most states, primary schools covered grade 1 to 4. In Berlin and Brandenburg they covered grades 1 to 6.
- Three forms of secondary schools with different leaving certificates and qualifications:

- The Hauptschule, which provided a basic general education which could lead to a vocational qualification or to further studies leading to a higher education entrance qualification. Hauptschulen normally covered the grades from the end of primary school to the end of compulsory education (grade 9 or 10 depending on the state).
- The Realschule, which provided a more extensive general education that lead to a vocational qualification or a higher education qualification. The standard Realschulen covered grades 5 to 10.
- The Gymnasium, which provided an education focussed on academic subjects. In most Federal States, some form of recommendation from primary school was required to be able to access a Gymnasium. The school leaving qualification (*Zeugnis der Allgemeinen Hochschulreife/ Abitur*) qualified its recipients to apply to university. The standard Gymnasium covered the grades from the end of primary school to grade 12 or 13.

The secondary school system has, however, been undergoing crucial changes in recent years in the majority of the Federal States, which are reflected upon in the following section. The video of a German lesson used in the International Classroom Lexicon Project was recorded in a Gymnasium.

Recent trends

Following the PISA shock after 2000, when German school students scored unexpectedly poorly on the first PISA assessment, there have been many changes to educational systems in Germany. The PISA results revealed that German school students had a much wider range of achievement than students in many other countries, and this was related to their attending different school forms. Comparative studies and trends of globalisation in education in general also had an impact on the German school structure which meant that today in most provinces in Germany the three-form school system is replaced by a two-form structure, the Gymnasium on the one side and a comprehensive school on the other side. At the same time, there was a trend to eliminate special schools for students with physical disabilities, which existed in former times as separate schools in Germany, and to integrate these students into the regular schools. The combination of these two trends means that classes are now more heterogeneous than in the late 1900s.

The reduction of the number of school forms was accomplished in many states by the elimination of *Hauptschulen* and *Realschulen*, and the creation of a combined school form. This school form is known by many different names across Germany. Mathematics is taught in all school forms, but different textbooks exist for different forms, and publishers offer textbooks specific to the curricula of the larger states as well. In 2019, ten of Germany's 16 Federal States offer now a system with secondary schools that cover different educational tracks that were traditionally assigned to *Hauptschule* and *Realschule*.

Another trend over recent years is an increase in the number of students who attempt and achieve a school programme leading to a school-leaving qualification that qualifies its recipients to apply to university (*Abitur*). To account for this trend, many of these newly created secondary schools also offer the option to work towards the *Abitur*, sometimes offering students one more year of schooling than a regular Gymnasium (i.e., offering the *Abitur* after a total of 13 years of schooling, whereas the *Gymnasium* in these Federal States offers the *Abitur* after 12 years of schooling).

Secondary school teacher education

As the focus of the lexicon is on secondary learner classrooms, we focus on the education of secondary teachers. Again, the Federal State structure leads to minor differences across states. The traditional path into teaching, which is still taken by the majority of teachers, consists of two steps: an academic part at a university (or, in some Federal States, a teaching college, *Pädagogische Hochschule*) and a more practice-oriented part of supervised teaching in school (the *Referendariat*).

A major characteristic of teachers in Germany is that every secondary school teacher qualifies to teach **two** subjects, which are not necessarily related in their content. This leads to five different parts in teacher education at university: subject courses for subject A, teaching-oriented courses (*Fachdidaktik*) for subject A, subject courses for subject B, *Fachdidaktik* for subject B, and general education courses. The subjects are regarded as equivalent. Academic preparation of teachers usually takes between 4 and 5 years (a 3-year Bachelor, and a 1- or 2-year Master's). While studying at university, teacher students are usually required to spend short time periods (typically between 6 weeks and 6 months) in school. During these *Schulpraktika*, the students are mainly encouraged to watch and analyse lessons, and sometimes to gather first experiences in teaching.

After successfully completing university, usually with a Master's degree, prospective teachers apply for a position to do their *Referendariat* in one of the Federal States. During the *Referendariat* which, depending on the Federal State, usually takes 18–24 months, teachers work in schools and undergo further preparation courses at the same time offered by the school board. While the first years of teacher education at university emphasise theoretical requirements for teaching, the *Referendariat* focuses on the practice of teaching. Future teachers are expected to learn from more experienced teachers, and their own lessons are often supervised by their instructor at the school and a teacher trainer from the schoolboard. At the end of the *Referendariat* (and in some Federal States also before the end), teachers receive a grade which either qualifies or disqualifies them for the job. Teachers who qualify then need to apply for a position in school.

In recent years, because of a huge lack of teachers there has been a growing trend in many Federal States to employ teachers who did not take the path outlined above, but who qualified for a different job (e.g. who hold a degree in engineering) to teach subjects for which there is a lack of teachers (e.g. physics). The proportion of these *Quereinsteiger* (cross-starters) is especially high when looking at the numbers of newly employed teachers, and it varies significantly between the Federal States with a peak in Saxony (46.6%) and no cross-starters in Bavaria, the Saarland, and Hessen (Klemm, 2019).

Methodology, the lexicon, and analyses

Developing the German Lexicon was an ongoing process involving successive cycles. In this section, we present the methodology used for creating the various versions of the lexicon. The German Lexicon team consists of three experienced mathematics teachers, two of them are now in positions mentoring new teachers, as well as three academic researchers, two experienced and one involved in doctoral research. All but one are native German speakers.

The initial lexicon

The starting point for the development of the German Lexicon was individual viewing by each team member of a video recording of a German mathematics lesson as well as two other videos from the other participating countries. The videos were assigned so that (aside from the German lesson) each one was viewed by one teacher and one researcher.

The focus was on recording events for which a descriptive term exists in German, and recording those terms, descriptions of their meanings, and references to examples of them in the videos. The team compared their individual lists of terms and discussed what terms to include and how to organise the terms.

In addition to compiling an initial list of words, descriptions, and examples from the videos, the research team also discussed possible ways to present the German Lexicon. Several terms were identified that fit within larger categories, for example, names of different phases of a lesson, which suggested other related terms, as well as a category of terms referring to the lesson structure. The presence of sets of terms that fit within larger categories suggested structuring the lexicon around such categories, but other terms did not fit this structure and so a system of cross-references was adopted instead. In Table 14.1, for example, the description of the term *Abschluss* includes cross-references to the term *Unterrichtsphase*, as well as to the term *roter Faden*. The latter applies to the overall lesson structure without being visible in a specific video example. The former term, *Unterrichtsphase*, points to the different lesson phases, *Abschluss* being one of them. The examples illustrate that the terms in the German Lexicon vary in their individual scope of what they describe.

This example also shows that the terms themselves were left in German. The original descriptions of the terms were in German, and they were then independently translated by two team members, who then negotiated the final English descriptions. In this translation process, it became clear that there would be issues in translating the terms themselves. For example, the term *roter Faden* translates literally as *red thread* but the sense of the term is *main theme* or perhaps better *leitmotif*. The translation *leitmotif* was rejected as it is not a commonly used term in English, but also because it has a distinctly different meaning in German. This is an example of the problem of false friends, which occurs also in translating German terms into English. For example, the term *Impuls* occurs in the German Lexicon and one is tempted to translate it into the similar English word *impulse,* but that does not have the same meaning. At this stage, the research team decided to avoid the problems with translating the terms themselves by not translating them.

The research team had at first been inclined to include any terms that any team member proposed, which resulted in a list of 105 terms. The team then reduced this list by

TABLE 14.1 An example from the initial lexicon.

Term	Description	Example
Abschluss	The end of a lesson-phase (see *Unterrichtsphase*), a period, or a unit, in which the unifying topic (see *Roter Faden*) is brought to a conclusion.	Beispiele DEU #1:18:52–1:21:13 (The teacher sums up the results of the lesson and highlights important parts) USA #1:37:20 - end (The students are given an exit task; they are allowed to leave as soon as they have completed this task)

independently indicating all terms that were considered important enough to include. Terms that were indicated by only one or no team member were removed and terms that all team members felt were important were marked as especially important for the evaluation phase (see below). Among the terms which were removed, there were for example terms for concrete objects used in German mathematics classrooms (*Geodreieck*), or assessment forms (*Portfolio, Lerntagebuch*). These terms were eliminated as we considered them as general pedagogic terms and wanted to focus more on terms typical for a German tradition and specific for mathematics education. *Geodreieck* was a term of debate as it qualified in the view of some team members both criteria, being typical for a German use in contrast to other countries where this instrument is not known, and being specific to mathematics teaching and learning. In the end, it was decided that the term does not describe a teaching activity and so it was excluded. Several synonyms were identified in the initial lexicon and merged, resulting in a lexicon of 78 terms, including six synonyms. Of these 14 were indicated as particularly significant, among these the aforementioned *Roter Faden*, but also terms such as *Kognitive Aktivierung* (referring to the alertness of students during a lesson), as well as *Fragend-entwickelndes Unterrichtsgespräch* (which refers to a method of questioning students during a plenary lesson phase, directed at a particular answer).

Local and national evaluations

The local evaluation was conducted in early 2016 with three focus groups of about four teachers, from Bremen and the surrounding region. Each focus group was moderated by a member of the research team, and each session was audio or video recorded. The evaluation focussed on the clarity of the descriptions (in German) as well as the currency of the terms in the communication of teachers and researchers. In the first phase, each individual teacher was presented with about 20 descriptions. The descriptions were distributed so that each description was read by at least one teacher, and the descriptions of the 14 significant terms by two teachers. The teachers were given two instructions:

- If you know a term to which this description fits, name it. If you have more than one idea, please underline the term you would most likely use.
- If you do not know a term to which this description fits, please make a cross [X].

In the second phase, the entire list of terms was presented to all teachers without descriptions. Individually, the teachers were instructed to:

- Go through the list. Delete all words that are not useful for describing mathematics teaching. If you know a listed word, but would use a synonym, please mark the word and indicate the synonym.
- Are there any other words that are specific or significant for mathematics teaching, but which are missing from this list? Please add these words (and if possible a suitable description) on an extra sheet.

In the third phase, a group discussion occurred, as a follow-up to the individual phases. The moderator went through the list in sequence and discussed the terms that were often deleted. The meanings the teachers had given to these terms were discussed, as well as other terms the teachers would assign to the term's description. Terms that none of the teachers had deleted were also discussed, to clarify if the teachers shared a common meaning for those terms, and possible synonyms they might also use.

The local evaluation resulted in the clarification of six descriptions, and the addition of ten new terms and descriptions. Terms were added if there was a reasonable consensus during the discussion that the terms are used by teachers. 17 terms were deleted, based on the criteria that none of the teachers marked the term as especially important, and more than one teacher marked the term as irrelevant. This process produced a lexicon of 75 terms and 69 descriptions (due to synonyms). Among the deleted terms in this phase, many focused on a meta-level of lessons, often looking at the orientation of a lesson or at the degree to which a lesson plan fits the needs of the learners or the topic. Examples for deleted terms are *Passung* (an adequate fit/interplay between various aspects of a lesson: cognitive demand, competencies, methods, content, groupings, and so on), *Verstehensorientierung/ Verfahrensorientierung* (orientation towards understanding vs. orientation towards procedures), *Prozessorientierung/ Produktorientiereung* (orientation towards working processes or towards a final product).

The terms in the lexicon produced through the local evaluation were included in a national online survey. Special emphasis was put on the terms which had been marked as highly relevant in the local evaluation. Six questionnaires were developed, and each participant saw one of these six, randomly selected. The questionnaire had three sections. In the first section basic demographic information was collected: teaching experience, role (e.g., teacher, head of school, probationary teacher), school level and type, and Federal State. In the second section, about 23 descriptions were given and terms matching them were asked for. Each description was included in two questionnaires. In the third section, 25 terms were given and the participants were asked:

- How frequently do you use this term? (5-point scale)
- How important is this term for mathematics teaching and learning? (5-point-scale)
- Which other terms would you rather use than the aforementioned?

Each term was included in two questionnaires, and terms were not included on the same questionnaires as their descriptions.

A total of 68 participants took part, 47 of them experienced teachers (mostly at secondary level), with the remainder being probationary teachers (five), board consultants (two), and researchers (seven) from across Germany, but concentrated in the area around Bremen.

Based on the results of the national evaluation, several terms were excluded as being unfamiliar to teachers, or perceived as irrelevant. Among these excluded terms are *Diagnostik* (uncovering learners'current state of learning by examining their working processes, actions, utterances, and/or products), *Natürliche Differenzierung* (providing tasks that allow students to work at their own level), and *Kleinschrittiges Vorgehen* (Degree of subdivision of a problem into smaller tasks). This produced a German Lexicon consisting of 65 main terms and 11 additional synonymous terms. An example is shown in Table 14.2. The entry for *Auswertung* includes a synonym, *Vergleich*.

Clarity checking and further revisions

The German Lexicon was then shared with other teams in the Lexicon project and the U.S. and Chinese teams were given the task of reading over the descriptions in the

TABLE 14.2 A sample of terms and descriptions following the national evaluation.

	Term	*Description*
4	Anschauungsmaterial	Objects, such as pictorial representations, through which content to be learned is made accessible.
5	Argumentieren, Begründen, Beweisen	A process competence ($\rightarrow$ *Prozessbezogene Kompetenz*) which describes a form of communicating in the mathematics classroom in which reasons are brought forward for the correctness or falsehood of a statement, or in which connections are established between mathematical concepts.
6	Äußere Differenzierung	Differentiation ($\rightarrow$ *Differenzierung*) in which an external authority divides the learners into groups.
7	Auswertung (*Vergleich*)	Comparative analysis of solution statements and thinking about tasks that were given as homework or within the lesson.

German Lexicon and commenting on the clarity of them. Shortly after they reported, members of the German team met with members of the French and Finnish teams for further discussions. The combined comments led to the revision of most of the descriptions, and the addition of English translations of the terms. To address the issue of literal translations differing from translations of the sense of the terms we followed a suggestion from the Chinese team and included both translations. Also in some cases, other synonyms were included with the translation. Table 14.3 shows these changes for three of the terms. The original idea of grouping the terms into categories was also reintroduced, but in the form of an additional index rather than a restructuring of the lexicon itself. This presentation is the current state of the German Lexicon, which we expect to continue to evolve as it is used.

Some specific remarks on the German Lexicon

During the process of translating the terms, several issues of interest became apparent. Some of these are illustrated by the words included in Table 14.3.

The problematique of literal translations

The English phrase *visual and manipulative materials* describes the nature of the materials called in German *Anschauungsmaterial*, but does not capture their purpose. The literal translation touches on this purpose, but *experience* does not capture the full depth of meaning of *Anschauung*, which conveys also elements of *intuition, opinion,* and *perception*. The difficulty of translating *Anschauung* is illustrated by Kant's famous aphorism, "Gedanken ohne Inhalt sind leer, Anschauungen ohne Begriffe sind blind." (1781, p. 51, 1787, p. 75, 1956, p. 95). In Meiklejohn's translation (Kant,1855), the phrase is translated as "Thoughts without content are void; intuitions without concepts, blind." (p. 46). Here *Anschauung* becomes *intuition*. However, this phrase is also rendered as "Percepts without concepts are blind, and concepts without percepts are empty." (e.g., Masih, 1994, p. 355). In this case, *Anschauung* is rendered as *concept*, as is the original German *Inhalt*, which is

TABLE 14.3 Example of terms from the current German Lexicon.

Term	Description	Examples and Non-examples
Anschauungsmaterial "Experience-material" (visual & manipulative materials)	Resources, such as visual representations, pictorial or concrete objects through which content to be learned is made accessible.	Example: Linking cubes, pictures, dynamic geometry software. A teacher uses the materials to show something to help students to see an idea.
Argumentieren, Begründen, Beweisen "argumentation, justifying, and proving" (reasoning)	A process competence (see *Prozessbezogene Kompetenz*) which describes a form of communicating in the mathematics classroom in which reasons are brought forward for the correctness or falsehood of a statement, or in which connections are established between mathematical concepts.	Example: Student explains a solution to $x - 2y = 3$; teacher emphasises "And can you explain why? Justify why it is this way?"
Äußere Differenzierung "external differentiation" (streaming [US: tracking, UK: setting])	Differentiation (see *Differenzierung*) in which learners are divided into schools or classes according to achievement level as well as other factors such as physical disabilities.	Example: In Germany after grade four or six the students are still streamed into different types of school according to their achievement level. *Non example:* Grouping, when students in one class are temporarily divided into different small groups for working on a specific task.

usually translated as *content.* All of these English words capture something of the meaning of *Anschauung*, but no direct equivalent exists.

In the case of *Argumentieren, Begründen, Beweisen*, the three words come directly from the national educational standards document (KMK, 2004), and correspond to the 2000 NCTM standard *Reasoning and Proof* (in the 1989 NCTM Standards, simply *Reasoning*). In this case, the literal translation *argumentation, justifying, and proving* is long, and the meaning and parallel with the NCTM Standards is better captured by the word *reasoning*, which lacks a simple equivalent in German.

The third term, *äußere Differenzierung*, corresponds closely in meaning to various terms used in the English speaking world, so three synonyms are provided. None of these, however, convey an important aspect of the German term, that the division is based on externally observable criteria such as prior school performance or a physical disability, not related to mathematical understanding of a particular topic. The literal translation and description help to add this nuance, and the example further reminds readers familiar with more inclusive school systems that differentiation in German schools is also done into different school forms.

Another difficulty occurs when a distinction is made in German that is not easily captured in English. For example, the example for *äußere Differenzierung* uses the word *grouping*, which is also the English equivalent given for two other terms, *Sozialformen* and *Binnendifferenzierung* (and also its synonym *innere Differenzierung*). As the word *grouping* is used in a range of ways, it is the closest translation of these two German terms, but with different meanings. Thus the literal translations, descriptions, and examples are important to further distinguish the difference between these terms.

The creation of compound words in German

Table 14.2 also includes an example of a peculiarity of German, the possibility of creating compound words of virtually unlimited length and complexity. The word *Anschauungsmaterial* is formed by combining two nouns *Anschauung* and *Material*. This new word derives its meanings from its components, but at the same time becomes a new entity. In contrast in English compounds remain phrases. "Visual aids" refers to a category of materials, but the decomposition "aids, that are visual" is easier. Another example is the word *Darstellungswechsel*. *Darstellung* means "representation" and *Wechsel* means "change" but there is more to *Darstellungswechsel* than just a change of the form of representation. In connecting *Darstellung* and *Wechsel* there seems to be incorporated a further element that the concept under observation stays invariant; only the form of representation is changed.

Depicting actual teacher language

Another issue that the German Lexicon team came across in their various evaluation phases while creating the German Lexicon is related to the fact that there are certain terms which, despite being familiar to mathematics education researchers or people working in the instruction of teachers, were ruled as "not used" or "barely used" for talking about the mathematics classroom by teachers. These include some of the terms which were deleted from the initial lexicon during the evaluation phases, such as *Diagnostik* (diagnostic), *Prozessorientierung, Produktorientierung, Verfahrensorientierung, Verstehensorientierung* (orientation towards processes, products, procedures, and understanding, respectively).

A unifying factor of these terms is their focus on the goals of teacher actions and teacher beliefs when planning a lesson. From a research perspective, it is desirable for teachers to reflect upon their teaching actions and their planning from these viewpoints of goals and beliefs. There are different potential origins of this lack of familiarity with the named terms. On the one hand, teachers do not often get to reflect on mathematics lessons once they successfully completed their teacher education. Collegial supervision scarcely happens, team teaching is rare, and there is little time during a typical school day to reflect upon lessons with others. If teachers spend very little time reflecting upon lessons, it is no wonder that their vocabulary scope for reflecting on lessons will be limited. On the other hand, the examples given do not relate to concrete actions seen in a lesson (and thus are hard to find in the lexicon videos) but rather to decisions made while planning a lesson.

These findings serve well to illustrate one of the German research team's struggles in the creation of our Lexicon. Whereas we agreed that it would be desirable for teachers to be aware of their underlying beliefs when planning a lesson, we still decided to exclude the abovementioned terms from the German lexicon, because they are not frequently used by teachers.

Lessons drawn from this process

While researching on the German Lexicon, our team had to deal with many different issues which provided beautiful opportunities for learning more, firstly about the mathematics classroom in Germany, but foremost about the actual language used by teachers. We found the next to endless possibility of creating new German words by linking nouns to be both, a blessing, and a curse. In our research team debates and in our local evaluation rounds with teachers we were sometimes unsure if we had just made up a new word or term (e.g. *Verlaufstransparenz*, transparency of a lesson's course; this word was part of our initial lexicon but got deleted in the process). We quarrelled over the inclusion of terms for concrete objects frequently used in the mathematics classroom (*Geodreieck*), we were happy to find some new terms for our definitions during the national evaluation (such as *Übergang* – transition – in addition to *Gelenkstelle* – joint position or hinge), and we enjoyed our struggle to find good translations of the terms we used, for it illustrated the need for a Lexicon.

The German Lexicon

Language is ever-developing, and so also in education new terms frequently enter teachers' vocabulary, while others are used less in turn. The German Lexicon we present is a snapshot of the language used by mathematics teachers in Germany. Due to large regional differences, we want to highlight that our initial list of words with which we started the process was created in the northwest of Germany. If the process had begun in a different German region, the outcome might have differed to a certain extent. Nevertheless, we are confident that the Lexicon we present depicts terms and definitions familiar to and used by most German mathematics teachers.

Next steps: Narratives

One outcome of the meetings with our French and Finnish colleagues was the writing of *narratives* in which a special effort is made to use the lexical terms to describe the lessons from the original videos. This idea was suggested by members of the French team, who had already exchanged narratives with the Czech team of the French and Czech videos, using both of their lexicons. These narratives show how the terms are actually used in practice to describe mathematics teaching, and so provide another kind of insight into the meaning of the terms. As Wittgenstein notes, "the meaning of a word is its use in the language" (1958, §43). A German narrative of the German lesson has been written and narratives of the French and Czech lessons, using the German Lexicon are in preparation.

These narratives allow for another mode of comparison of the lexicons, looking at how the same event is portrayed (or not) in narratives employing different lexicons. They also reveal additional issues related to translation. Each narrative is initially written in the language of the lexicon, and then translated to English. Readers of both the original language and English can then examine what nuances of the original are lost in the English version, and how these nuances might be conveyed in order to enrich international discourse.

Acknowledgements

Thanks to all the teachers who participated in the project, especially Stephanie Fraun, André Smole and the teacher in the video.

References

Blum, W. (2015). Quality teaching of mathematical modelling: What do we know, what can we do? In S.J. Cho (Ed.), *The Proceedings of the 12th International Congress on Mathematical Education* (pp. 73–96). Cham: Springer International Publishing.

Blum, W., Artigue, M., Mariotti, M. A., Sträßer, R., & Van den Heuvel-Panhuizen, M. (Eds.). (2019). *European Traditions in Didactics of Mathematics. ICME 13 Monographs*. New York: Springer.

Hefendehl-Hebeker, L., vom Hofe, R., Büchter, A., Humenberger, H., Schulz, A., & Wartha, S. (2019). Subject-matter didactics. In H.-N. Jahnke & L. Hefendehl-Hebeker (Eds.), *Traditions in German-speaking Mathematics Education Research. ICME 13 Monographs* (pp. 25–59). New York, NY: Springer.

Hudson, B., & Meyer, M. A. (Eds.). (2011). *Beyond Fragmentation: Didactics, Learning and Teaching in Europe*. Opladen, Germany: Barbara Budrich Publishers.

Jahnke, H.-N., & Hefendehl-Hebeker, L. (Eds.). (2019). *Traditions in German-speaking Mathematics Education Research. ICME 13 Monographs*. New York, NY: Springer.

Kaiser, G. (2017). *Proceedings of the 13th International Congress on Mathematical Education ICME-13*. Open Access http://creativecommons.org/licenses/by/4.0/Springer Open.

Kant, I. (1781/1787). *Kritik der reinen Vernunft*. Riga, Latvia: Johann Friedrich Hartknoch.

Kant, I. (1855). *Critique of Pure Reason* (trans. J. M. D. Meiklejohn). New York, NY: Prometheus Books.

Kant, I. (1956). *Kritik der reinen Vernunft*. R. Schmidt (Ed.). Hamburg, Germany: Felix Meiner Verlag.

Kirsch, A. (1977). Aspects of simplification in mathematics teaching. In H. Athen & H. Kunle (Eds.), *Proceedings of the Third International Congress on Mathematical Education* (pp. 98–120). Karlsruhe, Germany: Zentrum für Didaktik der Mathematik.

Kirsch, A. (1987). *Mathematik wirklich verstehen*. Köln, Germany: Aulis Deubner.

Kirsch, A. (2000). Aspects of simplification in mathematics teaching. In I. Westbury, S. Hopmann & K. Riquarts (Eds.), *Teaching as a Reflective Practice: The German Didaktik Tradition* (pp. 267–284). Mahwah, NJ: Erlbaum Associates.

Klemm, K. (2019). *Seiten- und Quereinsteiger_innen an Schulen in den 16 Bundesländern, Versuch einer Übersicht*. Berlin: Friedrich-Ebert-Stiftung. http://library.fes.de/pdf-files/studienfoerderung/15305.pdf

Klieme, E., Avenarius, H., Blum, W., Döbrich, P., Gruber, H., Prenzel, M., Reiss, K., Riquarts, K., Rost, J., Tenorth, H.-E., & Vollmer, H. J. (2003). *Zur Entwicklung nationaler Bildungsstandards. Eine Expertise*. Bonn, Germany: BMBF.

[KMK] Sekretariat der Ständigen Konferenz der Kultusminister der Länder in der Bundesrepublik Deutschland (2004). *Bildungsstandards im Fach Mathematik für den Mittleren Schulabschluss*. München, Germany: Luchterhand.

[KMK] Sekretariat der Ständigen Konferenz der Kultusminister der Länder in der Bundesrepublik Deutschland (2008). *Bildungsstandards der Kultusministerkonferenz. Erläuterungen zur Konzeption und Entwicklung*. Bonn, Germany. https://www.kmk.org/fileadmin/Dateien/veroeffentlichungen_beschluesse/2004/2004_12_16-Bildungsstandards-Konzeption-Entwicklung.pdf

[KMK & IQB] Sekretariat der Ständigen Konferenz der Kultusminister der Länder in der Bundesrepublik Deutschland [KMK] & Institut zur Qualitätsentwicklung im Bildungswesen [IQB]. (2010). *Konzeption der Kultusministerkonferenz zur Nutzung der Bildungsstandards für die Unterrichtsentwicklung*. Köln, Germany: Carl Link.

Lietzmann, W. (1923). *Methodik des mathematischen Unterrichts. 2. Teil: Didaktik der einzelnen Gebiete des mathematischen Unterrichts* (2. Auflage). Leipzig, Germany: Quelle & Mayer.

Masih, Y. (1994). *A Critical History of Western Philosophy*. Delhi: Motilal Banarsidass Publishers.

National Council of Teachers of Mathematics [NCTM] (1989). *Curriculum and Evaluation Standards for School Mathematics*. Reston, VA: Author.

National Council of Teachers of Mathematics [NCTM] (2000). *Principles and Standards for School Mathematics*. Reston, VA: Author.

Nührenbörger, M., Rösken-Winter, B., Link, M., Prediger, S., & Steinweg, A. S. (2019). Design science and design research: The significance of a subject-specific research approach. In H.-N. Jahnke, L. Hefendehl-Hebeker (Eds.), *Traditions in German-speaking Mathematics Education Research. ICME 13 Monographs* (pp. 61–89). New York, NY: Springer.

OECD (2019). *Education at a Glance 2019: OECD Indicators*. Paris: OECD Publishing. https://doi.org/10.1787/f8d7880d-en

Sträßer, R. (2019). The German speaking didactic tradition. In W. Blum, M. Artigue, M. A. Mariotti, R. Sträßer & M. Van den Heuvel-Panhuizen (Eds.), *European Traditions in Didactics of Mathematics. ICME 13 Monographs* (pp. 123–151). New York, NY: Springer.

Vorhölter, K., Greefrath, G., Borromeo Ferri, R., Leiß, D., & Schukajlow, S. (2019). Mathematical modelling. In H.-N. Jahnke & L. Hefendehl-Hebeker (Eds.), *Traditions in German-speaking Mathematics Education Research. ICME 13 Monographs* (pp. 91–114). New York, NY: Springer.

Wittgenstein, L. (1958). *Philosophical Investigations*. (trans. G. E. M. Anscombe.). Oxford, UK: Basil Blackwell.

Wittmann, E. C. (2004). Developing mathematics in a systematic process. In H. Fujita, Y. Hashimoto, B. R. Hodgson, P. Y. Lee, S. Lerman, & T. Sawada (Eds.), *Proceedings of the Ninth International Congress on Mathematical Education* (pp. 73–90). Dordrecht, Germany: Kluwer Academic Publishers.

15

GERMAN LEXICON

Jenny Christine Cramer, Christine Knipping, David A. Reid and Birte J. Specht

A German Lexicon

Based on the research described in *Chapter 14 The Current German Lexicon,* the 65 terms in the lexicon presented in this chapter are familiar to and used by German middle-school teachers. For each of the terms in the lexicon descriptive definitions were developed, capturing the way the term is used in practice. Most of these descriptions also include an example or two from the classroom. Where needed, a non-example, is also included to clarify the limits of the term's usage. These non-examples are practices that the term might be thought of illustrating, but that are not actually referred to by the term. The terms are presented in alphabetical order in the original German.

The first lexicon table (see Table 15.1) lists the original term (in German), a literal English translation, and a semantic English translation. If the literal translation and the semantic translation are the same, the literal translation is omitted. The second lexicon table (see Table 15.2) includes the operational definition of each term arranged into three columns. The first column lists the term. The second column gives a description of the term; these descriptions often include cross-references to other terms which are useful in understanding the relationships between terms. These are marked by (q.v.) if the cross-reference is to a term included in the description, or by (see *Term*) if the cross-referenced term is not otherwise included in the description. *Examples* and *Non-examples* from the classroom are found in column three.

Language is ever-developing, and so also in education new terms frequently enter teachers' vocabulary, while others are used less often. The German Lexicon in this chapter is a snapshot of the language used by mathematics teachers in Germany. It was arrived at through structured negotiations between researchers and the teaching community; progressive consolidation of this lexicon coincided with expansion in the accessed teaching community. Due to large regional differences, we want to highlight that we conducted our research in the northwest of Germany. If the research had been conducted in a different German region, the outcome might have differed somewhat. Nevertheless, we are confident that the Lexicon above depicts terms and definitions familiar to and used by most German mathematics teachers and could be used for the study and promotion of reflective practice of teachers.

TABLE 15.1 German Lexicon – Terms.

Abstrahieren "abstracting" (abstracting)	**Alltagsbezug** "Reference to everyday life" (contextualisation)	**Anschauungsmaterial** "Experience-material" (visual & manipulative materials)
Argumentieren, Begründen, Beweisen "argumentation, justifying and proving" (reasoning)	**Äußere Differenzierung** "external differentiation" (streaming [US: tracking, UK: setting])	**Bildungspläne, Rahmenpläne** "Educational plans" (curriculum [UK: syllabus])
Bildungsstandards Mathematik "Mathematics standards" (standards)	**Binnendifferenzierung/innere Differenzierung** "inner differentiation" (grouping)	**Diagnostik** "diagnostics" (formative assessment)
Didaktische Reduktion "didactic reduction" (focussing)	**Differenzierung** "differentiation" (differentiation)	**Einführung** "introduction" (introducing a new topic)
Einzelarbeit "individual work" (seat work)	**Erarbeiten** "working on something" (engaging)	**Fachsprache** "technical language" (technical terms)
Fehleranalyse "error analysis" (analysis of mistakes)	**Fehlerkultur** "error-culture" (embracing mistakes) ,	**Fragend-entwickelndes Unterrichtsgespräch** "question-developed teaching-discussion" (classroom discussion (Socratic dialogue))

(Continued)

TABLE 15.1 (*Continued*)

Frontalunterricht / Lehrerzentrierter Unterricht "frontal-teaching / teacher-centred teaching" (lecture)	**Gelenktes Unterrichtsgespräch** "guided teaching-discussion" (dialogical teaching)	**Gelenkstelle / Übergang** "join-place / crossing" (transition)	**Gruppenarbeit** (group work)
Handlungsorientierung "action-orientation" (action oriented teaching)	**Hausaufgabenkontrolle** "house-task-control" (checking homework)	**Hausaufgabenvergleich** "house-task-comparison" (comparing homework)	**Impuls** "impulse" (stimulating)
Individualisierung "individualisation" (individualising)	**Inhaltsbezogene Kompetenzen** (content related competencies)	**Klassenarbeit; [Klausur]** "class-work" (summative evaluation)	**Kleinschrittiges Vorgehen** "small-step action" (step by step progression)
Kognitive Aktivierung (cognitive activation)	**Kopfrechnen** "head-calculation" (mental calculation exercise)	**Lehrerrolle** "teacher's role" (teacher's position)	**Leitidee** "lead-ideas" (big ideas)
Lernumgebung "learn-surroundings" (learning environment)	**Methoden** (methods)	**Motivation** "motivation" (motivating)	**Partnerarbeit** "partner-work" (pair work)
Phasen [Unterrichtsphasen] ((lesson) phases)	**Prozessbezogene Kompetenzen** (process related competencies)	**Reflexion** "reflection" (review)	**Roter Faden** "red thread" (motif)
Sammeln "gathering" (bringing together)	**Schülerorientierter Unterricht** "student oriented teaching" (student centred teaching)	**Sicherung** "safeguarding" (consolidation)	**Sozialformen** "social-forms" (grouping)

(Stunden-) Einstieg "(lesson-) entry" (introduction to the content of the lesson)	**Stundenstruktur / Verlaufsplanung** "lesson structure / course planning" (structure of the lesson)	**Stundenverlauf / Unterrichtsverlauf** "lesson-course / teaching-course" (development of a lesson)	**Systematisieren / Strukturieren** (systematising, structuring)
Üben "exercising" (practicing)	**Unterrichtseinheit / Unterrichtssequenz** "teaching-unit / teaching- sequence" (teaching unit)	**Unterrichtsgespräch** "teaching-discussion" (dialogical teaching)	**Vereinbarungen** "agreements" (social norms)
Veranschaulichen "illustrating" (representing)	**Wiederholung** "repetition" (revision)	**Wochenplan** "week-plan" (weekly work plan)	**Zielorientierung** "goal-orientation" (goal oriented teaching)
Zusammmenfassung "together-setting" (synthesis)			

TABLE 15.2 German Lexicon – Terms with operational definitions.

Term	Description	Examples and Non-examples
Abstrahieren "abstracting" (abstracting)	Identifying an underlying feature from a specific case or a set of specific cases.	Example: Looking at different linear equations and their graphs and identifying what the coefficients in the equations mean in general. Non-example: *Students have to determine the side lengths of a rectangular frame made out of a 3 m long slat, where the longer sides are twice as long as the shorter ones. Here a connection between content taught and everyday life is expected and established. So applying already familiar mathematical structures in a context is the focus, rather than identifying these mathematical structures.*
Alltagsbezug "Reference to everyday life" (contextualisation)	Establishing a connection between content taught and everyday life.	Example: Students have to determine the side lengths of a rectangular frame made out of a 3 m long slat, where the longer sides are twice as long as the shorter ones. Here a connection between content taught and everyday life is expected and established. Non-example: *A captain owns 26 sheep and 10 goats. How old is the captain? (A modern version of the classical problem raised by Flaubert.* [*See:* Verschaffel, L.; Greer, B.; de Corte, E. (2000). Making sense of word problems, *Educational Studies in Mathematics, 42(2) (2000), pp. 211–213).*]

Term	Description	Examples and Non-examples
Anforderungsniveau "demand level" (difficulty level)	How demanding a task is meant to be. It can be influenced by different factors such as complexity or how obvious the solution is. It can be related to the task itself or the teacher's presentation of the task. The term itself does not specify what exactly is being addressed and outside observers of teaching.	Examples: (Complex) content can be taught in (graded) levels of difficulty/demand. For example, limits can be taught with a graphical focus on the process of approaching, or by focusing on strictly formal matters (ε-∂ definition, etc.) Problem types can have differing levels. By opening up problems, the "Anforderungsniveau" can rise (e.g. by omitting information). In a lesson, the problems given to the student groups differ in Anforderungsniveau in these ways: – Difficulty of the language used – Open versus guided tasks – Number of steps necessary for getting to a solution – How well known/ obvious the solution is (based on prior experience) – Amount of mathematical modelling required – Whether justifications/argumentations are requested
Anschauungsmaterial "Experience-material" (visual & manipulative materials)	Resources, such as visual representations, pictorial or concrete objects through which content to be learned is made accessible.	Example: Linking cubes, pictures, dynamic geometry software. A teacher uses the materials to show something to help students to see an idea.

(Continued)

TABLE 15.2 (*Continued*)

Term	Description	Examples and Non-examples
Argumentieren, Begründen, Beweisen "argumentation, justifying and proving" (reasoning)	A process competence (see *Prozessbezogene Kompetenz*) which describes a form of communicating in the mathematics classroom in which reasons are brought forward for the correctness or falsehood of a statement, or in which connections are established between mathematical concepts.	Example: Student explains a solution to $x - 2y = 3$; teacher emphasises "And can you explain why? Justify why it is this way?"
Äußere Differenzierung "external differentiation" (streaming [US: tracking, UK: setting])	Differentiation (see *Differenzierung*) in which learners are divided into schools or classes according to achievement level as well as other factors such as physical disabilities.	Example: In Germany after grade 4 or 6 the students are still streamed into different types of school according to their achievement level. Non-example: *Grouping, when students in one class are temporarily divided into different small groups for working on a specific task.*
Auswertung [Vergleich] "Evaluation [comparison]" (comparing results)	Comparative analysis of answers, solution methods, strategies, and approaches to tasks that were given as homework or within the lesson.	Examples: Students' individual solutions and approaches from a previous working phase are discussed. Discussion of homework; a second approach is considered A student's solution on the board is compared to the "cooking recipe." Students are asked to take a stand on his approach. Solutions for the different tasks arising from the group work are written on the board and discussed

Term	Description	Examples and Non-examples
Bildungspläne, Rahmenpläne "Educational plans" (curriculum [UK: syllabus])	State/Provincial frameworks for teaching mathematics. Normative guidelines that teachers must follow.	Example: For example, the curricula of Bremen and Lower Saxony can be found at: – https://www.lis.bremen.de (under Bildungspläne) – https://www.nibis.de (under Vorgaben: Curriculare Vorgaben) Non-example: The mathematics standards of The Standing Conference of the Ministers of Education and Cultural Affairs of the Länder in the Federal Republic of Germany; (see Bildungsstandards Mathematik)
Bildungsstandards Mathematik "Mathematics standards" (standards)	Normative guidelines for mathematics teaching in Germany. These influence but do not determine the Bildungspläne (q.v.).	Example: The mathematics standards of The Standing Conference of the Ministers of Education and Cultural Affairs of the Länder in the Federal Republic of Germany, see for example https://www.kmk.org/fileadmin/Dateien/veroeffentlichungen_beschluesse/2003/2003_12_04-Bildungsstandards-Mathe-Mittleren-SA.pdf Non-example: The federal curricula of the different provinces of Germany (see Bildungspläne).
Binnendifferenzierung / innere Differenzierung "inner differentiation" (grouping)	Differentiation in the classroom (see Differenzierung). In this form of differentiation the learners in a class are given modified tasks with different learning goals.	Examples: There are four equations to be solved, but the third and fourth are optional. Students choose whether to work on homework, or pay attention to the solutions being shown in class. The teacher arranges the students into groups according to a previously worked-out plan. They then receive tasks that differ in their level of demand (see Anspruchsniveau).

(Continued)

TABLE 15.2 (*Continued*)

Term	Description	Examples and Non-examples
Darstellungswechsel "Representation-change" (change of representation)	Working with a variety of representations of a concept or mathematical object and changing between representations. For example, a function can be represented by a graph, a list of ordered pairs, or a table of values, and connections can be drawn between these representations.	<u>Example:</u> A cuboid is first discussed verbally, and then the students look for a formula for the volume of a cuboid, then a sponge serves as a visualisation of a cuboid.
Diagnostik "diagnostics" (formative assessment)	Analysis and evaluation of learners' verbalisations or productions in order to gauge the knowledge/learning of one or more learners.	<u>Example:</u> A test given before or after a teaching sequence or a worksheet/notebook which the teacher reads during a teaching sequence to see individual learning progress.
Didaktische Reduktion "didactic reduction" (focussing)	Preparation/presentation of content with the aim of making the content more accessible to learners. In doing so, essential content relations are meant to become clearer or structures are supposed to be revealed and focused.	<u>Examples:</u> Choosing three different functions to work on the number of roots of quadratic functions. Choosing three functions with "easy" numbers and all possibilities (no root, one root, two roots), so that the students can focus on what is essential in this lesson without having a lot of functions to work on. *Non-example:* *A simplification.*

Term	Description	Examples and Non-examples
Differenzierung "differentiation" (differentiation)	Modification of the lesson to accommodate the diversity of learners in the class. This can be achieved in a variety of ways, for example, different tasks for different students.	Examples: Written hints given out during seat work. Prompts differing in difficulty (see *Impuls*) in case of learning barriers
Eigenverantwortliches / Selbstständiges Lernen "self-guided learning" (autonomous learning)	An approach to teaching in which learners must monitor their own progress and autonomously work through tasks.	Example: Working with a weekly plan (see *Wochenplan*)
Einführung "introduction" (introducing a new topic)	A phase in teaching when a new topic begins.	Examples: The teacher reminds the students of the scale as an image of equations known from a previous school year and uses it to introduce the idea of solving equations by making the same changes to each side Repetition of one-stage random experiments for the introduction of two-stage random experiments Non-example: *Reminding students what was already worked on; they have to explain how they know if a quadratic equation works*
Einzelarbeit "individual work" (seat work)	Learners working alone without interacting with their peers.	Examples: Pupils have to work on their own on equations and have to tell what type of relation it is. They are not allowed to work together but they can use their notes.

(Continued)

TABLE 15.2 (*Continued*)

Term	Description	Examples and Non-examples
Erarbeiten "working on something" (engaging)	Learners actively interact with the lesson content in order to understand it. This can be prompted by the teacher or the learners' peers, or it can occur without outside stimulation.	Example: Students are working on new tasks and trying to find out how to solve them Non-example: *Teacher motivating students.*
Erklärung des Vorgehens "explanation of action" (demonstration)	Explanation of the steps of a process or procedure.	Example: The teacher explains that one can solve a quadratic equation in normal form with a formula and which parts of the equation go where in the formula.
Fachsprache "technical language" (technical terms)	A means of communication specific to a discipline (e.g., mathematics), using specialised terms to describe actions and concepts. It includes symbols and agreed-upon ways of writing.	Example: Fraction, slope, linear equation are examples for technical terms in mathematics and mathematics classrooms. Non-example: *As described by Chevallard in his theory of didactic transposition, knowledge and technical language produced in mathematics as a discipline is transformed in teaching and not the same. For example: "equal fractions" is a term used in teaching, while in academic mathematics fractions are looked at as equivalent classes. Talking about equal fractions doesn't make any sense from an academic mathematical point of view.*
Fehleranalyse "error analysis" (analysis of mistakes)	Systematic discussion of an error or its source. This discussion with the class, is led by the teacher in order to establish a culture of embracing mistakes (see *Fehlerkultur*) in the classroom.	Example: A mistake within the homework is picked out and discussed.

Term	Description	Examples and Non-examples
Fehlerkultur "error-culture" (embracing mistakes)	A particular approach to dealing with errors, which includes the valuing of errors as learning opportunities (as a formative assessment).	Example: The teacher makes a mistake. A student notices it. The teacher is thankful and explains, why this mistake happened and that it is a classic error.
Förderung; ("Hilfe") (support (scaffolding))	Targeted support of individual learners or groups of learners' advancement.	Example: When a teacher introduces the addition of fractions, he/she might notice that some students have a limited understanding of basic operations and fractions. This might require of the teacher extra efforts, strategies, and a program for individual students. Non-example: *A hint or a piece of information given to students while they solve problems or revision of content together with the entire class.*
Fragend-entwickelndes Unterrichtsgespräch "question-developed teaching-discussion" (classroom discussion (Socratic dialogue))	Form of teacher-learner interaction in which the teacher poses specific questions to one learner or a group of learners in order to get to an intended answer or solution for a task. If the teacher guidance is very narrow, it is referred to as funnelling (see *Gelenktes Unterrichtsgespräch*).	Example: The teacher works through questions on a worksheet. The students answer.
Frontalunterricht / Lehrerzentrierter Unterricht "frontal-teaching / teacher-centred teaching" (lecture)	A teaching method in which the teacher tightly controls the situation and in which the information flow is from the teacher to the learners.	Example: The teacher lectures a content but raises once in a while questions to make sure that his students are still listening. Non-example: *Dialogical teaching, where the solution of a task is the focus and the engagement of students in this process is the intent.*

(Continued)

TABLE 15.2 (*Continued*)

Term	Description	Examples and Non-examples
Gelenktes Unterrichtsgespräch "guided teaching-discussion" (dialogical teaching)	A teaching method in which the teacher engages in dialogue with the learners in order to reach learning goals defined by the teacher. Funnelling is a very narrow version of this method.	Example: The teacher asks the students to compare what they did in the last lesson and what one student said just before. He then asks some more questions to push the students in the intended direction.
Gelenkstelle / Übergang "join-place / crossing" (transition)	A marker of the end of one phase of a lesson (see *Phasen*) and the beginning of the next one.	Example: Students find some interesting results to a question and the teacher makes a transition to widen the question or to go deeper into the subject.
Gruppenarbeit (group work)	Grouping (see *Sozialform*) in which learners work together in groups of more than two.	Examples: Students are split into five groups who work on different (similar) tasks. A problem is supposed to be solved in groups of four.
Handlungsorientierung "action-orientation" (action oriented teaching)	Teaching with special attention to the active participation of the learners. Teaching which engages the whole learner.	Example: Going out of the classroom to explore the intercept theorem (Thales' theorem) for determining a distance that cannot be measured directly, e.g. the height of a big building. Non-example: *Creating concrete objects together with the students in class, but never using these as manipulatives through which content can be learned.*
Hausaufgabenkontrolle "house-task-control" (checking homework)	Checking if the learners have completed their homework.	Example: First the teacher asks if anybody did the homework. Then he asks one student to write the homework on the blackboard, so that anybody can compare.

Term	Description	Examples and Non-examples
Hausaufgabenvergleich "house-task-comparison" (comparing homework)	Collective evaluation (see *Auswertung*) of homework.	Example: One student writes his homework on the blackboard. Then the students compare whether they did it the same way and discuss it.
Impuls "impulse" (stimulating)	Prompting using materials, a teaching act or similar means, that is supposed to have an effect on the learners, for example in getting them thinking (see *Kognitive Aktivierung*) This can also be done silently.	Example: Students calculate the angle sums of spontaneously drawn triangles and label them. Teacher: "So, couldn't we have foreseen this?" Non-example: Teacher says: "Think of Egypt!" with the goal to speak about triangles.
Individualisierung "individualisation" (individualising)	Teaching a class of students in which different materials, tasks etc. are used for each learner. Learners are allowed to move in their own pace.	Example: Students get different tasks or work to do according to their specific needs or capacities with the intend to help them to advance as good as possible. Non-example: Students who complete their work quickly are assigned extra work without taking into account their specific needs or capacities.
Inhaltsbezogene Kompetenzen (content related competencies)	Basic content strands in the standards for mathematics education (see *Bildungsstandards Mathematik*). They include: Number, Measure, Space and Shape, Functional relationships, and Data and Chance.	

(Continued)

TABLE 15.2 (Continued)

Term	Description	Examples and Non-examples
Klassenarbeit; [Klausur] "class-work" (summative evaluation)	A written test, usually given towards the end of a unit, to determine the level of learner achievement at that time. The results are shared with individual learners and the class.	Example: In German maths teaching at least four major tests of this kind are given to students in a year. Non-example: *Quizzes or mini-tests, as well as diagnostic assessment are not part of this.*
Kleinschrittiges Vorgehen "small-step action" (step by step progression)	A teaching style in which many small subtasks, steps in a longer procedure, are assigned.	Example: For example, when teaching students a proof of the irrationality of the square root of two, the proof is divided into little steps. First assuming that the square root is rational, then describing the root as a fraction, looking at its reduced form and so forth until finally producing a contradiction. Non-example: *A detailed focus on an aspect of a problem or students' engagement in coming up with a plan how to solve a problem.*
Kognitive Aktivierung (cognitive activation)	Engagement in an active and spirited interaction with the content, that can lead to connections between old and new knowledge.	Example: Students working by themselves on a mathematical problem, struggling with the mathematics, trying to find out and finally understanding how to do it. Non-example: *Using an algorithm to get a result.*
Kopfrechnen "head-calculation" (mental calculation exercise)	Calculation exercises that are meant to be done mentally, without using pencil and paper or a calculator.	Example: Learners in German grade 6 are supposed to be able to mentally calculate with simple fractions, such as 2/3 + ¼

Term	Description	Examples and Non-examples
Lehrerrolle "teacher's role" (teacher's position)	A role taken on by a teacher, which establishes her/his function and position in class. This can be achieved by different demeanours and interactions with the learners.	Example: The teacher reminds the students that now is not the time to have a chat but to listen and follow the lesson.
Leitideen "lead-ideas" (big ideas)	Important concepts in mathematics around which teaching can be organised, such as symmetry, continuity. Confusingly, in the Bildungsstandards (q.v.) the same term is used for Inhaltsbezogene Kompetenzen (q.v.).	Example: **Symmetry** or equivalence are examples of big ideas throughout a spiral curriculum. Non-example: Number and measurement are content standards, not big ideas.
Lernumgebung "learn-surroundings" (learning environment)	A situation in which learners are placed which has been designed to support their learning. It can be designed by the teacher or found in prepared materials (e.g. textbooks).	Example: Lessons as shown, planned and reflected on by by Jo Boaler and Cathy Humphreys in Connecting Mathematical Ideas: Middle School Cases of Teaching & Learning. Heinneman: Portsmouth, 2005.

(Continued)

TABLE 15.2 (*Continued*)

Term	Description	Examples and Non-examples
Methoden (methods)	Means of structuring learning and teaching. Examples include:	Examples: – Think-Pair-Share – Jigsaw groupings – Silent writing – Learner presentations – Placemat – Group puzzle – Stummes Schreibgespräch Think-Pair-Share: "first each of you on your own into your notebook, then check it with your neighbour" Division of the class into 6 student groups. These receive (similar) tasks. The procedure was previously captured on the board (see *Verfahrensorientierung*) Homework is given in which the students are supposed to further practice the lesson content.
Motivation "motivation" (motivating)	Engaging and preparing learners to participate in the lesson. A teacher action.	Example: The teacher brings along a picture with some relation to the content or explains its relevance or what impressed himself concerning this content with the aim to motivate the students. Non-example: *Using visual or manipulative materials with the aim to support understanding. Telling funny stories to make the students laugh.*

Term	Description	Examples and Non-examples
Partnerarbeit "partner-work" (pair work)	Grouping (see *Sozialform*) in which the learners work in pairs.	Examples: Students are supposed to work on a practice task in pairs. Students are asked to discuss with a neighbour A problem that was previously solved individually is now discussed with a partner.
Phasen [Unterrichtsphasen] ((lesson) phases)	Parts of the lesson separated by content and temporal structure. The first phase is usually a *Stundeneinstieg* (q.v.). These phases can be linked by the *Roter Faden* (q.v.).	Example: The teacher marks the beginning of a new lesson phase in which the students receive data tables. They then work on these tables until the comparison.
Prozessbezogene Kompetenzen (process related competencies)	Abilities laid out in the standards for mathematics education (see *Bildungsstandards Mathematik*) which do not relate to a specific content area. They include Modeling, Problem-solving, and Argumentation.	
Reflexion "reflection" (review)	Intensive consideration of content, often as a review of a recently learned concept.	Examples: Summing up and reflection on what the students learned about quadrilaterals, repetition of special properties of certain quadrilaterals and morphing of one quadrilateral into another, including necessary conditions. A clear presentation is developed. Repetition and reflection on what was elaborated lately. Teacher says this explicitly and emphasises that she wants to reflect on and highlight the interdependencies between things. Non-example: Describing something, drill and practice or repetition.

(Continued)

TABLE 15.2 (*Continued*)

Term	Description	Examples and Non-examples
Roter Faden "red thread" (motif)	Conceptual thread tying together a lesson or unit.	Example: In a problem based lesson, the solution of the problem.
Sammeln "gathering" (bringing together)	The teacher collecting and relating the learners' ideas related to a topic, in preparation for *Systematisieren* (q.v.). Often at the conclusion of an activity, but also includes brainstorming at the beginning of an activity.	Example: In the topic area of similarity, students draw triangles in different sizes with side ratios of 1:2:3. Following this activity, different dimensions are gathered on the board. (As a follow-up, the students would then measure angles and see that they are equal.)
Schülerorientierter Unterricht "student oriented teaching" (student centered teaching)	Teaching determined by the interests, questions and stimulus of the learners.	*Non-example:* *When students work by themselves on the context tasks around linear equation systems but are not following their own ideas or questions.*
Sicherung "safeguarding" (consolidation)	The recording of relevant findings and results, often in written form (poster, phrase, oral synthesis, presentation on blackboard by pupils or teacher)	Example: The teacher and one student write the results of the tasks done, on the blackboard.
Sozialformen "social-forms" (grouping)	Standard ways of organising learners to work together in the lesson. (see *Partnerarbeit, Einzelarbeit, Gruppenarbeit*)	Example: Examples for social forms are group work, pair work or seat work. *Non-example:* *A method (q.v.) is not a grouping nor is it a fixed differentiation in different classes.*

Term	Description	Examples and Non-examples
(Stunden-) Einstieg "(lesson-) entry" (introduction to the content of the lesson)	A marker of the beginning of the content related part of the lesson, in which the motif (see *Roter Faden* (q.v.)) of the lesson is taken up. It can occur in several different forms (see examples).	Examples: "informative entry" – The teacher gives information about the content of the lesson. "organisational entry" – The teacher gives information about the organisation of the lesson. "overview of the course of the lesson" – The teacher gives information about the content and/or the organisation of the lesson. "silent impulse" – The teacher shows a picture or a figure which has some relationship to the content of the lesson, which visualises the problem of the lesson. "repetition" – In the beginning of the lesson the teacher repeats the content of the last lesson or poses some questions about this. "outward guidance" – The teacher gives some information or asks some questions about the background of the problem to be worked on.
Stundenstruktur / Verlaufsplanung "lesson structure / course planning" (structure of the lesson)	How the content portion of the lesson is structured. Usually observable in the written lesson plan. Determines phases (q.v.).	Example: A structure of the lesson is to begin with an introduction to the content of the lesson followed by a phase of engaging in pair work, then bringing together in a dialogical teaching and finally making a synthesis.
Stundenverlauf / Unterrichtsverlauf "lesson-course / teaching-course" (development of a lesson)	The development of a lesson over time, the sequence of phases (q.v.) in a lesson.	Example: The development of the lesson is how it evolves in reality, so it can be the same as the structure of the lesson, but it also can vary because of some unexpected impulses or events.
Systematisieren / Strukturieren (systematising, structuring)	Organisation of content during the lesson. Follows *Sammeln* (q.v.). The teacher leads this stage of the lesson with the goal of structuring the ideas, results, and activities the students have worked on.	Examples: General propositions are established that are then worked out using concrete examples Students measure circumference and diameter of diverse round objects. In a whole-class discussion, results are compared. By looking at ratios or by visualisation, the regularity "The circumference is always 3.14 times as big as the diameter" can be found.

(Continued)

TABLE 15.2 (*Continued*)

Term	Description	Examples and Non-examples
Üben "exercising" (practicing)	Applying and testing (often newly) learned concepts by learners, to solidify and deepen understanding.	<u>Example:</u> After having worked out how to solve quadratic equations the students get some quadratic equations to solve. <u>Non-example:</u> *To try to solve quadratic equations without knowing how to do it.*
Unterrichtseinheit / Unterrichtssequenz "teaching-unit / teaching-sequence" (teaching unit)	A topic from a content area that is addressed over the course of several lessons, normally several lessons.	<u>Example:</u> A teaching unit can be a sequence of lessons about linear equations or about the Thales theorem and applications for example. <u>Non-example:</u> *Working just about a small part of a subject, for example to learn how to draw the graph of a linear function.*
Unterrichtsgespräch "teaching-discussion" (dialogical teaching)	A class discussion which can be openly moderated by the teacher or narrowly guided (see *Gelenktes Unterrichtsgespräch, Fragend-entwickelndes Unterrichtsgespräch*)	<u>Example:</u> Within the topic "slope," the students are supposed to report in a whole-class discussion on where they see everyday contexts for the topic.
Vereinbarungen "agreements" (social norms)	Conventions in the mathematics classroom.	<u>Examples:</u> One student has to come to the board and show what he did and a timer gives the time. Everybody has to look to the board and the student has to explain what he does.
Veranschaulichen "illustrating" (representing)	Making mathematical topics accessible, usually using concrete materials or pictorial representations. The action of using *Anschauungsmaterial* (q.v.)	<u>Examples:</u> Taking a cube and an octahedron to count the amount of edges, sides and vertices and to explore their connections. The students or the teacher draw a graph of a function to get an idea how it behaves going to infinity. <u>Non-example:</u> *A picture of Egypt when you are talking about triangles and pyramids.*

Term	Description	Examples and Non-examples
Wiederholung "repetition" (revision)	To pick up once more a topic that has already been encountered.	<u>Examples:</u> The lesson starts with a repetition of what was elaborated; teacher and students sum up the previous content together by the teacher asking questions and the students chorusing. Revision of what was taught before.
Wochenplan "week-plan" (weekly work plan)	A form of differentiation (see *Differenzierung*) in which each learner is given a listing of all that s/he should accomplish, either at school or at home, during the coming week. The learners can choose to some extent the order in which they accomplish tasks. Sometimes all students are given the same plan, but in other cases different plans are given to learners creating de facto groups.	<u>Example:</u> To do in week 17: – practice all you know about triangles in the following exercises: book, p. 18–19, n°3 – 5 + 9 + 11. – Read instructions on p. 20 in the book – Worksheet A or B – Workbook p. 22 with another student together – for fast students: additionally Workbook p. 23–24 <u>Non-example:</u> *Giving some exercises to do on just one day or just informing the students what will happen in this week and then the whole class works together at the same moment on the same tasks.*
Zielorientierung "goal-orientation" (goal oriented teaching)	An orientation towards reaching a goal in teaching.	<u>Example:</u> The teacher has the goal that the students learn how to draw a triangle with given side length or angles and chooses appropriate exercises and creates a good learning environment so that the students can reach the goal. He / she focuses on this subject and stresses other relating subjects at this moment only briefly. <u>Non-example:</u> *Working about a mathematical topic openly; letting the progress of the lesson be influenced by comments of the students or emerging ideas. Changing the progress of the lesson to follow a good question of a student.*
Zusammenfassung "together-setting" (synthesis)	Written or oral summary of fundamental content and results, by the teacher.	<u>Example:</u> After having worked out how to multiply with negative numbers the teacher explains the procedure in general and writes a sentence to copy on the board. <u>Non-example:</u> *The teacher explains a new topic and writes the procedure on the board.*

Acknowledgements

Thanks to all the teachers who participated in the project, especially Stephanie Fraun, André Smole and the teacher in the video.

Bibliography

Presentations and Papers

The work of the German national team has been presented locally and internationally at the following meetings and conferences:

- European Conference on Educational Research (ECER) 2017
- Conference for Didactics in Mathematics in Heidelberg 2016

A selection of publications from the German research team include:

Cramer, J., & Knipping, C. (2016) Das "Lexicon"-Projekt: Weltweite Begriffssysteme zur Beschreibung von Mathematikunterricht. In *Beiträge zum Mathematikunterricht 2016: Vorträge auf der 50. Tagung für Didaktik der Mathematik vom 07.03.2016 bis 11.03.2016 in Heidelberg* (pp. 217–220). Münster: WTM, Verlag für wissenschaftliche Texte und Medien.
Krieger, I. (2018) *Erstellung und Vergleich von Narrativen Über Mathematikstunden Verschiedener Länder In Anlehnung an das Lexicon-Project*. Thesis. Carl von Ossietzky University, Oldenburg.

16

THE EVOLVING NATURE OF THE JAPANESE LEXICON IN A TRADITION OF LESSON STUDY

Yoshinori Shimizu, Yuka Funahashi, Hayato Hanazono and Shogo Murata

Introduction

Becoming a teacher in the Japanese educational system means not only to finish a teacher preparation course and to pass an examination to be recruited, but also to learn more about subject matters, students' learning, and teachers' key roles in teaching and learning informally with and from their colleagues. In other words, beginning teachers in Japan are expected to continue these learnings through informal interactions with their experienced colleagues, in addition to a formal educational system. Experienced teachers are also expected to develop their teaching competence gradually and continuously throughout their careers. From this perspective, learning to teach in the classroom is regarded as a life-long process, which is closely tied to participating in social and cultural activities related to the community of teachers.

Lesson study (*jugyo kenkyu*), which is a common element in Japanese educational practice, takes place in various contexts (Fernandez & Yoshida, 2004; Huang & Shimizu, 2016; Lewis & Tsuchida, 1998; Shimizu, 2010, 2014; Stigler & Hiebert, 1999). Lesson study can be characterised, in particular, as an approach to professional development whereby a group of teachers collaborate to improve classroom teaching by studying the subject matter they teach, how students solve problems in the classroom and learn the subject, and to what extent they might change their instruction according to the particular theme set by the group of teachers participating, or by their school, in relation to the emphases in the national curriculum guidelines.

The Japanese tradition of Lesson study has created a teaching community in which observation and discussion of mathematics lessons is an integral part of professional practice. In the discourse of lesson study particular pedagogical terms and phrases can be found to have specific significance to describe particular roles of teachers and students (O'Keefe, Clarke, & Xu, 2006; Shimizu, 2006). In this context, identifying and unpacking the pedagogical vocabulary of teachers will provide an opportunity to understand the nature of classroom mathematics teaching and then provide suggestions for both the improvement of classroom teaching and professional development in both pre-service and in-service teacher education.

This chapter tries to address the question: *What are the terms and phrases that Japanese teachers use to describe the phenomena of the middle-school mathematics classroom?* The research reported in this chapter explored the constituent elements and structural characteristics of the Japanese Lexicon. The study aimed to create a Japanese Lexicon, as a part of The International Classroom Lexicon Project which is being undertaken in Australia, Chile, China, the Czech Republic, Finland, France, Germany, Japan, Korea, and the United States of America to document and compare the naming systems in the participating ten countries (Mesiti & Clarke, 2018). Documenting and comparing teaching vocabulary will provide an opportunity to understand the nature of classroom mathematics teaching in these countries and facilitate international comparative research. Creating a Japanese Lexicon in the long tradition of lesson study provides an idiosyncratic approach to the purpose of the Lexicon Project – that is, to identify how mathematics teachers in different countries see the teaching-learning process and the terminology used by professional educators.

Theoretical background and the context for creating a Japanese Lexicon

The theoretical position adopted by this project is that our experience, engagement, and reflection are shaped in various degrees by the language that is available (Sapir, 1949). In this position, we can see and talk only about those things that we can name and in this project we examine this normative role of language in relation to classroom practice and research.

There is a greater recognition that mathematics classrooms need to be considered as cultural and social environments in which individuals participate, and that teaching and learning activities taking place in these environments should be studied as such (e.g. Cobb & Hodge, 2011; Lerman, 2006; Seeger, Voigt & Washescio, 1998). Seminal works by Bishop (1988a, 1988b) have directed researchers to the pivotal role that culture plays in both teaching and learning of mathematics. Säljö (2010) emphasised the importance of Vygotsky's work on the role of cultural tools in supporting learning. He drew attention to ways in which historical aspects of culture support student learning: "[Cultural tools] are the products of the development of practices in society over time" (Säljö, 2010, p. 499). They support communication between teacher and learner, and they also support a child thinking alone using cultural tools such as language, mathematical symbols, and/or concrete artifacts in doing so. Thus, learning is supported through access to the knowledge, tools, rules, and ways of thinking a culture has previously developed.

Moreover, it is the case for the practice of lesson study that cultural tools are the products of the development of practices in society over time. The Japanese tradition of Lesson Study has created a teaching community in which discussion of mathematics lessons is an integral part of professional practice (Fernandez & Yoshida, 2004; Huang & Shimizu, 2016). Social practices, such as lesson study, have been argued to be discursively constituted where people become part of practices as practices become part of them (Lerman, 2002).

Japanese lesson study has a function of turning practitioner knowledge into professional knowledge (Hiebert, Gallimore, & Stigler, 2002). Participants in lesson study generate practitioner knowledge that links with particular practice in a detailed and specific manner. That knowledge is then integrated during the cycles of lesson study by discussing and reflecting on the actual teaching that is public and sharable. This provides an opportunity for lexical items to be shared by groups of teachers and by the broader community of teachers. Hiebert and Stigler (2017) argue that a focus on teaching, not on teachers, can shape a coordinated system for improving classroom instruction. For the case of Japanese lesson study, in contrast to the case of lesson study in the United States, four elements contribute to sustaining lesson study as a system for improvement: shared learning goals for students, widely used curricula, assessments with useable feedback for teachers, and professional development to enculturate teachers into the system for improvement. Talking directly about teaching they have just observed, the participant teachers in lesson study can have opportunities to share value-laden terms as they work towards improvement of teaching. Developing and sharing pedagogical terms and phrases through lesson study provides a mechanism for producing value-laden lexicon items.

The process of constructing the Japanese Lexicon

The first stage: Identifying "popular" terms in lesson study

In lesson study, an outside expert is often invited as an advisor who facilitates the post-lesson discussion and makes comments on the possible improvement of a lesson from a broader viewpoint (Fernandez & Yoshida, 2004; Shimizu, 1999, 2009). The expert may be an experienced teacher, a supervisor at the local board of education, a principal of a different school, or a professor from the nearby university. In some cases, the group of teachers may meet with the expert several times prior to conducting the research lesson to discuss issues such as reshaping the goal of the lesson, clarifying the rationale of a particular task to be presented in the classroom, expanding the range of anticipating students' responses to a task, and so on. In this context, the outside expert can be a collaborator who shares responsibility for the quality of a lesson with the teachers, not just an outside authority directing the team of teachers.

The first author has been deeply involved in lesson study of various forms and in various contexts over the past decades. Teachers engaged in lesson study have made numerous suggestions for improving classroom teaching in mathematics (Shimizu, 2009). The activity of lesson study includes careful planning and implementing the research lesson as a core of the whole activity, followed by post-lesson discussion and reflection by participants. In the discourse of teachers in planning, implementing, and reflecting on lessons, particular pedagogical terms and phrases are often used in the contexts of examining classroom instruction (Shimizu, 1999). With this in mind, the Japanese national team members first identified a group of terms that are "popular" and quite often used by groups of teachers. These included:

- *Hatsumon: Hatsumon* means asking a key question for provoking students' thinking at a particular point in a lesson. At the beginning of the lesson, for example, the teacher may ask a question to probe or promote students' understanding of a problem. On the other hand, in a whole-class discussion the teacher may ask, for example, about the connections among the proposed approaches to a problem or the efficiency and applicability of each approach.
- *Kikan-shido* during structured problem-solving time when students work on a problem on their own: *Kikan-shido* means "instruction at students' desk" and includes a

purposeful scanning by the teacher of students' problem-solving. The teacher moves about the classroom, monitoring students' activities silently, doing two important activities that are closely tied to the whole-class discussion that will follow. First, he or she assesses the progress of students' problem-solving. In some cases, the teacher suggests a direction for students to follow or gives hints to the students for approaching the problem. Second, he or she will make a mental note of several students who made the expected approaches or other important approaches to the problem. They will be asked to present their solutions later. Thus, in this period of the purposeful scanning, the teacher considers questions like "Which solution methods should I have students present first?" or "How can I direct the discussion towards an integration of students' ideas?" Some of the answers to such questions are to be prepared in the planning phase but some are not.

- *Neriage* in whole-class discussions: This is a term for describing the dynamic and collaborative nature of a whole-class discussion in the lesson. The term *neriage* in Japanese refers to "kneading up" or "polishing up". In the context of teaching, the term works as a metaphor for the process of polishing students' ideas and getting an integrated mathematical idea through a whole-class discussion. Japanese teachers regard *neriage* as critical for the success or failure of the entire lessons. Based on his or her observations during *kikan-shido*, the teacher carefully calls on students, asking them to present their methods of solving the problem on the chalkboard. The teacher encourages students to find the mathematical connections among alternative solution methods and leads a discussion on them.

- *Matome* at the final summing up phase of lessons: *Matome* in Japanese means summing up. Japanese teachers think that this stage is indispensable to any successful lesson. At the *matome* stage, the teacher tends to make a final and careful comment on students' work in terms of mathematical sophistication. In general, the whole-class discussion is reviewed briefly and what the students have learned is summarised by the teacher.

- *Bansho* as providing a bird's-eye view of an entire lesson: For a research lesson, teachers carefully plan and implement how to organise writing on the chalkboard. *Bansho* means writing on the chalkboard in front of the class. The chalkboard may be divided into a few parts, such as the problem for today, students' alternative solutions, and the summary of what the class learned. In some cases, teachers may use a smaller white board, which will be incorporated into the large picture, to invite students to write their ideas to present to their classmates later. *Bansho* can provide the "afterword," a birds-eye view of the entire lesson. It is referred to quite often during the post-lesson discussion.

Through the discussion of these popular terms, the team came up with the following eight terms to be included in a short survey to mathematics teachers at lower and senior high schools and elementary school teachers whose professional focus is on mathematics ($n = 156$); *bansho* (board writing), *kadai-settei* (setting a task), *hatsumon* (key questioning), *jiriki-kaiketsu* (problem-solving on their own), *mitori* (monitoring students' learning), *neriage* (kneading up), *matome* (summing up), and *furikaeri* (looking back). Teachers were asked to rate the extent to which they use these terms by choosing one of the following options: (1) never, (2) not really, (3) sometimes, or (4) very often.

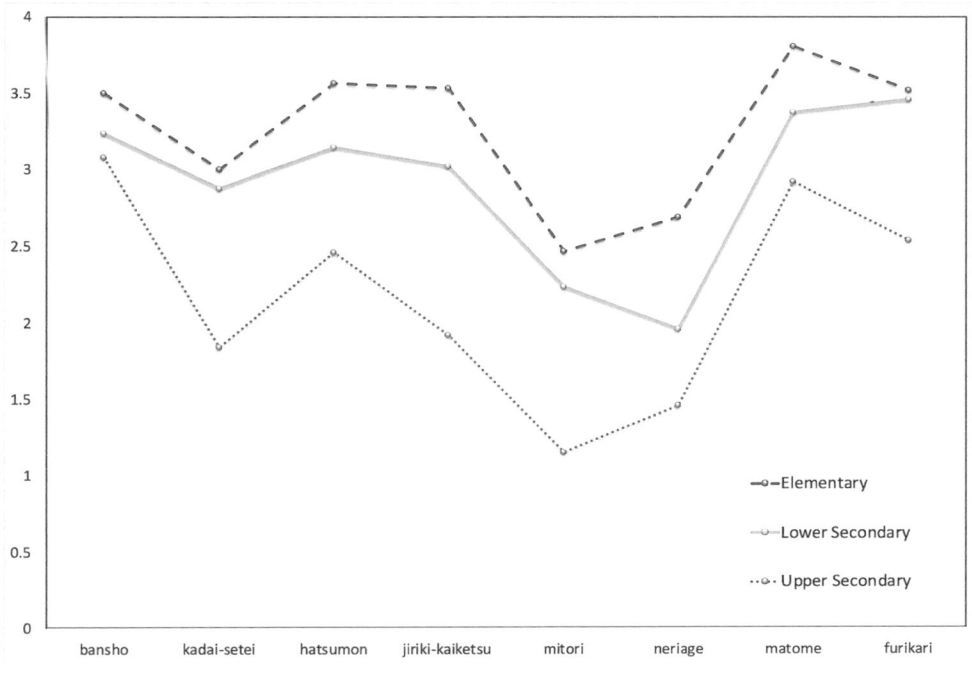

FIGURE 16.1 How often do you use this term?

The results, displayed in Figure 16.1 reveal that there are differences in the ratings by teachers among items, and between the school levels of teachers. As Figure 16.1 shows, in particular, items like *bansho, hatsumon, jiriki-kaiketsu, matome,* and *furikaeri* got higher rates than the others. We also noticed that differences can be attributed to school settings, as well as by the way pre-service teachers are prepared. We hypothesise that these differences can be explained by the engagement with the practice of Lesson Study in different school settings. In other words, the Elementary teachers use these popular terms more often than teachers in other settings because of their increased engagement in Lesson Study. Nishimura et al. (2013), in their survey of more than 1100 schools, found that 96% of elementary schools, 92% of lower secondary schools, and 82% of upper secondary schools engaged in the practice of observing "research" lessons. However, engagement in discussion about these lessons revealed larger differences: 94% of elementary schools engaged teachers in a subsequent discussion, this dropped to 89% for lower secondary schools and 78% for upper secondary schools.

Teachers were also invited to write their own familiar terms. Major additional terms proposed by the teachers included: *meate, nerai, manabiai, mitooshi, group-gakushu, riyuu, setsumei, happyo,* and *oshieai.* These terms and their meanings are displayed in Table 16.1.

Through the first stage of selecting and collecting pedagogical terms and phrases, six categories of terms and phrases emerged: general phases of lessons, structured problem-solving phases, organisation of lessons, general teaching strategy, explicit reference to the values related to mathematics, and other terms. Some of these categories are related directly to "teaching mathematics through problem solving."

The second stage: Video viewing for the draft lexicon

TABLE 16.1 Additional key terms proposed by the teachers surveyed.

Japanese Terms	Meaning in English	Responses
meate	Goal of the lesson	22 (15.8)*
nerai	Task to be achieved	11 (7.9)
manabiai	Learning together	10 (7.2)
mitooshi	Outlook, Perspective	7 (5.0)
group-gakushu	Group work	6 (4.3)
riyuu	Reason	6 (4.3)
setsumei	Explaining	6 (4.3)
happyo	Presentation	5 (3.6)
oshieai	Teaching each other	4 (2.9)

* Numbers in parentheses show percentages of respondents.

The Japanese Lexicon team members took part in a video viewing process to identify the terms and phrases used by Japanese school mathematics teachers. The Japanese team consisted of two researchers, two experienced teachers, and two doctoral students. The Japanese Lexicon was constructed by watching video material and reviewing time-stamped transcripts and classroom support materials for one lesson of mathematics in an eighth-grade classroom in a public school in Ibaraki prefecture, and by classifying a common set of video records of mathematics lessons drawn from all countries participating in the Lexicon Project, in order to identify the terms in our local language that in combination constitute our national pedagogical lexicon. We use these terms to discuss, analyse, reflect upon, and theorise about the mathematics classroom.

Data generation was undertaken simultaneously in Japan and other eight participating countries. Each participating team contributed a videotaped lesson which was included in a stimulus package. The stimulus material was viewed by team members in each country to identify the well-established pedagogical terms or phrases of used in the communities. These terms are supplemented with the clearest possible operational definitions in English, describing both the form and function of each named term. The combined classroom video material from the participating countries then becomes a source of multiple high definition video exemplars of each of the named terms.

We created a draft lexicon of 76 terms and produced extensive descriptions of each one by viewing videos. As mentioned earlier, the Japanese draft lexicon items were initially organised into six categories that were later formalised as follows:

> General Phases of Lessons (16 terms), including the Introductory phase (6 terms), the Main phase (7 terms), and the Closing phase (3 terms);
> Structured Problem-Solving Phase (28 terms);
> Values Related to Mathematics (7 terms);

TABLE 16.2 Organisation of lexicon items and entries (for *kikan-shido*).

Category: General Phases of Lesson			
Lexicon Items (in Japanese)	*Description (in Japanese)*	*Example*	*Non-Example/ Comments*
机間指導 **kikan-shido**	子どもが問題を個々に、またはグループで解決している間に、教師が観察して教室内を歩き回りながら、ヒントを与えたり、指導して誤った考えを修正したりする.	The teacher walks between students' desks, watching students works and making it sure that "a better deal" can include "the same" chance.	Walking between the desks to distribute a worksheet is not the case.
Lexicon Items (in English)	*Description (in English)*	*Time*	
instruction between desks	The teacher moves about the classroom, monitoring students' activities and giving some help.	JP: 00:08:36:00–00:16:13:14	

General Teaching Strategy (15 terms);
Lesson Planning (3 terms); and
Context/Environment (2 terms).

Each lexicon item in Japanese is organised with entries of a description in original language and an English approximation together with example and non-example (in some cases) from video clips. An example term together with its accompanying description and examples is presented in Table 16.2.

The third stage: National validation by conducting an electronic survey

An electronic survey was conducted for a national validation to examine how familiar the terms were to the mathematics teachers in Japan. General demographic data was collected about the teachers, this included: age (5-year intervals from 20), gender (female, male, prefer not to say, other), living regions (among six), employment status, school type, and current job (teacher, teacher educator, researcher, administrator, other). Years of teaching experience was also asked, using 5-year intervals. The survey was completed by 101 teachers and of those, 72% were male and 26% were female. The teaching experience and the age distribution of the respondents are well- balanced.

As shown in Table 16.3 overall, some terms were very familiar to the respondents, although some terms were somewhat less frequently in use (e.g. *label*). Based on the responses by teachers, some terms were excluded from the lexicon and some terms were added. In this case, the term *label* was excluded because the percentage of familiarity when including results for "quite often," "often" and "sometimes" did not surpass the threshold of 50%.

TABLE 16.3 A sample of responses about familiarity of terms.

How often do you hear about the term?					
Candidate Terms	Quite Often	Often	Sometimes	Not Really	Never
label (labelling)	9%	8%	27%	28%	29%
kimari (social norms for learning)	13%	23%	20%	30%	13%
tougou (integration)	13%	18%	34%	26%	10%
goto wo ikasu (capitalising on an error)	18%	36%	31%	12%	4%
minomawari (connecting maths real world)	30%	44%	18%	4%	3%
yosa (merits)	33%	39%	20%	3%	4%

Organisation of lexicon terms and entries

A total of 70 lexicon terms or phrases are identified as the Japanese Lexicon. Characteristic of the relationships among terms within the Japanese Lexicon were the multi-layered intentions of Japanese teachers as these were inferred by the Japanese team and represented in the constituent elements of the Lexicon. Similarly, distinctive were the values implicit in many of the identified terms. The Lexicon study has provided insight into the naming system employed by mathematics teachers in Japan in relation to their classroom practice by documenting and interpreting the constructs that are well-known and understood and are used in discussions with others.

Contrasts between "familiarity," "occurrence," and "my use" or "use by others"

The results of the lexicon survey showed some interesting tendencies among teachers' responses. Tables 16.4 and 16.5 give the teachers' responses to questions about:

- Familiarity, "How often do you hear about the term?"
- Occurrence: "In your experience does the term refer to something that occurs frequently in the classroom?"
- My Use: "Do you use this term in conversation with your colleagues"
- Use (by others): "Is this term widely used by other teachers?"

While teachers tend to report their familiarity of *matome* (see Table 16.4) with high rates of 74% for "quite often" and 23% for "often," their views of occurrence tended to be lower than that for familiarity (57% for "quite often" and 26% for "often"). There was not a large difference between personal use of the term *matome* in contrast with use by others.

TABLE 16.4 Lexicon survey results about the item: *Matome.*

Matome					
	Quite Often	*Often*	*Sometimes*	*Not Really*	*Never*
Familiarity	74%	23%	2%	1%	0%
Occurrence	57%	26%	14%	3%	0%
My Use	57%	22%	16%	2%	3%
Use (by others)	53%	18%	22%	5%	2%

TABLE 16.5 Lexicon survey results about the item: *Kikan-shido.*

Kikan-shido					
	Quite Often	*Often*	*Sometimes*	*Not Really*	*Never*
Familiarity	80%	16%	3%	1%	0%
Occurrence	55%	29%	11%	4%	1%
My Use	57%	22%	16%	2%	3%
Use (by others)	53%	18%	22%	5%	2%

Other lexicon items show a similar pattern. The results for *kikan-shido* are given in Table 16.5. While most teachers responded they are very familiar with the term (80% for "quite often" and 16% for "often"), their responses regarding occurrence tended to be lower than for familiarity (55% for "quite often" and 29% for "often"). For the term *kikan-shido* there was little difference between personal use and use by others.

Discussion

Evolving nature of lexicon

The Japanese Lexicon can evolve and be updated to reflect changing trends and foci in the practice of lesson study, the revision of the National Course of Study (which is conducted every ten years), and the composition of teachers in Japanese society.

The Japanese team has noted in the research literature evidence of the "evolution" in the lexicon towards terms referring to more explicit intervention by teachers, particularly when "teaching between desks" (*kikan-shido*) (O'Keefe, Clarke, & Xu, 2006). In the last decades, the Japanese term for "teaching between desks" has changed to include *kikan-junshi* (monitoring between desks), *kikan-shido* (teaching between desks), and *kikan-shien* (supporting between desks).

Lewis and Tsuchida (1998) discuss that key elements for lesson study include a shared, frugal curriculum, collaboration among teachers, self-critical reflection, and stability of educational policy. Hiebert and Stigler (2017) also point out that lesson study in Japan is a system for improving teaching that is supported by well-defined, shared learning goals, a curriculum aligned with these goals, assessment that provide information teachers need to improve instruction, and professional development that socialises teachers into the practice of improving teaching. The Japanese Lexicon provides a key resource to support these processes.

In the current national curriculum standards (MEXT, 2010), classroom activities such as communication, discussion, explaining, and writing are strongly valued and emphasised as "activities with languages" in all the subject areas. Ongoing lesson studies in public schools often focus on introducing peer dialogues, small-group discussions, writing, and so on, in addition to the study of certain subject matters. Also, the revision of national curriculum standards introducing change in the scope and sequence of mathematical content influences the choice of themes to be pursued through lesson study. For example, in the new curriculum the introduction of an early conceptualisation of a common fraction is taught in second grade, while in the former national curriculum standards the concept of a common fraction was introduced in fourth grade. Teaching common fractions might therefore be a "hot topic" to be examined through lesson study at the elementary level. Learning to cope with such new visions in mathematics curriculum, teaching methods, and general emphases in education is one of the major aims for teachers who are involved in lesson study. Also, in the revised national curriculum standards (MEXT, 2017), "integration of mathematical ideas" is explicitly described in the goal of teaching mathematics. Then the team *tougou* (integration) appeared with relatively low familiarity may become popular term.

Labeling practical wisdoms to share

Discussing lessons in a teaching community created through the tradition of lesson study provides focus for the improvement of teaching and learning. Each lexicon term/phrase is situated in the context of professional practice with the goal of improving teaching and learning. It should be noted that particular lexicon terms and phrases shared by the teachers emerged in the process of analyses. For example, the phrase "HA, KA, SE," which stands for the importance of generalisability, quickness, simplicity, and exactness of mathematical methods, was often found in the comments made by the teachers. The phrase is directly related to the value of mathematics and teachers use the phrase when they discuss the efficiency of mathematical methods.

It is important to note that these lexical terms and phrases are used in the discourse in particular contexts embedded in a whole system for describing a particular style of teaching. Structured problem-solving is often mentioned to describe the Japanese system with an emphasis on students' thinking on problems posed. Japanese mathematics teachers often organise an entire lesson by posing just a few problems with a focus on students' various solutions to them. Educating teachers about lesson plans includes understanding of key pedagogical terms.

Lesson plans as vehicles for sharing pedagogical terms

One of the most distinctive results to emerge from the documentation and analysis of the Japanese Lexicon was the recognition of the central importance of the lesson plan as a critical vehicle for teachers' use of their professional lexicon. The establishment of a nationally standardised format for lesson planning was seen to have a normative effect on both the pedagogical options available to Japanese teachers and the substance and structure of the Japanese Lexicon.

Educating teachers about lesson plans includes understanding of key pedagogical terms. By using and listening to these terms, pre-service as well as in-service teachers gradually become members of the community of teachers. The informal aspects found in the process of teacher education support the formal systems of teacher education programs in Japan.

In the process of a lesson study, lesson plans are used as "vehicles" with which teachers can learn and communicate about the topic to be taught, anticipated students' approaches to the problem presented, and important teachers' roles in lessons.

Concluding remarks

For more than a decade, educators and researchers in the field of mathematics education have been interested in lesson study as a promising source of ideas for improving education. For a Japanese mathematics educator who has been deeply involved in lesson study for more than two decades in local contexts, this "movement" has provided an opportunity for reflecting on how lesson study as a cultural activity works as a system embedded in the entire society as well as in the local community of teachers with shared values and beliefs. The Japanese Lexicon can be a window with which teachers and researchers can reflect on what is valued in describing lessons and compare these descriptions with the Lexicons from other countries.

Further studies are needed to analyse differences in familiarity with terms and phrases among groups of teachers of different ages and years of teaching experience, to explore the Lexicon in the practice of lesson study in more detail, and to look at an alignment of the Lexicon with the changing emphases of any new national courses of study.

References

Bishop, A. J. (1988a). *Mathematical enculturation. A cultural perspective on mathematics education*. Dordrecht, The Netherlands: Kluwer Academic Publishers.

Bishop, A. J. (1988b). *Mathematics education and culture*. Dordrecht, The Netherlands: Kluwer Academic Publishers.

Cobb, P., & Hodge, L.L. (2011). Culture, identity, and equity in the mathematics classroom. In E. Yackel, K. Gravemeijer, & A. Sfard (Eds.), *A journey in mathematics education research: Insights from the work of Paul Cobb* (pp. 179–195). New York, NY: Springer.

Fernandez, C., & Yoshida, M. (2004). *Lesson Study: A Japanese approach to improving Mathematics teaching and learning*. Malwah, NJ: Lawrence Erlbaum.

Hiebert, J., & Gallimore, R., & Stigler, J.W. (2002). A knowledge base for the teaching profession: What would it look like and how can we get one? *Educational Researcher*, 31(5), 3–15.

Hiebert, J., & Stigler, J. W. (2017). Teaching versus teachers as a lever for change: Comparing a Japanese and a U.S. perspective on improving instruction. *Educational Researcher*, 46(4), 169–176.

Huang, R., & Shimizu, Y. (Eds.). (2016). Improving teaching, developing teachers and teacher educators, and linking theory and practice through lesson study in mathematics: an international perspective, *A Special Issue of ZDM Mathematics Education*, 48(4), 393–409.

Lerman, S. (2002). Cultural, discursive psychology: A sociocultural approach to studying the teaching and learning of mathematics. In C. Kieran, E. Forman & A. Sfard (Eds.), *Learning discourse: Discursive approaches to research in mathematics education* (pp. 87–113). Dordrecht: Kluwer Academic Press.

Lerman, S. (2006). Cultural psychology, anthropology and sociology: The developing "strong" social turn. In J. Maasz & W. Schloeglmann (Eds.), *New mathematics education research and practice* (pp.171–188). Rotterdam: Sense Publisher.

Lewis, C., & Tsuchida, I. (1998). A lesson is like a swiftly flowing river. *American Educator*, 22(4), 12–17, 50–52.

Mesiti, C., & Clarke, D. (2018). The professional, pedagogical language of mathematics teachers: A cultural artefact of significant value to the mathematics community. In E. Bergqvist, M. Österholm, C. Granberg, & L. Sumpter (Eds.), *Delight in mathematics education* (Vol. 3, pp. 379–386). Umeå, Sweden: PME.

Ministry of Education, Culture, Sports, Science and Technology (MEXT) (2010). *National curriculum standards for kindergarten, elementary school, lower and upper secondary school*. Tokyo: The Ministry [in Japanese].

Ministry of Education, Culture, Sports, Science and Technology (MEXT) (2017). *National curriculum standards for kindergarten, elementary school and lower secondary school*. Tokyo: The Ministry [in Japanese].

Nishimura, K., Matsuda, N., Ohta, S., Takahashi, A., Nakamura, K., & Fuji, T. (2013). A survey on implementation of Research Lesson in Mathematics in Japan. Mathematics Education: A Journal of Japanese Society of *Mathematical Education*, 95(6), 2–11 [in Japanese].

O'Keefe, C., Clarke, D. J., & Xu, L. (2006). Kikan-Shido: Between desks instruction. In D.J. Clarke, J. Emanuelsson, E. Jablonka, & I.A.C. Mok (Eds.), *Making connections: Comparing mathematics classrooms around the world* (pp. 73–105). Rotterdam, The Netherlands: Sense Publishers.

Säljö, R. (2010). Learning in a sociocultural perspective. In P. Peterson, E. Baker, & B. McGaw (Eds.), *International encyclopedia of education* (3rd edition.) (pp. 498–502). Amsterdam, The Netherlands: Elsevier.

Sapir, E. (1949). *Selected writings on language, culture and personality*. Berkeley: University of California Press.

Seeger, F., Voigt, J., & Washescio, U. (Eds.). (1998). *The culture of the mathematics classroom*. New York, NY: Cambridge University Press.

Shimizu, Y. (1999). Aspects of mathematics teacher education in Japan: Focusing on teachers' role. *Journal of Mathematics Teacher Education*, 2(1), 107–116.

Shimizu, Y. (2006) How do you conclude today's lesson? The form and functions of "matome" in mathematics lessons. In D. Clarke, J. Emanuelsson, E. Jablonka, & I.A.C. Mok (Eds.). *Making connections: Comparing mathematics classrooms around the World* (pp. 127–146). Rotterdam, The Netherlands: Sense Publishers.

Shimizu, Y. (2009). Characterizing exemplary mathematics instruction in Japanese classrooms from the learner's perspective. *ZDM-The International Journal of Mathematics Education*, 41(3), 311–318.

Shimizu, Y. (2010). Mathematics teachers as learners: Professional development of mathematics teachers in Japan. In F.K. Leung & Y. Li (Eds.), *Reforms and issues in school mathematics in East Asia: Sharing and understanding mathematics education policies and practices* (pp. 169–180). Rotterdam, The Netherlands: Sense Publishers.

Shimizu, Y. (2014). Lesson study in mathematics education. In S. Lerman (Ed.), *Encyclopedia of Mathematics education* (pp. 358–360,). New York, NY: Springer.

Stigler, J. W., & Hiebert, J. (1999). *The teaching gap: Best ideas from the world's teachers for improving education in the classroom*. New York, NY: Free Press.

17

JAPANESE LEXICON

Yoshinori Shimizu, Yuka Funahashi, Hayato Hanazono and Shogo Murata

A Japanese Lexicon

The Japanese Lexicon of 70 terms is considered familiar by Japanese teachers in the mathematics education community. The terms of the validated lexicon are organised into six categories: General Phases of Lessons (16 terms); Structured Problem Solving Phase (28 terms); Values Related to Mathematics (7 terms); General Teaching Strategy (15 terms); Lesson Planning (3 terms); and Context/Environment (2 terms).

In this chapter the terms are presented in the original language (Japanese including the romanisation) as well as a closest English translation (in brackets). Operational definitions were developed for each of the terms and these include a general description of the classroom phenomena together with classroom illustrations. The terms are organised alphabetically by the romanisation.

The first lexicon table (see Table 17.1) lists the terms of the lexicons in Japanese, the romanisation, and the closest English translation. The second lexicon table (see Table 17.2) includes the operational definition of each term arranged into three columns. The first column lists the term itself; the second column gives a description and in some cases a note or comment (in italics). Examples are found in column three.

The terms of the Japanese Lexicon have been organised into six categories (see Table 17.3):

1. General Phases of Lessons (16 terms), including the Introductory phase (6 terms), the Main phase (7 terms) and the Closing phase (3 terms)
2. Structured Problem Solving Phase (28 terms)
3. Values Related to Mathematics (7 terms)
4. General Teaching Strategy (15 terms)
5. Lesson Planning (3 terms)
6. Context/Environment (2 terms)

TABLE 17.1 Japanese Lexicon – Terms.

板書 **bansho** (board writing)	板書計画 **bansho keikaku** (planning board writing)	導入 **dounyu** (introduction)	振り返り **furikaeri** (looking back)
学級経営 **gakkyu-keiei** (classroom management)	学習課題 **gakushuu-kadai** (learning tasks)	学習感想 **gakushuu-kansou** (journal writing)	誤答を活かす **goto wo ikasu** (capitalising on an error)
グループ学習 **group-gakusyuu** (group work)	具体物の利用 **gutaibutsu no riyou** (using teaching materials)	始めの挨拶 **hajime no aisatsu** (greetings at the start of a lesson)	はかせ **hakase** (ha-ka-se)
反例 **hanrei** (use of a counterexample)	発表 **happyou** (presentation of student's solution)	発問 **hatsumon** (key questioning)	発展 **hatten** (extension)
比較検討 **hikaku-kentou** (comparing)	ヒントカード **hint card** (cards with some hints)	褒める **homeru** (praising students)	一斉討議 **issei tougi** (whole class discussion)
自力解決 **jiriki kaiketsu** (problem solving on their own)	授業の山場 **jugyo no yamaba** (key moment in the lesson)	課題設定 **kadai-settei** (setting a task)	活用 **katsuyou** (functional use)
キーワード **ki-wado** (key words)	机間巡視 **kikan-junshi** (walking between desks for purposeful scanning)	机間指導 **kikan-shido** (instruction between desks)	既習事項の掲示 **kishuujiko no keiji** (use of posters for scaffolding)
個に応じる指導 **koni-oujiru-shido** (differentiation)	根拠・理由 **konkyo/riyuu** (justification)	教科書の参照 **kyoukasyo no sannsyou** (referring to textbooks)	共通点と相違点 **kyoutsuten to souiten** (similarities and differences)
共有 **kyouyuu** (sharing)	教材研究 **kyouzai-kenkyu** (analysing teaching materials)	教材の提示 **kyouzai no teiji** (presenting the instructional materials)	学びのきまり **manabi no kimari** (social norms for learning in the classroom)
学び合い **manabiai** (learning together)	まとめ **matome** (summing up)	めあて **meate** (sharing the goal)	身の回り **minomawari** (connecting mathematics to the real world)

見通し **mitooshi** (outlook or perspective)	みとり **mitori** (monitoring students when working)	問題条件の確認 **mondai jyoukenn no kakunin** (clarifying conditions of the problem)	問題解決型授業 **mondai kaiketsugata jyugyo** (structured problem solving)
問題提示 **mondai-teiji** (posing a problem)	ねらい **nerai** (clarifying the aim of activities)	練り上げ **neriage** "kneading up" (comparing and discussing alternative solutions and ideas)	ノート **nohto** "note" (note-taking)
教え合い **oshieai** (teaching each other)	終わりの挨拶 **owari no aisatsu** (closing greetings)	ペア学習 **pea-gakusyuu** (pair work)	練習 **renshu** (exercise)
生徒指導 **seito-shido** (management of student behaviour)	説明 **setsumei** (explanation)	式 **shiki/** 式化 **shikika** (use of literal symbols)	指名 **shimei** (naming)
小集団指導 **sho-shudan-shido** (teaching small groups)	集団思考 **shudan-shikou** (collective thinking)	宿題の確認・答え合わせ **syukudai no kakuninn, kotae awase** (checking homework)	宿題の指示 **syukudai no shiji** (assigning homework)
多様な考え **tayou na kangae** (inventing alternative ideas)	適用・応用 **tekiyou/ouyou** (application)	統合 **tougou** (integration)	机の配置 **tsukue no haichi** (physical arrangement of desks)
ワークシート **wakushito** (worksheet)	予告 **yokoku** (looking forward/ preview)	よさ **yosa** (merits)	予想 **yosou** (conjecturing/ anticipating)
予想される子どもの反応 **yosousareru kodomo no hannou** (anticipated student responses)	前時の振り返り **zenji no furikaeri** (review of the previous lesson)		

TABLE 17.2 Japanese Lexicon – Terms with operational definitions.

Term	Description	Examples
板書 **bansho** (board writing)	The teacher writes a problem of key concepts on the blackboard in front of the classroom. Students also present their idea by writing on the blackboard. *By making a board writing plan, the teacher can see the flow of the lesson.*	The chalkboard may be divided into a few parts, such as the problem for today, students' alternative solutions, and the summary of what the class has learned. In some cases, teachers may use a smaller white board, which will be incorporated into the large picture, to invite students to write their ideas to present to their classmates later.
板書計画 **bansho keikaku** (planning board writing)	The teacher writes problems, looks for solving a problem, students' solutions, and summary on the blackboard during lesson. The teacher plans layout and colour scheme of these descriptions beforehand. *The teacher believes that organising the chalkboard is a key ingredient to organising students' thinking and understanding.*	The teacher decides to use the space of the chalkboard to include the problem for the today's lesson for the upper left, students' alternative solution methods in the middle, and some remarks and summary of what students have learned to the right.
導入 **dounyu** (introduction)	The teacher tells a story related to the topic to be learned in the lesson. *This term is used both in a very broader sense of introductory phase of lesson in a particular action to involve students to the topic today by telling some stories related today's topic.*	"Today, we are going to think about something on functions that you have learned in the elementary school."
振り返り **furikaeri** (looking back)	The teacher and students reflect on what the class learned in the today's lesson. *Through this activity, teacher sheds light on the important ideas or mathematical contents.*	The teacher asks students to say what the class learned in today's lesson.
学級経営 **gakkyu-keiei** (classroom management)	In the long run, teacher try to establish social norms for learning in classroom.	The teacher sets the goal of the class and have a class meeting to share the rules in the classroom.

Term	Description	Examples
学習課題 **gakushuu-kadai** (learning tasks)	From a problem or a situation presented at the beginning, teacher tries to focus on the particular aspect of mathematical tasks to be tackled by the students. *Tasks here are those facilitate students' thinking.*	Given a problem within daily life, a task of finding two variables and their relation was formulated and then a simultaneous linear equation is found.
学習感想 **gakushuu-kansou** (journal writing)	At the final phase of the lesson teacher may ask students to write their feeling and thinking as a journal writing.	The teacher asks students, "What did you understand from today's lesson?" The teacher also asks students to write down points of what they understood, what they noticed, further questions they had, and so on.
誤答を活かす **goto wo ikasu** (capitalising on an error)	The teacher presents an error found in students' answer to facilitate a discussion on the topic and to learn from it. *This is not for the teacher to tell the answer is wrong but the intention is to clarify.*	The teacher may present a typical error in number sentence such as $0.2 \times 0.3 = 0.6$ for clarifying the meaning of place value.
グループ学習 **group-gakusyuu** (group work)	Students are divided into different groups and the students in each group work together and discuss on the tasks or solutions. *This form of activity is mostly taken before comparing and discussing alternative solutions and ideas in a whole class. Number of students in a group is larger than two.*	Students are divided into groups of four and working on the problem together.
具体物の利用 **gutaibutsu no riyou** (using teaching materials)	The teacher uses materials in order to explain a problem, a solution, a mathematical idea and so on.	Teacher asks students to manipulate marbles to find a particular number pattern to find sum of triangular numbers.
始めの挨拶 **hajime no aisatsu** (greeting at the start of a lesson)	At the start of a lesson, students stand-up and then both teacher and students make a bow.	This activity is a formal declaration of the beginning of a lesson.

(Continued)

TABLE 17.2 (*Continued*)

Term	Description	Examples
はかせ **hakase** (ha-ka-se)	The teacher and students evaluate the solution methods in particular their quickness, simplicity, and correctness.	At the end of the lesson, the teacher asks students to reflect on what was good about dealing with the problem mathematically.
反例 **hanrei** (use of a counterexample)	Giving a counterexample to show a proposition does not hold. *The context for the use of counterexample, to show something is not hold, is the key for this item.*	One student said "4 ÷ 2 = 2, 8 ÷ 4 = 2. So, I think that the quotient of dividing an even number by an even number will be an even number at any time." Then another student said "But if we calculate 6 ÷ 2 it will be 3, but 3 is not an even number."
発表 **happyou** (presentation of student's solution)	A student's written and verbal presentation of a solution in the front of the classroom. The solution might be presented on the board, on a document camera, as a poster, and so on.	The students present their solutions and the teacher accepts and/or elaborates on them.
発問 **hatsumon** (key questioning)	The teacher asks a thought-provoking key question in the entire lesson.	At the beginning of the lesson, teacher asks a question to probe or promote students' understanding of the problem. In a whole class discussion the teacher asks, for example, about the connections among the proposed approaches to the problem or the efficiency and applicability of each approach.
発展 **hatten** (extension)	An extended problem is presented or mathematical extension to the problem solved in the lesson is discussed.	Expanding the range of numbers in an expression (e.g. expanding from integers to decimal numbers), or formulating and solving an extended problem.
比較検討 **hikaku-kentou** (comparing)	Alternative solution methods to the problem are presented by students and compared in a whole class setting. *If there are no alternative solution method presented by the students, the teacher themselves present methods that hel she want students to learn.*	Two students who had used different approaches (the addition and subtraction methods and the substitution method) were asked to present their solutions on the blackboard, and whole class discussion focused on the key ideas included in each of the solution.

Term	Description	Examples
ヒントカード **hint card** (cards with some hints)	The teacher gives cards with some hints to students who have difficulty to solve a problem when students solve it individually. *The intention is not providing a procedure for solving the problem, but giving a hint to think about.*	The teacher gives students a card with the question, "What if you add or subtract equations to solve simultaneous linear equations?"
褒める **homeru** (praising students)	The teacher praises students on their idea, formal/ informal presentation, attitude and so on even if it is incorrect. *The teacher praises students either from a mathematical point of views or with an educational intention.*	The teacher praises a student's idea in front of the classroom or gently talks only to that student.
一斉討議 **issei tougi** (whole class discussion)	The teacher and students have a discussion as a whole class.	The teacher stands at the front of the classroom and orchestrates a discussion while the students sit at their desks.
自力解決 **jiriki kaiketsu** (problem solving on their own)	Students solve the problem on their own, sometimes alone and sometimes in a group of two to four.	The teacher moves about the classroom, monitoring students' activities silently. The teacher then walks around the classroom while the students are engaged on the problem to monitor to what extent students understand the problem or how many students have a difficulty.
授業の山場 **jugyo no yamaba** (key moment in the lesson)	The teacher may reflect on afterwards what at which point was the key moment, the climactic point, of the entire lesson. *A climax may not be observable in some case and it may be difficult to operationalise how to identify the climax. But it is still key term for discussing the lesson.*	The teacher discusses alternative solution methods to identify a more sophisticated one.

(Continued)

TABLE 17.2 (*Continued*)

Term	Description	Examples
課題設定 **kadai-settei** (setting a task)	The teacher intentionally presents a situation and then the class finds a problem to solve for the situation.	The situation is "Put water into a tank. The following table shows various amounts of time and the corresponding depths of water accumulated in the tank." The problem to solve in the situation is "Explore the relationship between time and depth."
活用 **katsuyou** (functional use)	The teacher asks students to apply the knowledge to various contexts including real world situations. *Solving a similar problem in the same context to the previous one is not the case. It is necessary for Katsuyou to formulate the situation into a mathematical problem and to interpret the results of mathematical solution for the situation.*	Using knowledge of linear function, students find the temperature at the 6th station on Mt. Fuji (altitude 2500 m) when the temperature at Kawaguchi Lake Weather Center at the base of Mt. Fuji (altitude 860 m) is 23.3 °C.
キーワード **ki-wado** (key words)	The teacher mentions to students the key words for what students are learning or what they have learned. *The teacher usually writes the key words on a blackboard so that the students can be notified so that it can become one of the elements to look back on.*	The teacher pointed to several expressions written on the blackboard and says, "And, these are called linear functions. Because they are all shown as a linear equation, they are called linear functions. Take a note to it somewhere in your notebook."
机間巡視 **kikan-junshi** (walking between desks for purposeful scanning)	The teacher moves about the classroom, monitoring students' activities silently.	The teacher walks around the classroom while the students are engaged in the problem to monitor to what extent the students understand the problem or how many students have a difficulty.
机間指導 **kikan-shido** (instruction between desks)	The teacher moves about the classroom, monitoring students' activities silently, and give some hints or helps individually. *This activity is very important for the teacher to plan how to conduct the following whole class discussion.*	The teacher assesses the progress of students' problem solving. In some cases, the teacher suggests a direction for students to follow or gives hints to the students for approaching the problem. Second, he will make a mental note of several students who made the expected approaches or other important approaches to the problem.

Term	Description	Examples
既習事項の掲示 **kishuujiko no keiji** (use of posters for scaffolding)	The teacher displays posters to the wall of the classroom to show what they have learned recently. *The content of posters is something related to the topic to be learned and not a mere repeat of the previous topic.*	The teacher displays a poster with the methods for constructing angle-bisector to invite students to find the drawing a perpendicula line.
個に応じる指導 **koni-oujiru-shido** (differentiation)	The teacher provides different material or points of entry for students who are at different levels. The teacher may give a hint to the struggling students to give a more difficult problem to advanced students.	The teacher asks "Are you all done? I see some of you need some more time, Am I right? I will give you a minute or two. So, I want you, who are through, to do the basic exercise or a chapter review. I think you will have enough time to work on at least one problem. "
根拠・理由 **konkyo/riyuu** (justification)	The teacher and students refer to the reasons for what they found	Students are invited to state not only procedures for solving a problem but also the reasons why they thought to use these.
教科書の参照 **kyoukasyo no sannsyou** (referring to textbooks)	The teacher asks students to refer to the textbook page in order to confirm a general expression or an explanation in mathematics.	The teacher said, "The definition of DAINYU-HOU (the way of substitution), I mean the explanation, well, I'll read the sentence on page forty-four, the two sentences before question six. Uh, we have removed Y by substituting number one into number two. This way of answering, the way of leaving only one kind of letters by substitution is called DAINYU-HOU."
共通点と相違点 **kyoutsuten to souiten** (similarities and differences)	The teacher has students look for similarities/differences between several students' idea or new learned content and content already learned.	The addition and subtraction methods and the substitution method for solving simultaneous linear equations can be compared to find a commonality of eliminating one variable at a time.
共有 **kyouyuu** (sharing)	The teacher invites students to present their ideas to be shared in the classroom.	Two students who had used different approaches (the addition and subtraction methods and the substitution method) were asked to present their solutions with writing and explaining on the blackboard.

(Continued)

TABLE 17.2 (*Continued*)

Term	Description	Examples
教材研究 **kyouzai-kenkyu** (analysing teaching materials)	Analysing educational goal of a particular lesson, mathematical background of a topic taught, teaching materials from children's psychological aspects.	The teacher may choose an appropriate real-life situation for teaching linear function and examine the use of numerical value used in the situation.
教材の提示 **kyouzai no teiji** (presenting the instructional materials)	The teacher presents the teaching materials in front of the classroom.	The teacher presents instructional materials on the blackboard in the form of a poster that contains the problem for today. Or, teacher demonstrates throwing a dice on the desk for asking students to think about the probability.
学びのきまり **manabi no kimari** (social norms for learning in the classroom)	The teacher refers to the rules for the students to follow in the lesson. *Kimari can be written on the poster to be shared or can be implicit as a hidden curriculum.*	The teacher explicitly refers to a rule in classroom such as, "listening to the presentation by classmates carefully."
学び合い **manabiai** (learning together)	The teacher assigns a time period for the students to learn together with their neighbours. *The term implies an opportunity to learn from each other.*	Students share their own solution methods or mathematical ideas with their neighbours.
まとめ **matome** (summing up)	The teacher talks with the whole class to highlight and to summarise the main point of the lesson. What the students have discussed in the lesson is reviewed briefly and what they have learned during the lesson is highlighted and summarised by the teacher in the whole class setting. *In some cases, the teacher may ask the students to summarise what they learned in the lesson in their own words.*	Both what the students have engaged in the lesson and the points of whole class discussion are reviewed briefly and what the students have learned is summarised by the teacher.

Term	Description	Examples
めあて **meate** (sharing the goal)	After setting the task for the lesson teacher and students explicitly confirm the goal for today's lesson in relation to the problem to be solved.	The teacher establishes the goal for today's lesson in accordance with the nature of new mathematical content. The teacher said, "In Lesson 2, we looked at the table from the horizontal relationship, such as two times and three times. … Today, I want you to find out the vertical relationship." Then, the teacher wrote on the blackboard "Let us find the fixed number that does not change by examining the table vertically."
身の回り **minomawari** (connecting mathematics to the real world)	The teacher mentions the application of what students learned in the lesson to a problem in the real-world context.	For teaching the concept of an expected value, the teacher starts with the story of buying a lottery ticket.
見通し **mitooshi** (outlook or perspective)	A few students are invited by teachers to present their outlook for the forthcoming activities to do. *The teacher encourages students to develop their own views of how to get the answer and invites them to share their views and approaches on how to explore the task.*	The teacher asks the students, "What is the difference from the other problem we have solved yesterday?," "What have we been learning?," "Do you have any idea to solve this problem?"
みとり **mitori** (monitoring students when working)	The teacher usually walks around the classroom while the students are engaging on the problem to monitor to what extent the students understand the problem or how many students have a difficulty.	The teacher observes carefully to have children specifically write how to solve the problem and why they did so.
問題条件の確認 **mondai jyoukenn no kakunin** (clarifying conditions of the problem)	The teacher and students discuss the conditions and assumptions to be considered during problem solving. Through this activity, students capture the problem situation formulated into a mathematical problem.	After getting two solutions to a quadratic equation which comes from the original problem situation, the teacher asks, "Which solution fits well with the situation?"

(*Continued*)

TABLE 17.2 (Continued)

Term	Description	Examples
問題解決型授業 **mondai kaiketsugata jyugyo** (structured problem solving)	A type of lesson which is structured in order for learners to learn knowledge, skills, mode of thought and attitude through experiencing the problem–solving individually and in group. This type of lesson is usually composed of phases.	For example, the teacher organises a lesson that involves: proposing the problem, problem solving on their own, comparing and discussing alternative solutions and ideas, summing-up, and practising.
問題提示 **mondai-teiji** (posing a problem)	The teacher presents the problem for today's lesson.	The teacher poses a task for the day that involves finding a relationship.
ねらい **nerai** (clarifying the aim of activities)	The teacher and students discuss the aim of activities in relation to the goal.	The teacher organises the classroom interaction sequentially by providing the students with the mathematical task and organising activity toward what she/he wants to them to learn.
練り上げ **neriage** "kneading up" (comparing and discussing alternative solutions and ideas)	Students' alternative ideas are presented on the chalkboard, to be compared with each other with oral explanations. *A mere sequential presentations of students' solution methods does not apply to this activity. This is a term for describing the dynamic and collaborative nature of a whole class discussion in the lesson*	For this activity, the teacher carefully calls on students, asking them to present their methods of solving the problem on the chalkboard, selecting the students in a particular order. The teacher's role is not to point out the best solution but to guide the discussion by the students towards an integrated idea and a new mathematical concept.
ノート **nohto** "note" (note-taking)	Students take note of their own ideas, classmates' ideas, descriptions written on the board, teacher's comments.	Students write down problem for the day and their solution to it. They make use of what is written in the notebook in the next lesson. Notebooks are their history of what they have learned.

Term	Description	Examples
教え合い **oshieai** (teaching each other)	One student in a pair teaches another student. Students share their own solution methods or mathematical ideas.	The pair of students work together and then one of the two teaches another a correct solution method.
終わりの挨拶 **owari no aisatsu** (closing greetings)	Students stand-up and then both teacher and students bow. *This activity is a formal declaration of the end of a lesson.*	
ペア学習 **pea-gakusyuu** (pair work)	As a particular form of group worked, students work in groups of two.	Students are divided into pairs and working on the problem together or sharing their own ideas.
練習 **renshu** (exercise)	The teacher asks students to work on a group of tasks to which what they had learned can be applied.	"So, let's try and do some exercises in the textbook. Okay, try to do exercise one in page fifty-seven."
生徒指導 **seito-hido** (management of student behaviour)	The teacher organises the class with particular rules for behaviour. *The teacher may just prohibit something in the classroom but having a student to reflect on them is the key for seito-shido.*	The teacher may give a direction to a particular student or group and warn them in various forms.
説明 **setsumei** (explanation)	Students explain mathematical reasons for a proposition or what they found.	Trying to explain their ideas in a way everyone can understand. Speaking in front of the blackboard, with their face and body looking toward the class.
式 **shiki/** 式化 **shikika** (use of literal symbols)	The teacher and students use mathematical expressions to represent a relationship between variables and explain the meanings of them.	Students are working on a mathematical modelling problem and making a table to find the relationship between two variables to find an equation as a model for the real situation.

(Continued)

TABLE 17.2 (*Continued*)

Term	Description	Examples
指名 **shimei** (naming)	The teacher invites a particular student to answer the question or to present her idea.	The teacher says, "As I'm looking at your answers of those who have started this question, I've found that there are two different ways for answering. So, I'll ask some representatives to come up in the front and to write these two ways down. Ken, go up in the front and answer it. The left one. [To Nao] Can you answer it with this way, please?"
小集団指導 **sho-shudan-shido** (teaching small groups)	The teacher provides different material or points of entry for small group of students who have similar difficulties.	The teacher rearranges the classes into several small groups according to the degree of mastery of the subject at the time of mathematics classroom.
集団思考 **shudan-shikou** (collective thinking)	The teacher and students interactively communicate and share their ideas and thoughts in a whole class discussion mode with public talks.	After teacher and students clarify and share the condition of the problem to be solved, they start working on it with communicating interactively and exchanging ideas or strategies to find the solution to the problem.
宿題の確認・答え合わせ **syukudai no kakuninn, kotae awase** (checking homework)	The teacher invites students to answer the questions in the homework. Typically, this activity is done in the beginning of a lesson.	"I gave you homework yesterday. Um, page forty-two, question three, number one. ...Today, I will ask someone who has done this homework to show his or her answer. I mean, I'll ask them to write the answer on the blackboard."
宿題の指示 **syukudai no shiji** (assigning homework)	The teacher assigns the students homework.	"I want you all to try the similar problem, which I will tell you now, and solve it by yourselves. It is homework. ... Number one under question three on page forty-two. I want you to work on it yourself."

Term	Description	Examples
多様な考え **tayou na kangae** (inventing alternative ideas)	The teacher explicitly asks students to find an alternative solution method once they find one solution method.	The problem for today's lesson: Let's think about how to calculate 1.2×3 1) Change multiplication into addition $1.2 \times 3 = 1.2 + 1.2 + 1.2 = 3.6$ 2) Calculate in terms of 0.1 as the unit $1.2 \times 3 \downarrow 0.1$ as the unit $12 \times 3 = 36 \downarrow 0.1$ as the unit 3.6. 3) Calculate by using the property of multiplication $1.2 \times 3 = \downarrow \times 10 \downarrow \times 10 \; 12 \quad \times 3 = 36 \; 36 \div 10 = 3.6$
適用・応用 **tekiyou/ouyou** (application)	The teacher asks students to apply the knowledge to various contexts including real world situations.	After learning simultaneous equations, students try to solve the following task: One group used 13 eggs to make a cupcake with pudding. I decided to make 30 pieces of pudding and cupcakes together using 13 eggs. From one egg you can make two for a pudding and three for a cupcake. How many eggs should we use to make a cupcake with pudding?
統合 **tougou** (integration)	Several solution methods or students' ideas are integrated from a mathematical point of view.	Natural numbers 2, 4, 6... can be integrated as even number. They can be integrated further in to common fractions, by representing them as $\frac{2}{1}, \frac{4}{1}, \frac{6}{1}$
机の配置 **tsukue no haichi** (physical arrangement of desks)	Teachers talk about the physical arrangement of desks in discussing on the appropriate learning environment.	"I'd like to divide you into groups but I guess some of these rows aren't quite straight, so, well, try facing your desk with someone near you, maybe someone in the next row."
ワークシート **wakushito** (worksheet)	A problem situation or mathematical tasks are printed on a paper.	The teacher passes out worksheets saying "I will distribute a worksheet to you."

(Continued)

TABLE 17.2 (*Continued*)

Term	Description	Examples
予告 **yokoku** (looking forward/preview)	The teacher informs students of the forthcoming topic related to the topic they have learned.	"In the next class I'd like you to examine the number of steps."
よさ **yosa** (merits)	The teacher and students evaluate the solution methods to the problem in terms of mathematical values such as efficiency, simplicity, clearness, and exactness.	After comparing different solutions which were presented by two students, the teacher said "This one is much easier, right? I think you can know which point was difficult as you compare the difficult way and the easier one."
予想 **yosou** (conjecturing/anticipating)	A few students are invited by the teacher to present what they anticipate for the answer to the problem or their conjecture.	The student is investigating what he can say about the sums of three consecutive natural numbers. If the numbers are 1, 2, and 3, then $1 + 2 + 3 = 6$. If the numbers are 2, 3, and 4, then $2 + 3 + 4 = 9$. If the numbers are 3, 4, and 5, then $3 + 4 + 5 = 12$. Based on his investigation above, he made the following prediction. "The sum of 3 consecutive natural numbers will be a multiple of 3."
予想される子ども の反応 **yosousareru kodomo no hannou** (anticipated student responses)	The teacher anticipates students' responses to problem or to his/her "Hatsumon." These anticipated responses are used for lesson planning.	In planning a lesson, the teacher anticipates examples of students' responses to the problem to be presented and prepares for the support in cases. Student responses are aligned with the sophistication of solution methods.
前時の振り返り **zenji no furikaeri** (review of the previous lesson)	The teacher reminds the students of what they have learned in the previous lesson. *Just looking back on what happened in the classroom is not the case. Referring to the previous topic is key.*	The teacher may refer to the previous page in the text.

TABLE 17.3 Japanese Lexicon – Terms by category.

General Phases of Lessons: Introductory	greetings at the start of a lesson; introduction; review of the previous lesson; presenting the instructional materials; learning tasks; key questioning
Main	comparing and discussing alternative solutions and ideas; group work; pair work; learning together; teaching each other; capitalising on an error; key moment in the lesson
Closing	looking back; summing up; closing greetings
Structured Problem Solving	structured problem solving; posing a problem; setting a task; clarifying conditions of the problem; sharing the goal; clarifying the aim of activities; outlook or perspective; conjecturing/anticipating; problem solving on their own; walking between desks for purposeful scanning; instruction between desks; monitoring students when working; differentiation; teaching small groups; presentation of student's solution; inventing alternative ideas; whole class discussion; key words; *comparing and discussing alternative solutions and ideas*; comparing; similarities and differences; integration; sharing; exercise; application; extension; functional use; looking forward/preview
Values Related to Mathematics	merits; ha-ka-se; use of literal symbols; connecting mathematics to the real world; explanation; justification; use of a counterexample
General Teaching Strategy	classroom management; management of student behaviour; social norms for learning in the classroom; naming; collective thinking; using teaching materials; board writing; referring to textbooks; note-taking; journal writing; worksheet; cards with some hints; assigning homework; checking homework; praising students
Lesson Planning	analysing teaching materials; anticipated student responses; planning board writing
Context/ Environment	physical arrangement of desks; use of posters for scaffolding

Note: Terms in italics are in more than one category.

The Japanese Lexicon is the product of intense work by the research members of the national team, as well as other teachers and members of the mathematical community. As discussed in *Chapter 16*, the Japanese Lexicon reflects the vocabulary of a community whose traditions include purposeful engagement in observation and discussion of classroom teaching. We hope that this collection of terms will be useful for the preparation and professional development of teachers and teacher candidates, will aid in the promotion of reflection, and provide the academic community a tool for comparison to support insight into their own professional vocabulary.

Acknowledgements

This work was funded by two grants from Japan Society for the Promotion of Science (19KK0056 and 24653266). A special note of gratitude to the experienced teacher who invited us into his classroom to videotape a single lesson of year eight Mathematics.

Bibliography

Presentations and Papers

The work of the Japanese national team has been presented locally and internationally at the following meetings and conferences:

- American Educational Research Association (AERA) Annual Meeting 2017, 2018
- European Association for Learning and Instruction (EARLI) Biennial Conference 2017
- The Second International Symposium on Mathematics Education (sponsored by ACME) 2019

A selection of peer-reviewed conference publications from the Japanese research team include:

Shimizu, Y. (2017). *The lexicon project: Identifying and unpacking the technical vocabulary of Japanese mathematics teachers. The Future of Mathematics Education and Mathematics Education for the Future: The Proceedings of 2017 NIMS & KSME International Workshop on Mathematics Education* (117–122). Daejeon, South Korea: NIMS.

Shimizu, Y. (2018). *Constructing lexicon of mathematics lessons: A focus on descriptive pedagogical terms. The 51st Annual Research Conference of Japanese Society of Mathematics Education.* Okayama, Japan [in Japanese].

18

IDENTIFYING AND DOCUMENTING KOREAN MIDDLE-SCHOOL MATHEMATICS CLASSROOM PRACTICES

Hee-jeong Kim and Hyungmi Cho

Introduction

When describing classroom phenomena or teaching-learning interactions in non-English speaking countries, it is necessary to increase the clarity of the meaning of pedagogical terms used locally. It is also important to compare the English-translated terms with the term in the original language (Clarke, 2013). Based on the need to re-conceptualise and redefine country-specific pedagogical terms, international researchers from Chile, China, the Czech Republic, Finland, France, Germany, Japan, the United States, and Australia had been conducting international collaborative research as part of The International Classroom Lexicon Project. Korea was the most recent addition to the international team and joined because it deemed as important the documentation of Korean teachers' professional language. This documentation is not only relevant for Korean teachers' professional learning and observation processes (Sherin, Jacobs, & Philipp, 2011) to improve classroom practice, but also for the investigation of the cultural mathematics classroom practices used in Korea.

Language that professional mathematics teachers use in their community describes how they view mathematics classrooms culturally as a place for practice (Stigler & Hiebert, 1999). Sfard (2008) defined discourse as a form of communication, involving people who are able to participate in the given discourse. Given this, it is important to understand how discourses are portrayed in a community and how these differ from one community to another. Although the timely significance of documenting and identifying the lexicon of teachers has already been discussed in the introduction to this book, research on the Korean teacher's lexicon, to describe classroom practice and activities, is scarce. Therefore, inspired by the aims and goals of The International Classroom Lexicon Project, the Korean researchers decided to participate. This chapter presents the results of the study of Korean pedagogical terms of mathematics teachers.

The context of the Korean educational system

The Korean educational system

In Korea, compulsory education begins at age 7 and comprises 6 years of elementary school and 3 years of middle school, for a total of 9 years. After compulsory education, there are another 3 years of the high-school system, followed by 2–3 years of college or 4–6 years of the university education system. There are four different types of high schools: general academic schools, specialised high schools (e.g. vocational schools, alternative schools), special-purpose high schools (e.g. foreign language schools, science schools, gifted schools, art schools, etc.), and autonomous high schools (Figure 18.1).

The Korean educational system is strongly centralised and adopts the national curriculum, which has been revised multiple times. Traditionally, the Korean national curriculum has pursued school curriculum standardisation and has been strictly controlled by government (So, 2019). However, the educational reform that occurred in the 1990s has given schools and its teachers more autonomy. The most recently revised national curriculum of Korea, the 2015 Revised National Curriculum, also suggests educational reforms including a competency-based curriculum and the implementation of an exam-free semester or year in middle school (Ministry of Education, 2015).

The process of becoming a mathematics teacher in Korea

To become a mathematics teacher in Korea, a person must possess a teacher's certificate, obtained after graduating either from a 4-year university's department of mathematics education (including for students who enrolled in a department of pure mathematics taking curriculum for teaching profession) or a 2-year program in a graduate school of education after completing a bachelor's degree, and meet the teacher qualification criteria. Courses that count for the teacher qualification criteria are mainly determined by the Ministry of Education, and the curriculum in teacher education programmes is operated within the standards set by the Ministry of Education (2019). The curriculum includes the study of mathematics educational theories (such as the theory of

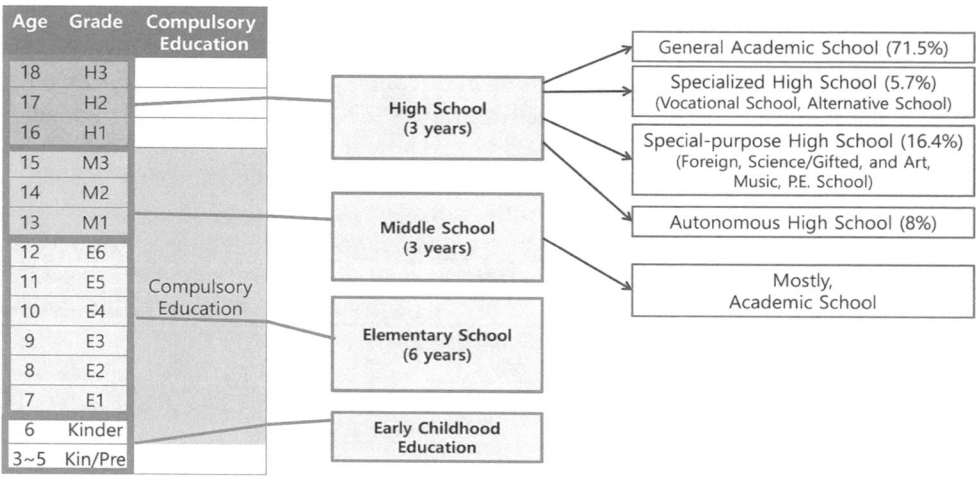

FIGURE 18.1 Korean educational system.

mathematics education, the theory and practice of school mathematics curriculum and school mathematics evaluation; research on the mathematics curriculum and teaching), teaching practicum, general educational theory, and college-level pure mathematics. By completing these theoretical-rich subjects and the teaching practicum, students can obtain teacher qualification. However, to become a teacher in national or public schools, they must pass the national employment exam held once a year, which is fiercely competitive and is very difficult to pass.[1]

The questions in this national employment exam are based on the principles of various educational and mathematics educational theories and college-level pure mathematics. The intention of questions about the mathematics educational theories in the exam are to evaluate how pre-service teachers understand theories of secondary students' mathematical learning and theories of teaching mathematics, and how they might apply those theories into mathematics classroom where they would teach. The exam also evaluates pre-service teachers' understanding of the Korean national curriculum, which strictly guide curriculum of Korean public schools. This national teacher employment exam system may influence Korean pedagogical naming system.

Traditional Korean classroom practices and school culture

Compressed modernity is a feature of the history of Korean education, similar to education in other East Asian countries (Chang, 2010; Lee, 2012; Sato, 2011; Tsuneyoshi, 2004). According to Chang (2010), compressed modernity refers to the situation in a country in which the political, economic, and social changes occur in a complex and condensed manner in a very short time period. Korea experienced this compressed modernity and Korean education is also influenced by this extreme change. Secondary education enrollment rates have risen sharply from 30% to as much as 90% in Korea from 1965 to the mid-1980s (Koo, 1998). This compressed modernisation brings about high social mobility and therefore leads to extreme competition in college entrance examinations. Lee (2012), an educational researcher, points out that a teacher-focused lesson in the overcrowded classroom has been the most common form of teaching in Korea because it is most economically efficient. In a teacher-focused lesson, the traditional method where a teacher explains and the student notes down what the teacher says is employed. Lee (2010), a mathematics educational researcher, explains that the mathematics classroom also employs teacher-focused lessons for the same reason that teachers are aware of the effects of external factors such as competition and success in the teaching environment. She also argues that Korean mathematics teachers continually explore the substance and values of mathematical knowledge to further improve their teaching and to recover humanism from compressed modernity. This classroom culture was represented in the Learner's Perspective Study (Clarke, Keitel, & Shimizu, 2006).

Recently, Korean education has been reformed, leading to several progressive school-change projects in Korea during the last two decades. These school change movements have become a catalyst for nationwide school transformation by making changes in pedagogy (Sung & Lee, 2017). One of the recent school change movements is called the *Hyukshin* schools, which means the "reforming of schools." Teachers in *Hyukshin* schools change their teaching practices to follow a more democratic classroom culture with a student-centred pedagogical practice while respecting the students' autonomy and authority. This change has led to a nationwide classroom culture change by adding collaboration to classroom teaching, as well as the sharing of beliefs and pedagogical strategies in the teacher learning community. This collaborative culture, a change from the traditionally competitive culture, has rapidly spread throughout the country, resulting in changes to mathematics classroom practices in Korean schools. Currently, these two different social values and pedagogical strategies are mixed. These traditional and reformed perspectives to the Korean classroom culture may provide context for the Korean mathematics classroom practice and could also influence Korean pedagogical naming systems.

The research methodology

Research contexts and overarching goals

In order to identify and organise mathematics classroom pedagogical terms, it was critical to look at and name them from the perspective of classroom mathematics teachers. It was also important that researchers had the ability to identify the complex and multifaceted aspects of classroom practices. Thus, following the project-wide protocol of The International Classroom Lexicon Project the Korean research team comprised of researchers, with international research experience and expertise in analysing classroom practices, and, most importantly, middle-/high-school mathematics teachers who could help identify pedagogical terms and analyse mathematics classroom practices from the point of view of field experts. The Korean research team consisted of two PhD researchers and three secondary school mathematics teachers as well as two research assistants who were involved with the more technical aspects of collecting video material.

The process of identifying the Korean lexicon

As a late addition to the international project the Korean research team was guided by the protocols established by the larger international team (these are discussed in the introduction to this book). However, while the other teams were able to access a stimulus package of videos from around the world, the Korean team used stimulus material from nine Korean mathematics lessons ranging from 7th to 8th grade. As with the other countries, the research team was not limited to what they saw in the lessons, pedagogical terms describing any other classroom phenomena were also included in the lexicon.

In the Korean case, two researchers and three professional teachers participated in identifying terms while watching the video stimulus (two teachers were high-school teachers, and the other a middle-school teacher). All three teachers had more than ten years of teaching experience in middle and high schools and had earned a master's degree in mathematics education. Once the team had almost finalised the terms, the descriptions of each term were reviewed by a professor of Korean language education. After revising the lexicon lists and descriptions of each term according to the professor's advice, the team finalised the terms and translated them into English.

Building an initial version of a Korean classroom lexicon

The building of the first version of the pedagogical terms had successive cycles which included the watching of nine video stimuli, building terms, reflecting, and revising. While the International Classroom Lexicon Team intended to build each country's classroom lexicon with a focus on naming pedagogical phenomena, the Korean lexicon, similar to lexicons of other countries, does not include physical tools used in classrooms such as chalk and calculators, nor does it include mathematical content terms such as algebra and functions. At the beginning of the process of building the classroom lexicon, the three professional teachers and two PhD researchers identified pedagogical terms while watching video stimuli at regular weekly meetings. When there was disagreement between the researchers and mathematics teachers in identifying pedagogical terms, the terms were constructed based solely on the opinions of professional mathematics teachers. This is because the primary goal of this research is to understand what kinds of terms are being used among professional teachers in teacher communities.

The initial version of the Korean lexicon included pedagogical terms, their operational definitions, and their corresponding key characteristics. The initial terms were categorised into six categories: lesson preparation, lesson structure, instruction-learning activity, assessment, improving teaching practice, and explicit strategy related to mathematics (see the initial category of Korean lexicon: Cho & Kim, 2018; Kim & Cho, 2018). The initial pedagogical terms were reviewed by two mathematics teachers who did not participate in building the original terms. They also reviewed the categories and suggested more pedagogical terms where possible and appropriate. The revised lexicon included terms suggested by educational policy research, such as *process-focused assessment*, and other terms such as *reframing* which are used in teacher professional communities. The review and revision of the Korean lexicon were conducted multiple times. After these multiple revisions, as soon as the internal members decided to proceed with a national confirmation, the lexicon survey was distributed to national Korean teachers.

National confirmation of the Korean classroom lexicon

The revised initial Korean lexicon included 112 pedagogical terms. These terms were organised into three different surveys in order to increase the response rate from the national Korean teachers. This was done because the teachers advised us that their workloads were very high, and the response rate might be low when all the terms were included in only one survey. To this end, each survey consisted of 30–35 terms, and it took about 20 minutes to complete one survey. Each survey contained the following statements and questions:

- How familiar you are with terms related to teaching and learning activities? To what extent do the following terms appear in your conversations with colleagues?
- To what degree do the following teaching and learning activities occur in your classroom?

The surveys were distributed to about 200 teachers nationwide, of which 147 completed at least one survey, setting the response rate at about 79%. Among the 112 terms, those that were identified as less frequently used in conversations with colleagues and less frequently occurring in classroom settings were selected and discussed with the teacher advisory group. 105 terms were confirmed by the national survey, but at the end, 103 terms were finalised from consulting Korean language experts. The finalised 103 terms were organised into 8 categories. Some of the deleted terms included the following: contests for improving classroom teaching, formative assessment sheet, gifted class, interpreting contextualised problems, reframing, and sense-making, among others. These terms were deleted because the national survey results showed that they neither appear in conversations with colleagues nor in teachers' own teaching. The eight categories in the revised version were: instruction-learning activity (35 terms), explicit strategy related to mathematics instruction-learning activity (18 terms), improving instruction-learning practice (8 terms), lesson format (9 terms), assessment (14 terms), lesson structure (6 terms), lesson preparation (3 terms), and others (10 terms). These categories were grouped based on common features of terms after the terms were identified. Consequently, these categories were inductive groupings of the identified pedagogical terms and were used as a framework for viewing the pedagogical phenomena occurring in current Korean mathematics classrooms. While the complete list of the terms in each category is provided in the next chapter of the book, some examples featured in the categories are introduced here.

The category of *instruction-learning activity* comprises 35 terms that relate to teaching actions, student actions, and teacher–student interactions. Traditionally, in Korea, when one says "lesson" (수업, *sueob*), it includes both teaching and learning rather than specifically focusing on just one aspect of either teaching or learning. In the professional community of teachers, teachers pertain to the phenomena of teaching and learning in the classroom as an instruction-learning activity (교수학습활동, *gyosu-hagseub hawldong*). Because of this, the team has named this category as such. The terms *calling attention, introducing learning goals, guiding activity sheets, motivating,* and *thought-provoking questioning* are examples representing teaching actions. The terms *discussing, peer-mentoring, problem-solving, answering, exploring activity, making arguments,* and *listening* are examples representing student actions. The actors of some terms, such as *questioning, listening,* and *summarising core concepts,* can be either teacher or students. That is, in certain situations, students can *question* a teacher or another student. A teacher may *listen* to students' explanations, and other students may also carefully *listen* to what another student argues or explains in relation to the mathematical concepts.

The category of *explicit strategy related to mathematics instruction-learning activity* comprises 18 terms. In this category, the terms are also related to *teaching actions, student actions,* and *teacher–student interactions* but more specifically associated with mathematics as a subject. Concepts and principles focused on *learning, connecting math concepts with previous lessons, investigating mathematical properties,* and *mathematical communications* are the terms in this category.

The category of *improving instruction-learning practice* was added because of a certain phenomenon pointed out by teachers while watching the video stimulus. Some of the classroom videos were recorded as part of a research lesson that opened the teacher's classroom to other teachers, principals, and other researchers; this is part of a traditional school-based supervision culture in Korea (see Pang, 2016). Several terms identified by the professional teachers on the research team while watching videos were related to this culture: *action research, classroom observation, instructional supervision, peer-supervision, research lesson, researching instruction, self-supervision,* and *teachers' council in schools* (8 total terms).

The other categories are lesson format (9 terms; e.g. *collaborative learning, flipped learning, lecture style lesson, learner-centred lesson*), assessment (14 terms; e.g. *attitude scores, formative assessment, group scores, process-focused assessment*), lesson structure (6 terms; e.g. *introductory phase of lesson, lesson finale, main phase of lesson, small group activity*), lesson preparation (3 terms; e.g. *instruction-learning process plan, reconstituting textbook, researching lesson materials*), and others (10 terms; e.g. *alignment of lesson and assessment, mathematical competency*).

The professional language of teachers

The complete lexicon is listed in the next chapter. Some examples of more familiar pedagogical terms which are used more frequently in the teaching context are presented here.

Pedagogical terms of Korean teachers used in conversations with colleagues and in practice

For each category of terms, the average degree to which a pedagogical term appeared in a conversation of teachers with colleagues as compared to that in their own teaching resulted in no significant differences on a 5-point Likert scale, in which 1 is not familiar (seldom talk about the term with colleagues) and 5 is highly familiar (very often talk about the term while having conversation with colleagues) (see Table 18.1). The scores represent the average in each category.

The terms that often appear in conversations with colleagues (i.e. familiarity) are those related to *instruction-learning activity* and *assessment*. The most frequently used terms in teaching practice are *lesson structure* and *instruction-learning activity*. On the other hand, the terms in the category of *classroom improvement* resulted in the lowest average in both familiarity and use in teaching practice.

The top 20 pedagogical terms that appear in conversations with colleagues (i.e. familiarity) are shown in Table 18.2. These represent the terms with the highest average scores of teachers responding to each term.

Many of these terms are in the categories of *instruction-learning activity* and *assessment*. Although the terms in the *instruction-learning activity* category correspond to the highest rate of all the 105 pedagogical terms (34.29%), 11 of the top 20 pedagogical terms fell into this category in terms of the "familiarity" survey question. That is, teachers most frequently use professional language related to *instruction-learning activity*. In addition, 6 terms related to *assessment* (*class participation, problems, performance assessment, midterm or final exam, essay-type questions, process-focused assessment*) also appeared in the

TABLE 18.1 Categories of Korean pedagogical terms with national confirmation.

Category of Korean pedagogical terms	Familiarity	Classroom practice and use
Instruction-learning activity	3.25	3.58
Explicit strategy related to mathematics instruction-learning activity	2.75	2.75
Improving instruction-learning practice	2.65	2.16
Lesson format	2.84	2.79
Assessment	3.25	3.19
Lesson structure	3.11	3.91
Lesson preparation	3.21	3.33
Others	3.05	2.68

TABLE 18.2 The top 20 Korean pedagogical terms corresponding to familiarity.

Ranking	Pedagogical terms	Average	Category
1	Problem-solving	4.5	Instruction-learning activity
2	Class participation	4.48	Assessment
3	Problems	4.22	Assessment
4	Questioning	4.13	Instruction-learning activity
5	Guiding activity sheet	4.08	Instruction-learning activity
6	Performance assessment	4.03	Assessment
7	Writing on a blackboard	4.00	Instruction-learning activity
8	Motivating	3.94	Instruction-learning activity
9	Thought-provoking questioning	3.93	Instruction-learning activity
10	Driving student presentation	3.85	Instruction-learning activity
11	Researching lesson materials	3.75	Lesson preparation
12	Explaining	3.72	Instruction-learning activity
13	Midterm/Final exam	3.69	Assessment
14	Encouraging	3.67	Instruction-learning activity
15	Circuit guiding	3.65	Instruction-learning activity
16	Student presentation lesson	3.65	Lesson format
17	Alignment of curriculum, instruction, and assessment	3.63	Others
18	Process-focused assessment	3.61	Assessment
19	Wrapping-up activity sheet	3.56	Instruction-learning activity
20	Mathematical essay problems	3.54	Assessment

top 20 terms, and 40% of the terms (6 of 15) in the *assessment* category were frequently used in conversations with colleagues. This shows that Korean teachers mainly focus on instructional practices and assessments, which may be a result of the sensitive climate of assessment in Korean society (Lee, 2010).

The top 20 terms from the results of the survey regarding usage in teaching practice are shown in Table 18.3. These terms are categorised into *instruction-learning activity* (10 terms), *lesson structure* (4 terms), and *lesson format* (4 terms). The results indicate that Korean teachers have a high regard for classroom structure and overall lesson format, and they try to implement these in their own teaching. These terms related to lesson structure are also very important in Korean mathematics classrooms. Korean teachers are taught that a lesson should be coherent during each of its phases (*introductory phase of lesson, main phase of lesson, lesson finale*), and it should be organised similar to the plot of a play

TABLE 18.3 The top 20 Korean pedagogical terms used in teachers' own teaching practice.

Ranking	Pedagogical terms	Average	Category
1	Explaining	4.50	Instruction-learning activity
2	Main phase of lesson	4.41	Lesson structure
3	Problem-solving	4.39	Instruction-learning activity
4	Introductory phase of lesson	4.34	Lesson structure
5	Lesson finale	4.31	Lesson structure
6	Calling attention	4.25	Instruction-learning activity
7	Reminding prior lesson	4.25	Explicit strategy related to mathematics instruction-learning activity
8	Thought-provoking questioning	4.22	Instruction-learning activity
9	Writing on a blackboard	4.22	Instruction-learning activity
10	Technology-enhanced lesson	4.22	Lesson format
11	Multimedia lesson	4.18	Lesson format
12	Informing about next lesson	4.16	Instruction-learning activity
13	Motivating	4.16	Instruction-learning activity
14	Lecture-style class	4.15	Lesson format
15	Questioning	4.11	Instruction-learning activity
16	Class participation	4.11	Assessment
17	Interdisciplinary integration lesson	4.11	Lesson format
18	Responding	4.09	Instruction-learning activity
19	Listening	4.06	Instruction-learning activity
20	Lesson finale	4.03	Lesson structure

with an introduction, action, climax, and conclusion. Korean teachers also consider *lesson format* in their teaching as well as *instruction-learning activity*, and most of the lesson structure that they use in their own teaching are *technology-enhanced lesson, multimedia lesson, lecture class,* and *convergence lesson.*

By comparing these two survey results, we can note that there are differences between "familiarity" and "usage in practice." In other words, pedagogical terms such as *problems, performance assessment,* and *process-focused assessment*, which were found to occur relatively frequently in conversations with colleagues, showed a relatively low frequency (out of the top 100 terms) in usage in teachers' own teaching practice. Terms such as *technology-enhanced lesson* and *convergent lesson* were examined as terms that occur frequently in usage in the teachers' own teaching practice but rarely in their conversations with colleagues. In particular, *convergence lesson* was ranked 112th for conversations with colleagues. *Convergence lesson* was actually emphasised recently in the Korean 2015 Revised National Curriculum, but is not frequently discussed, suggesting that although teachers might attempt to implement convergent classes it is not a discussion topic with colleagues.

Discussion

Korean mathematics teacher education and classroom lexicon

From our investigation, Korean pedagogical terms can be characterised by the appearance of terms that are made and used based on the national teacher education programme and a teacher's academic background. Some terms, such as *action research, problem-solving, discovery learning, mathematical modelling*, and *peer-mentoring*, can be characterised as originating from a foreign language and translated into Korean or as an English expression used in its original language. The pedagogical terms related to academic background may be derived from the teacher education programme in Korea, which emphasises theoretical exploration centred on mathematics educational psychology, rather than the structural aspects of the Korean language. As described in an earlier section, in order to become a mathematics teacher in Korea, theoretical learning related to mathematics instruction-learning theory, which includes works by Piaget, Bruner, Dienes, Schoenfeld, and Freudenthal, is emphasised. Certain terms used by teachers seem to come from these works.

Additionally, the terms in the *lesson structure* and *lesson preparation* categories are related to what teachers have learned as pre-service teachers in the teacher education programme. In the teacher education programme, pre-service teachers not only study mathematics education theory in depth but also engage in teaching practice. They spend many hours in the design of lessons, demo-teaching, and teaching reflection. Throughout this process, terms related to structural lesson planning, such as *introductory phase of lesson, main phase of lesson, lesson finale, motivating, reminding prior lesson*, and *connecting math concepts with previous lessons*, are embedded in Korean mathematics teachers' lexicon.

Gaps in policy and practice

Most of the terms identified by teachers refer to how mathematics teachers see and name the pedagogical phenomena occurring in mathematics classrooms spontaneously. However some terms such as *process-focused assessment, convergence lesson*, and *alignment of curriculum-lesson-assessment*, were influenced by national educational policy. These nationally generated terms are in accordance with educational policy and direction, which are transmitted to the lead teachers through national- or school district-level teacher training. Teachers attempt to implement them in their own teaching after training, but it may take a long time to reach an agreed meaning for the proposed pedagogical terms at the national level. In particular, in the early stages of educational policy, practical implementation methods are often lacking, so pedagogical practice is not well implemented. For example, the term *process-focused assessment* was used in the 2015 Revised National Curriculum, but individual teachers interpreted it differently, so it was used differently depending on the school site (Shin, Ahn & Kim, 2017) when the 2015 Revised National Curriculum was launched. To narrow the gap between the terms initiated by educational policy and those from classroom practice, teacher professional development programmes were created and applied to integrate these educational policy terms into classroom practices. However, the surveys conducted in this study show that these programmes seem to be insufficient for teachers to implement the terms in classrooms. Further on-going analysis of these terms with long-term perspectives is needed to track the terms that represent and name pedagogical practices in Korea.

Unnamed instruction-learning activities in Korea

We also found that there are several terms that teachers could not name, although they occur frequently in mathematics classrooms in Korea. First, it was hard to name the phenomena of teachers supporting or facilitating students' learning by navigating among students to check, monitor, and pay attention to their learning. This is a common phenomenon in Korea and is vital in teaching activities. Currently, teachers have named it *circuit guidance*, but we all agreed that this term is not sufficient to describe the particular phenomenon. This term, *circuit guidance,* seemed to be created and used at the beginning when small group activities were considered important pedagogical strategies. Additionally, this term has been used to indicate the situation that teachers or principals move between schools to teach subjects or manage schools. Therefore, our research team and advisory expert teachers wanted to name it to refer to the instructional activities that include instruction as a means of scaffolding or facilitating in recent learning theories that promote and support small groups or individual students. However, this proved difficult, and it was not sufficient to use the existing term, *circuit guidance*; we all agree that it is necessary to use a new term.

Second, there was no term to describe an activity in which a teacher reframes or rewords what was presented or explained in his/her language in mathematically clear terms. These instructional activities are not intended to be repeated by the teacher but to improve the quality of communication and the clear expression of the mathematical ideas or content of the student's presentation. This is found in the video stimulus, but there was no pedagogical term to name it. This activity was named by one of the expert teachers as *reverting (reframing)* in Korean as mentioned by Park (2013), but the term itself is not familiar to national teachers. This activity is conducted by the teacher with the intention of improving the quality of mathematical communication if students cannot precisely express their opinions in their own words. Therefore, this is vital and must be named to help the teachers improve their teaching.

In research regarding classroom discourse, especially in English, we can name it as *reframing,* but the purpose of this study is not to generate pedagogical terms related to classroom phenomena from the researchers' point of view but to identify and organise terms named from the professional teachers' point of view. Thus, we have attempted to avoid directly suggesting research-oriented terms at the moment. However, teachers are aware of the need to name these activities and expect that teachers in their community gradually suggest terms to refer to them. In this way, it is possible to create terms that can be used in the classroom and to refer to classroom activities with terms with clear and agreed-upon meanings in the teacher community in Korea.

Concluding remarks

Explaining the phenomenon of teaching and learning in the classroom is deeply related to the terms and language that researchers and/or teachers use. The key structures and phenomena are reliant upon how terms are clearly being used in everyday conversations as well as in practice. Expanding the vocabulary resources of teachers and researchers who discuss pedagogical activities in the classroom improves the human ability to decipher the real meanings of words (Bruner, 1996). Discourse about the terms from other languages or cultures will also allow us to reflect on our own language and discourse. This also allows for alternative ways of seeing and interpreting words used in the classroom. Thus, building the Korean classroom lexicon through the International Classroom Lexicon Project can reconcile both insider and outsider perspectives of classroom practice within our own culture and school system. Even though there are several challenging issues that need further analysis, this study already provides new insights into the pedagogical traditions that need to be explored by comparing them with the use of languages in other countries – in turn fostering sensitivity to the terms used.

Acknowledgements

This project was supported by the 2018 Hongik University Research Fund.

Note

1 In 2017, the number of national mathematics teachers was 9263, which was over 4 times higher than the number of private mathematics teachers, 2004. The national employment exams, which are the only way to become a public school teacher, are set to select the number of teachers according to teacher supply and demand. The first examination round is conducted using the same questions nationwide according to strict evaluation procedures, and candidates are selected to continue to the next examination round. The final selection of successful candidates is made after the second exam, which consists of a teaching demonstration, interview, and essay. The national exam is fiercely competitive, varies from year to year, and is very difficult to pass; in 2019, the number of exam fails to passes was 20-to-1 in Seoul (Seoul Metropolitan of Education, 2019).

References

Bruner, J. (1996). *The culture of education.* Cambridge, MA: Harvard University Press.

Chang, K. (2010). *South Korea under compressed modernity: Familial political economy in transition.* London & New York: Routledge.

Cho, H. & Kim, H-j. (2018). Identifying and documenting pedagogical terms in Korean mathematics classroom. *Journal of Korea Society Educational Studies in Mathematics: School Mathematics.* 20(3), 463–481.

Clarke, D. J. (2013). Contingent conceptions of accomplished practice: The cultural specificity of discourse in and about the mathematics classroom. *ZDM*, 45(1), 21–33.

Clarke, D. J., Keitel, C., & Shimizu, Y. (Eds.). (2006). *Mathematics classrooms in twelve countries: The insider's perspective.* Rotterdam, NL: Sense Publishers.

Kim, H-j. & Cho, H. (2018). *The international classroom lexicon project: Identifying and documenting Korean middle school mathematics classroom practices.* In E. Bergqvist, M. Österholm, C. Granberg, & L. Sumpter (Eds). *Proceedings of the 42nd Conference of the International Group for the Psychology of Mathematics Education* Vol. 5 (p. 86). Umeå, Sweden: PME.

Koo, S. (1998). Demographic transition, education and economic growth in East Asian countries. In H. S. Rowen (Ed.). *Behind East Asian growth: The political and social foundations of prosperity* (pp. 234–262). London: Routledge.

Lee, K. H. (2010). Searching for Korean perspective on mathematics education through discussion on mathematical modelling. *Journal of Educational Research in Mathematics*, 20(3), 221–239.

Lee, Y. (2012). An investigation into the possibility and practicality of an East Asian model of education. *Korean Journal of Comparative Education*, 22(5), 1–32.

Ministry of Education (2015). 2015 Revised National Curriculum in Mathematics.

Ministry of Education (2019). 2019 Handbook of Teacher Qualification Examination.

Pang, J. (2016). Improving mathematics instruction and supporting teacher learning in Korea through lesson study using five practices. *ZDM*, 48(4), 471–483.

Park, H. (2013). *Teachers grow by lessons (in Korean)*. Seoul: Mamedrim.

Sato, M. (2011). Imagining neo-liberalism and the hidden realities of the politics of reform: Teachers and students in a globalized Japan. In D. B. Willis & J. Rappleye (Eds.), *Reimagining Japanese education: Borders, transfers, circulations, and the comparative* (pp. 225–246). Oxford: Symposium Books Ltd.

Seoul Metropolitan of Education (2019). *2010 Secondary School Teacher Examination Status (Final)*. Retrieved from http://www.sen.go.kr/web/services/bbs/bbsView.action?bbsBean.bbsCd=23&bbsBean.bbsSeq=433

Sfard, A. (2008). *Thinking as communicating: Human development, the growth of discourses, and mathematizing*. Cambridge, UK: Cambridge University Press.

Sherin, M. G., Jacobs, V. R., & Philipp, R. A. (2011). *Mathematics teacher noticing: Seeing Through teachers' eyes*. New York: Routledge.

Shin, H., Ahn, S., & Kim, Y. (2017). A policy analysis on the process-based evaluation: Focusing on middle school teachers in Seoul. *The Journal of Curriculum and Evaluation*, 20(2), 135–162.

So, K. H. (2019). A review on the historical changes and recent reform trends in South Korean National Curriculum. *The SNU Journal of Educational Research*, 28(1), 87–103.

Stigler, J. W., & Hiebert, J. (1999). *The teaching gap: Best ideas from the world's teachers for improving education in the classroom*. New York: Free Press.

Sung, Y. K., & Lee, Y. (2017). Is the United States losing its status as a reference point for educational policy in the age of global comparison? The case of South Korea. *Oxford Review of Education*, 43(2), 162–179.

Tsuneyoshi, R. (2004). The new Japanese educational reforms and the achievement "crisis" debate. *Educational Policy*, 18, 364–394.

19

KOREAN LEXICON

Hee-jeong Kim and Hyungmi Cho

A Korean classroom lexicon

The lexicon presented in this chapter consists of 103 terms, and these terms are organised into eight categories. Operational definitions, for each of the terms, were developed to include a description of the classroom practice as well as examples from a classroom. For each term, the Korean term is presented, then the romanisation of the Korean term follows, and finally, an English translation of the term is given. The romanisation of the Korean terms is based on the rules of the Revised Romanisation of Korean, the standard pronunciation of the National Academy of the Korean Language. The terms are listed alphabetically by romanisation.

The first lexicon table (see Table 19.1) lists the terms of Korean lexicon with English-translated terms. The second lexicon table (see Table 19.2) lists the terms (column one) of the Korean lexicon with operational definitions (column two) and examples (column three).

The lexicon was organised into eight categories based on common features of terms, inspired by professional teachers (see Table 19.3). The categories are as follows: *Instruction-learning activity* (35 terms), *Explicit strategy related to mathematics instruction-learning activity* (18 terms), *Improving instruction-learning practice* (8 terms), *Lesson format* (9 terms), *Assessment* (14 terms), *Lesson structure* (6 terms), *Lesson preparation* (3 terms), and *Others* (10 terms).

The lexicon documented in this chapter reflects the terms Korean teachers use to name pedagogical phenomena of their mathematics classroom. We agree that the language and pedagogical terms are still in the process of evolving, but documenting the current Korean pedagogical terms provides insights into which our community members – teachers, teacher educators, policymakers, and researchers – become aware the characteristics of the terms that are made and used by mathematics teachers. It also provides fruitful insights into pedagogical traditions and ways to develop their professional noticing. We hope that this lexicon research supports respect of the practice of classrooms in each culture and assists in the elaboration of theories of teaching and learning in mathematics classrooms.

TABLE 19.1 Korean Lexicon – Terms.

아이디어 연결짓기 **aidieo yeongyeoljisgi** (making connections among ideas)	받아적기 **bad-ajeoggi** (writing down)	발견학습 **balgyeonhagseub** (discovery learning)	발문 **balmun** (thought-provoking questioning)
발표수업 **balpyo sueob** (presentation focused lesson)	발표자 선정하기 **balpyoja seonjeonghagi** (selecting presenters)	발표유도 **balpyoyudo** (driving student presentation)	방과후 학교 **bang-gwahu haggyo** (after school class)
보상하기 **bosanghagi** (rewarding; including individual rewarding, group rewarding)	채점하기 **chaejeomhagi** (scoring and checking answers)	차시예고 **chasi yego** (informing about next lesson)	답변하기 **dabbyeon hagi** (responding)
도입 **doib** (introductory phase of lesson)	동기유발 **dong-gi yubal** (motivating)	동료장학 **donglyo janghag** (peer-supervision)	동료평가 **donglyo pyeong-ga** (peer-assessment)
EBS 수학 **EBS suhag** (EBS math)	개별학습 **gaebyeol hagseub** (individual learning)	개별학습지 풀기 **gaebyeolhagseubji pulgi** (solving individual worksheet)	개념 및 원리 중심 학습 **gaenyeom mich wonli jungsim hagseub** (concepts and principles focused learning)
강의식 수업 **gang-uisig sueob** (lecture style class)	가꾸로 수업 **geokkulo sueb** (flipped classroom)	공학도구 활용 수업 **gonghagdogu hwal-yong sueob** (technology enhanced class)	격려하기 **gyeoglyeohagi** (encouraging)
경청하기 **gyeongcheong hagi** (listening)	과정중심평가 **gwajeong jungsim pyeong-ga** (process-focused assessment)	과제평가 **gwajepyeong-ga** (assignment assessment)	귀납적으로 탐구하기 **gwinabjeog-euro tamguhagi** (inductive investigating)

(Continued)

TABLE 19.1 (*Continued*)

교과서 재구성 **gyogwaseo jaeguseong** (reconstituting textbook)	교재연구 **gyojaeyeongu** (researching lesson materials)	교사협의회 **gyosa hyeob-uihoe** (teachers' council in school)	교사연수 **gyosa yeonsu** (teacher training)
교수학습 과정안 **gyosuhagseub gwajeong-an** (lesson plan)	교육과정 수업 평가의 일체화 **gyoyuggwajeong sueob pyeong-gaui ilchehwa** (alignment of curriculum, instruction and assessment)	학생활동중심 수업 **hagsaenghwaldong jungsim sueob** (student activity-centered class)	학생이 칠판에 풀이적기 **hagsaeng-i chilpan-e pul-ijeoggi** (students' writing solutions on blackboard)
학생 사고 수집하기 **hagsaeng sago sujibhagi** (collecting student thinking)	학습자 중심 수업 **hagseubja jungsim sueob** (learner-centered class)	학습목표 재확인 **hagseubmogpyo jaehwag-in** (reconfirming learning objectives)	학습목표 제시 **hagseubmogpyo jesi** (introducing and engaging in lesson goals)
핵심정리 **haegsimjeongli** (summarising core concepts)	핵심정리 같이 읽기 **haegsimjeongli gat-I ilg-gi** (reading core theorem together)	호명하기 **homyeonghagi** (calling students)	형성 평가 **hyeongseong pyeong-ga** (formative assessment)
활동지 안내 **hwaldongji annae** (guiding activity sheet)	활동지/학습지 정리하기 **hwaldongji/hagseubji jeonglihagi** (wrapping-up activity sheet or worksheet)	협력 학습 **hyblyeog hagseub** (collaborative learning)	이해 확인하기 **ihae hwag-inhagi** (checking comprehension)
인성 교육 **inseong gyoyug** (humanising education or character education)	자기장학 **jagijanghag** (self-supervision)	자기평가 **jagipyeong-ga** (self-assessment)	전체학습 **jeonchehagseub** (whole group lesson)

전개 **jeongae** (main phase of lesson)	정당화하기 **jeongdanghwa hagi** (justifying)	전시 학습 개념과의 연결 **jeonsi hagseub gaenyeongwaui yeongyeol** (connecting math concepts with previous lessons)	전시학습 상기 **jeonsihagseub sang-gi** (reminding prior lesson)
질문 **jilmun** (questions)	주의환기 **jooui hwangi** (recalling attention)	중간기말고사 **jung-gan/gimal gosa** (midterm/final exam)	말하기 평가 **malhagi pyeong-ga** (evaluating verbal communication)
멀티미디어 수업 **meoltimidieo sueob** (multimedia lesson)	모둠활동 **modumhwaldong** (small group activity)	문제 **munje** (problem)	문제 풀기 **munje pulgi** (problem-solving)
문제읽기 **munjeilg-gi** (reading problems)	문제유형 제시 **munjeyuhyeong jesi** (providing types of problems)	난이도 제시 **nan-ido jesi** (informing difficulty of problems)	논의하기 **non-uihagi** (discussing)
논술형 **nonsulhyeong** (mathematical essay problems)	노트필기 **noteupilgi** (note-taking)	판서 **panseo** (writing on blackboard)	상황설명 **sanghwang seolmyeong** (explaining contexts of problems)
설명하기 **seolmyeonghagi** (explaining)	선행학습 **seonhaeng hagseub** (accelerated learning)	스토리텔링 수학 **seutolitelling suhag** (storytelling mathematics or contextually rich mathematics)	시험범위 제시 **siheombeom-wi jesi** (providing a range of exam)
실행연구 **silhaeng yeongu** (action research)	실생활 맥락 도입 **silsaenghwal maeglag doib** (connecting to real life contexts)	소집단 점수 **sojibdan jeomsu** (group scores)	수업 **sueob** (lesson)

(Continued)

TABLE 19.1 (*Continued*)

수업 참여도 **sueob cham-yeodo** (class participation)	수업 참관 **sueob chamgwan** (classroom observation)	수업 장학 **sueob janghag** (instructional supervision)	수업 평가 일체화 **sueob pyeong-ga ilchehwa** (alignment of instruction and assessment)
수업의 마무리 **sueob-ui mamuri** (lesson finale; concluding phase of lesson)	수업연구 **sueob-yeongu** (researching instruction)	수행평가 **suhaeng pyeong-ga** (performance assessment)	수학 동아리 **suhag dong-ali** (math club)
수학 교과역량 **suhag gyogwa yeoglyang** (mathematical competency)	수학의 활용 제시 **suhag-ui hwal-yong jesi** (providing application of mathematics)	수학 용어 정의하기 **suhag yong-eo jeong-uihagi** (defining mathematical terms)	수학과 진로 지도 **suhaggwa jinlo jido** (career guiding related to mathematics)
수학적 모델링 **suhagjeog modelling** (mathematical modeling)	수학적 성질 탐구 **suhagjeog seongjil tamgu** (investigating mathematical properties)	수학적 의사소통 **suhagjeog uisasotong** (mathematical communication)	순회지도 **sunhoejido** (circuit instruction or circuit guiding)
탐구활동 **tamgu hwaldong** (exploration activity)	탐구한 성질 정리하기 **tamguhan seongjil jeonglihagi** (organising inquiry features)	태도 점수 **taedo jeomsu** (attitude scores)	토론하기 **tolonhagi** (making arguments)
또래교수/동료 멘토링 **ttolaegyosu/donglyo mentoling** (peer-instruction/peer-mentoring)	연구수업 **yeongusueob** (research lesson)	융복합 수업 **yungboghab sueob** (interdisciplinary integration lesson)	

TABLE 19.2 Korean Lexicon – Terms with operational definitions.

Term	Description	Examples
아이디어 연결짓기 **aidieo yeongyeoljisgi** "idea connecting" (making connections among ideas)	The activity of connecting various mathematical ideas raised in class.	Linking Pascal's triangle with the number (i.e. combination) of n to m methods in any order.
받아적기 **bad-ajeoggi** "dictating" (writing down)	Students write down in a way that teachers' or other students' explanations, talks or solutions written in a blackboard.	
발견학습 **balgyeonhagseub** "discovering learning" (discovery learning)	Terms similar to the concept presented by Bruner or Freudenthal, where students discover mathematical concepts or principles on their own.	"There are several primary functions and their graphs. Let's look at the features between the graphs and explore what a and b mean in $y = ax + b$."
발문 **balmun** (thought-provoking questioning)	Thought-provoking questions are that teacher asks to develop students' (mathematical) thinking. To answer this question, students need to consider various aspects of mathematics.	"What mathematical evidence did the university researchers use to predict the results of global warming research?"
발표수업 **balpyo sueob** "presentation lesson" (presentation focused lesson)	A lesson mainly with which students present or "discuss" their (or small groups') findings.	A class in which students are asked to present their research and exploration of materials.
발표자 선정하기 **balpyoja seonjeonghagi** (selecting presenters)	Teacher selects which student will share the results of small group activities or individual works in front of the whole class.	
발표유도 **balpyoyudo** (driving student presentation)	Teacher moves to encourage students to present their works during the lesson.	
방과후 학교 **bang-gwahu haggyo** "after school" (after school class)	A supplementary class of learning or various activities conducted inside the school after the regular classes of the school.	A teacher's class after regular classes as a part of school curriculum.
보상하기 **bosanghagi** "rewarding" (rewarding; including individual rewarding, group rewarding)	Rewarding students according to their "class participation" in order to increase student participation. It includes individual rewards and small group rewards.	Teacher: Let's applaud this group because they solved this problem well, asking and informing each other what they don't know.

(Continued)

TABLE 19.2 (*Continued*)

Term	Description	Examples
채점하기 **chaejeomhagi** (scoring and checking answers)	Students score their own or peer's problem-solving after students have answered or solved a problem so that they can check their own achievement.	An activity of grading together the questions students have solved, such as "Now, what is y for number one?"
차시예고 **chasi yego** "informing next lesson" (informing about next lesson)	Teacher briefly introduces what they will learn in the next lesson related to what they learned today.	"Next time, we will learn how to draw a graph with the concept of slope that we learned today."
답변하기 **dabbyeon hagi** "answering" (responding)	Students answer and respond to the teacher's "thought-provoking question", and this response can be the starting point or basis for mathematical discussion	In the activity of drawing a graph on a coordinate plane, S: "I'm going to draw a straight line, and I'm going to put the baobab tree as the reference point." T: "What's the point of reference? What do you think?" S: "When we have the reference point, we can know how far it is from the point."
도입 **doib** "introduction" (introductory phase of lesson)	At the beginning of the class, teacher introduces a main lesson briefly with motivational activities. Lesson introduction includes the activities of motivating students to engage in the learning of the lesson, providing real-life contexts of the concepts, etc.	(After doing activities related to the concepts of algebraic expressions) You really think I am a wizard? There must be something. Right? There's got to be a way. It's going to be hard to explain how it works before learning about lesson unit 3. But after we study hard in Unit 3, you can figure out how it works."
동기유발 **dong-gi yubal** "stimulating of motivation" (motivating)	Teacher's attempts to increase the concentration of instruction by stimulating students' curiosity and interests in a (non-mathematical or mathematical) context related to the math content being studied. In the introduction of textbooks or lessons, it is mainly embodied as a problem in the context of real life.	Teacher: This summer, I am suffering from a more severe drought than ever before. People are very interested in meteorological forecasts. How does the weather center predict the meteorological forecast? When learning linear functions, it can cause students to interest in the contents with the context of everyday life that can be linked to linear functions.

Term	Description	Examples
동료장학 **donglyo janghag** (peer-supervision)	To provide scholarship among teachers for the purpose of improving educational activities. Usually, teacher work on researching on instruction-learning activity or on improving the teaching method with fellow teachers of the same year class assignment.	Observing peers" classroom and discuss what they noticed and how to improve instruciton-learning, teaching methods, or educational effects.
동료평가 **donglyo pyeong-ga** (peer-assessment)	Evaluation and feedback between students.	Teacher: In each group, you suggested different methods as a way to suggest the government's carbon dioxide regulations to cope with climate change. Let's change each other's group and analyse how they model and solve the problem in the other group, and evaluate and discuss their strengths and weaknesses.
EBS 수학 **EBS suhag** (EBS math)	EBS is a public education broadcasting station in Korea. EBS Math is equipped with materials related to mathematics learning. It includes videos of lectures taught by EBS instructors, various math activity materials, and mathematics history materials, which are available for free to teachers and students.	
개별학습 **gaebyeol hagseub** (individual learning)	A class structure in which students work individually and silently in class.	T: "Let's solve this problem silently for about 5 minutes" Then, students work individually,
개별학습지 풀기 **gaebyeolhagseubji pulgi** (solving individual worksheet)	Activities for students to solve problems on a worksheet individually.	Teacher: That's all I'm going to talk about today, Now, individually solve the problems on page 72.
개념 및 원리 중심 학습 **gaenyeom mich wonli jungsim hagseub** (concepts and principles focused learning)	Teaching and learning activities in which the teacher faithfully explains mathematical concepts or allows students to explore and construct their own concepts and principles.	Overall class activities such as presenting newspaper articles analysing heat wave damage caused by greenhouse gases and students predicting linear function and linking them to algebraic expressions (methods).

(*Continued*)

TABLE 19.2 (*Continued*)

Term	Description	Examples
강의식 수업 **gang-uisig sueob** (lecture style class)	A class in which the teacher's lectures are central.	A class in which teachers continue to explain and lead the whole class.
거꾸로 수업 **geokkulo sueb** "backwards lesson" (flipped classroom)	A new form of instructional-learning that provides students with pre-learning and offline classes through short videos of 10 minutes online, and provides students-centered exploratory classes during class.	Students watch and learn short videos before class so that they construct their knowledge before class. During class, they discuss what they learned and questions that they have.
공학도구 활용 수업 **gonghagdogu hwal-yong sueob** "technology using lesson" (technology enhanced class)	Technology enhanced class refers to classes using exploratory geometrical software or application related to learning math, ICT, VR, etc.	Showing the process of parabola being made using geogebra software.
격려하기 **gyeoglyeohagi** (encouraging)	Teacher moves that give students positive feedbacks so that they don't lose interest in learning mathematics.	When students say the wrong answer, not just say "It's not correct" Teachers find out why the wrong answer came out, give students positive feedback without having a negative perception of math.
경청하기 **gyeongcheong hagi** "active and careful listening" (listening)	Active listening to understand the explanations of teacher or fellow students' arguments.	Listening to other students' arguments or solutions during a small group works or whole class discussion.
과정중심평가 **gwajeong jungsim pyeong-ga** (process-focused assessment)	Contrary to the assessment of learning outcomes, it is a form of continuous diagnosis and feedback that helps students learn.	A series of diagnostic-assessment-feedback courses in which teachers provide appropriate feedback by observing the misconception of students in the class. For example "I will explain again because there are so many students who are confused about the concept of a gradient"
과제평가 **gwajepyeong-ga** (assignment assessment)	Evaluating students' homework or their work on tasks in class.	Students form groups and perform tasks, then teacher scores performed tasks.

Term	Description	Examples
귀납적으로 탐구하기 **gwinabjeog-euro tamguhagi** "inductive investigating" (inductive investigating)	Exploring mathematical concepts and principles in an inductive manner through activities that consist of similar mathematical structures before the definition of a mathematical term or concept is presented.	Inductive guessing or understanding of the mathematical concepts (combination) associated with the shortest-paths through a simple example.
교과서 재구성 **gyogwaseo jaeguseong** "textbook reorganising" (reconstituting textbook)	To alter or modify the exploratory content by individual teachers to better support students learning in classroom, not to follow the order of learning presented in the textbook.	In the textbook first presented the concept of a linear function, the teacher reconstructed it and introduced the concept of slope first using the slope of the ski slope, etc., then s/he provided the concept of a linear function.
교재연구 **gyojaeyeongu** "lesson materials research" (researching lesson materials)	To study textbooks and teaching materials, taking in to account the teaching and learning process in order to improve students' learning.	Research to find out and study which materials are appropriate and more necessary for students to learn concepts more effectively.
교사협의회 **gyosa hyeob-uihoe** "teachers' council" (teachers' council in school)	A group of teachers to discuss together to solve problems or practice together. The issues discussed here are classroom educational activities, exams, problems that occur at school. They share a common responsibility. Usually teachers who teach same subject have a council in middle/high schools.	Same grade teachers come together to discuss the exam of mid-term and final.
교사연수 **gyosa yeonsu** (teacher training)	The process of learning and researching by teachers to develop their expertise. Training or professional development refers to "study and polish of learning", which indicates that the teacher is a person who studies and studies the teacher. In order to develop the teacher's expertise, teacher training can be called "the process of the teacher's research and professional development"	
교수학습 과정안 **gyosuhagseub gwajeong-an** "instruction-learning process plan" (lesson plan)	A teacher's plan for teaching-learning a lesson and preparing students worksheets. The result of predicting students' reaction and thinking experiments about teaching and learning situation (classroom situation).	Lesson plan, which is recording the results of the teaching experiments containing the principles of the composition of the lesson and the expected responses of students, and the teacher's statements.

(Continued)

TABLE 19.2 (*Continued*)

Term	Description	Examples
교육과정 수업 평가의 일체화 **gyoyuggwajeong sueob pyeong-gaui ilchehwa** "curriculum lesson assessment integrating" (alignment of curriculum, instruction and assessment)	Currently, the Korean curriculum is revised every 3–5 years as the format of revision. Making consistency the standards of achievement in curriculum documents with the content of lessons and their assessments.	Consistent with the standards of achievement in curriculum documents with the content of lessons and their assessments.
학생활동중심 수업 **hagsaenghwaldong jungsim sueob** (student activity-centered class)	A class similar to a learner-centered or student-participatory class but focused more on students' engaging in math activities.	Class that focuses on the activity of finding the area by dividing the given graph directly to know the concept of integral
학생이 칠판에 풀이적기 **hagsaeng-i chilpan-e pul-ijeoggi** (students' writing solutions on blackboard)	An activity in which one or more students come up and write their own solutions to a given problem.	Showing problems, "graph $y = -2x + 1, y = 3x-1$" on the screen, two students come out and graph each of them on the blackboard.
학생 사고 수집하기 **hagsaeng sago sujibhagi** (collecting student thinking)	Pedagogical moves that collect mathematical thinking and misconceptions that students have.	Teacher: What is an irrational number? Student 1: Root 2 Teacher: What else is an irrational number? Student 2: some point some some Teacher: Again? Is there anyone else who can explain the irrational numbers? Student 3: a number that is not rational. The teacher writes the students' responses on the blackboard and uses them for teaching.
학습자 중심 수업 **hagseubja jungsim sueob** (learner-centered class)	A class in which the learner wishes to discover, understand, or organise his/her learning content on his/her own. This term is used in terms that contrast with the class in the teacher-centered or delivery style lesson.	A class in which learners have ownership and accountability for their own learning, by organising and understanding knowledge through peer mentoring, discussion learning, exploring concepts.

Term	Description	Examples
학습목표 재확인 **hagseubmogpyo jaehwag-in** "learning goals reconfirming" (reconfirming learning objectives)	Teacher restates learning goals or reconfirms whether students know learning goals during a lesson so that students can focus on the learning objectives of the lesson and their achievements of the learning goals.	"We learned the geometric meaning last time. While it's important to know the concept, it's also important to use it. So we're going to practice how to apply what we've learned on these graphs today. So, think about what you need to learn today, think about what your goals are. "Or, "Today, we are going to know how to make connections between algebraic expression and graphs of this function that we learned yesterday. Remind that it was our goal of today's lesson."
학습목표 제시 **hagseubmogpyo jesi** "learning objectives presenting" (introducing and engaging in lesson goals)	Activities that teacher guides or helps students recognise learning goals that is, what the learner should be able to do after completing the activity or assignment.	Teacher: "Today's learning goals is that "you can explain the slope of a graph of linear functions. After investigating the graphs on the activity sheet, let's study how the slope appears in graphs and algebraic expressions."
핵심정리 **haegsimjeongli** (summarising core concepts)	To summarise the main points or key points of what is learned in the current class	Teacher: "Today, we've learned about prime factorisation. To sum up, a prime number is a number that has only one and itself as a divisor among natural numbers greater than one. Prime factorisation is the product of natural numbers as a product of prime numbers.
핵심정리 같이 읽기 **haegsimjeongli gat-I ilg-gi** (reading core theorem together)	Reading aloud the contents of the textbook in class together.	Teacher: (to the whole classroom) The content we learned today is at the bottom of page 72. Shall we read it together? "Students: (seeing the textbook) Of the natural numbers a prime number is said ~~"

(Continued)

TABLE 19.2 (*Continued*)

Term	Description	Examples
호명하기 **homyeonghagi** "calling students" name" (calling students)	Teacher's pedagogical move that calls for a student's name or student's number, and then asks him/her for an answer to a presentation or question. It is also behavior to call attention to.	To call students' names directly to make them solve a problem. For example,"Soomin, Seonghee and Wonseok, let's come out and solve the problem".
형성 평가 **hyeongseong pyeong-ga** (formative assessment)	A brief assessment at the end of the lesson to make sure students understand what they have learned.	Teacher: Today, we have learned how to draw a graph with slope and y-intercept. Now let's check whether you achieved the learning goals.
활동지 안내 **hwaldongji annae** (guiding activity sheet)	**Activity sheets** are learning materials that teachers use in class to restructure textbooks to help students construct mathematical concepts. **Guiding activity sheet** is teacher's guiding or explanation of how the contents of the activity sheets are related to the learning objectives and what activities should be done with the activity sheet.	Teacher: The content of this activity sheet is the slope of the linear function. Think about how the graph of the function will change with the change in slope and try to work it out for 10 minutes with your friends.
활동지/학습지 정리하기 **hwaldongji/hagseubji jeonglihagi** (wrapping-up activity sheet or worksheet)	Summarising or organising the mathematical concepts or contents of the activity sheets or worksheets students used in class.	
협력 학습 **hyblyeog hagseub** (collaborative learning)	Learning in which students work with each other around the same goal to achieve their learning objectives	"On your desk, there is the graph on B4 size paper. We're going to interpret this graph in groups. I learned about the geometric meaning and I have to write two things about the characteristics. Let's group together an interpretation of the graph and present it." Collaborating with each other to reach the same goal efficiently to do group assignments in order to achieve the same goal effectively.

Term	*Description*	*Examples*
이해 확인하기 **ihae hwag-inhagi** "understanding confirming" (checking comprehension)	The teacher checks the knowledge or understanding of students through various forms such as questions and quizzes.	"Student A? Do you understand what the teacher said earlier? Can you tell the difference between the combination and the sequence?
인성 교육 **inseong gyoyug** "humanism education" (humanising education or character education)	Education emphasises humanising or human characters such as caring, communication and collaborative competence among students; the educational vision that is emphasised to avoid knowledge-based schooling and to promote a well-rounded person.	Teacher: What's more important than knowing it alone is to get to know it with friends. Ask each other questions you don't know and explain what you know well.
자기장학 **jagijanghag** "self-supervising" (self-supervision)	Teacher reflects own practice for his or her development, or to attend a lecture at a research institution or conferences to develop his or her expertise.	
자기평가 **jagipyeong-ga** (self-assessment)	To look back and reflect their own learning.	Making a self-assessment, as in "Today, we have learned about the linear function, shall we check how much we understand ourselves?"
전체학습 **jeonchehagseub** "whole group learning" (whole group lesson)	A class structure consisting of a teacher's explanation, a full discussion, or learning activities for the entire class.	Teachers explains math concepts or lead discussion to the whole class. T: "Look at this relationship and the expression of the relationship. This relationship in which only one value of y is determined when the x value is determined. When there's this relationship between the two variables, we call it a function."
전개 **jeongae** "main body of lesson" (main phase of lesson)	A lesson structure which refers to key instructional-learning activities to achieve learning objectives.	Based on the contents related to the learning goals, the classes are conducted using conceptual descriptions, group activities, personal problem-solving activities, and presentation activities.

(Continued)

TABLE 19.2 (*Continued*)

Term	Description	Examples
정당화하기 **jeongdanghwa hagi** "justifying" (justifying)	Activity that claims that a mathematical proposition, mathematical idea, or owns' mathematical opinion is true or false	
전시 학습 개념과의 연결 **jeonsi hagseub gaenyeomgwaui yeongyeol** "previous lesson concepts connection" (connecting math concepts with previous lessons')	To link the concept of mathematics in the current lesson with the previously learned concepts.	
전시학습 상기 **jeonsihagseub sang-gi** "prior learning reminding" (reminding prior lesson)	Reminding students of concepts or content learned before and related to this lesson. This is usually seen at the beginning of the lesson.	Teacher The first time, you took a point on the coordinate plane, you drew a graph, and then you drew it with a parallel shift, right Today we're going to learn about the third way.
질문 **jilmun** (questions)	Questions the teacher asks students to find out if they know mathematical facts, as opposed to thought-provoking questioning. The answers to these questions are often procedural explanations or short answers.	The question students ask to see if they know the mathematical facts, such as "What is the next value?"
주의환기 **jooui hwangi** (recalling attention)	Teacher's pedagogical moves to make students' attention on the teacher or on tasks.	In a noisy classroom, teacher rings a bell at the corner of a blackboard. The noisy students stop talking and focus on the teacher. Or, after the group activities, the teacher "one, two, three" and the students concentrate on the class with three applause.
중간/기말고사 **jung-gan/gimal gosa** (midterm/final exam)	In order to evaluate students' learning in a school, students take examinations in whole subjects for almost a week. Subject Teachers of the school make problems and grade the examinations together.	An assessment conducted to assess learning levels and overall learning of almost all subjects in the mid-term and end-of-term of a semester.

Term	Description	Examples
말하기 평가 **malhagi pyeong-ga** "speech evaluation" (evaluating verbal communication)	Assessing student's presentation or verbal explanation of math-related content (e.g, math problem-solving, math posters, etc.).	After writing a statistics poster, teacher assesses the process of explaining the purpose, data collection method, interpret the results, etc., and evaluates whether the expression is mathematically fluent or accurate or logical, etc.
멀티미디어 수업 **meoltimidieo sueob** (multimedia lesson)	It refers to classes that utilise multimedia instructional-learning materials such as videos	Teaching using video related to mathematical concepts such as EBS Math
모둠활동 **modumhwaldong** (small group activity)	A lesson structure in which two or more students explore, discuss, and/or solve problems in a group.	"If you look at your desks, there are three graphs on B4 sized paper. Since we have nine of them, we're going to take one of them and interpret the graph. Let's discuss it among the group members and try to interpret the graph, thinking about the geometric implications that we learned last time." Students explore graphs in a small group.
문제 **munje** "problems" (problem)	A refined form of question for the exercise of learned concepts and procedures to check students' achievement. It may include multiple-choice questions, essay types of questions.	
문제 풀기 **munje pulgi** (problem-solving)	After learning a concept, solving similar problems requiring using same skill and concept. These days, teachers have refrained from doing **this (problem-solving)** in the lesson.	To solve a problem such as "arrange y = 3x + 1, y = −2x + 4, y = x-2 in order of the large slope."
문제읽기 **munjeilg-gi** (reading problems)	An activity that reads the tasks or problems presented together.	"Let's read problem number 2, page 72 together."
문제유형 제시 **munjeyuhyeong jesi** (providing types of problems)	Informing test problems by categorising similar types of problems.	
난이도 제시 **nan-ido jesi** "providing a degree of difficulty" (informing difficulty of problems)	Informing the difficulty of problems and concepts by the teacher.	To be able to provide how difficult problems would be, such as "What we're going to learn today is a little bit deeper than what we learned last time."

(*Continued*)

TABLE 19.2 (*Continued*)

Term	Description	Examples
논의하기 **non-uihagi** (discussing)	Activities (speaking) that share a variety of ideas and ways to solve math problems, activities, and tasks.	Sharing the problem-solving process, and sharing the strengths and weaknesses of the various problem-solving processes.
논술형 **nonsulhyeong** "essay style" (mathematical essay problems)	Commonly referred to as a form of question that is not multiple choice or short answer.	In case of a middle-school exam, an essay type question such as, "Write your answer whether the outer and inner points of a triangle are the same, and write your reason as well."
노트필기 **noteupilgi** "note-taking" (note-taking)	Students take notes summarises, problem-solving or other contents learned during lesson on their notebook. Usually performed by students within the time allotted by the teacher in class.	Teacher Write down the contents of the blackboard in your notes, and summarise the gradient of the linear function graph in your notes.
판서 **panseo** (writing on blackboard)	Writing an idea and an additional explanation on the board for showing to the whole class. Teacher usually contemplate and plan how to record mathematical concepts and explanations effectively to the whole class.	"Let's figure out the slope in terms of the definition of slope. What's the definition of slope? [Write it on the blackboard on one corner] Let's draw this graph, and that graph on the blackboard. [Draw each graph in an organised way]" These records on the blackboard are written for all students to see well.
상황설명 **sanghwang seolmyeong** "situation explanation" (explaining contexts of problems)	Describing the outside contexts of mathematical concepts that relate to the problem situation.	"We need to find out when Chul-soo and Min-su can meet. But now Chul-soo is walking, and Min-su is riding the bus." It has nothing to do with solving the problem, but it's about explaining the external situation of the problem.

Term	Description	Examples
설명하기 **seolmyeonghagi** (explaining)	An activity in which a teacher explicitly tells students the meaning of a concept and/or strategies of problem-solving; or students tells how oneself solves a problem.	Explaining the definition of the slope by comparing the skewed extent of two graphs and providing examples such as, T: "when you draw a graph of y = 2x + 1, y = 3x + 1, you can find that when the value of x increases by 1, the value of y increases by 2 in the graph of y = 2x + 1. And when the value of x increases by 1, the value of y increases by 3 in the graph of y = 3x + 1. At this point, the greater the increase in y, the faster the slope becomes."
선행학습 **seonhaeng hagseub** "preceding learning' (accelerated learning)	Preceding learning the contents before they learn in school especially in cram schools.	Before having a lesson in school, learning the contents from a private educational institute or private tutoring in advance.
스토리텔링 수학 **seutolitelling suhag** "storytelling math" (storytelling mathematics or contextually rich mathematics)	A mathematical learning activity that provides narratives to students in the process of mathematics learning through stories, in order to provide them with a practical context in mathematics learning.	"I'm going to find out about the log today. Why was the log created? Sixteenth-century mathematician Napier wondered if there was a simple way for people to calculate large numbers. It's the logarithm that was invented." Before learning math concepts, it's about telling relevant stories and drawing attention.
시험범위 제시 **siheombeom-wi jesi** "a range of exam providing" (providing a range of exam)	Presenting a range of learning points to be included in various tests, such as midterms and final exams that the teacher constitutes.	"This midterm exam will be covering from the front to 120p on textbook"
실행연구 **silhaeng yeongu** (action research)	Classes to perform for the purpose of study of effective of teaching methods or measurement of educational effectiveness about one's teaching.	

(Continued)

TABLE 19.2 (*Continued*)

Term	Description	Examples
실생활 맥락 도입 **silsaenghwal maeglag doib** "real life context introducing" (connecting to real life contexts)	Guiding students by connecting the concept of mathematics to the context of real life.	"The parabolic path from which the window flies becomes a secondary function. So, if you can draw a secondary function, you can get the distance to the point where the window flew, right?" Explaining when concepts are used in everyday life, as shown in "and guiding concepts.
소집단 점수 **sojibdan jeomsu** (group scores)	When students work in small group, a teacher scores and quantifies the excellence of small group activities by criteria.	"Group 1 and 3 got them scores because of their good activity."
수업 **sueob** "instruction-learning" (lesson)	**Lesson** commonly includes both teaching activities and learning activities. It is used as "teachers **do(conduct)** lessons" or "students **receive** lessons" in Korea.	
수업 참여도 **sueob cham-yeodo** "class participation rate" (class participation)	The degree to which student represents the level of participation in the class, including presentations, questions (as students do), activeness in group activities.	
수업 참관 **sueob chamgwan** "lesson observation" (classroom observation)	Observing classroom for the purpose of lesson analysis, lesson improvement, and/or research. Types of classroom Observation include the purposes of supervision guidance, observing curricular experts' lesson, teacher aspirants or practitioners, parents or local personnel, and research.	
수업 장학 **sueob janghag** "lesson supervision" (instructional supervision)	Professional technical advising activities to effectively achieve the learning objectives of lesson. Instructional advice aims to improve teaching practice and solve problems related to pedagogy with the aim of improving student learning.	
수업 평가 일체화 **sueob pyeong-ga ilchehwa** "lesson assessment integrating" (alignment of instruction and assessment)	To make the consistent between lessons and evaluations. Occasionally, for the sake of discernment of evaluations, there were cases where evaluation items were very difficult with the content that was not taught. Refraining from this and encouraging teachers to make assessment items from the content of the lessons.	After learning the similarity ratio, to make consistency between the lesson with the contents of an assessment by providing problems that can be solved using the ratio of similarity.

Term	Description	Examples
수업연구 **sueob-yeongu** "instruction researching" (researching instruction)	Researching instruction includes all practical action studies that relate to the pedagogical principles, instruction-learning activities, and curriculum in order to improve instruction. This researching instruction is observed by school administrators and supervisors in school district.	
수업의 마무리 **sueob-ui mamuri** "lesson finale" (lesson finale; concluding phase of lesson)	A lesson structure that teacher and students reflect and summarise mathematical concepts on what students have learned at the end of the lesson. The teacher commonly assigns time for the **lesson finale** in the lesson plan so that students explicitly know what they have learned today and what they will learn in the next lesson related to today's lesson.	Teacher What we wanted to learn today was to use the differential meaning and the geometric meaning of the average rate of change to mathematically characterise the graph. So, we interpreted each graph, right? So, we had the time to find the differential meaning and the geometric meaning of the average rate of change and to present it mathematically.
수행평가 **suhaeng pyeong-ga** (performance assessment)	One of the encouraging evaluation methods, in contrast to the paper-pencil based assessment, which is a result-based or knowledge-based assessment that evaluates the students' progress and results.	Follow the instructions to carry out the survey activities and write a report based on the survey results and your analysis.
수학 동아리 **suhag dong-ali** (math club)	A form of community activity in which students participate autonomously without grade or class classification in each school, mainly playing math games, exploratory activities related to mathematics and organising math booth at science festival.	
수학교과역량 **suhag gyogwa yeoglyang** "mathematical subject competency" (mathematical competency)	Subject-specific competencies refers to the ability to perform in terms of knowledge, skills, values, and attitudes that are specific to the learner's ability to show when he/she learns a particular subject. Mathematical competencies refer to the ability to perform mathematics including the intellectual aspects, skills, values, and attitudes necessary to learn mathematics.	"With this project, we can develop our information processing skills by using our creative fusion capabilities and engineering tools."

(*Continued*)

TABLE 19.2 (*Continued*)

Term	Description	Examples
수학의 활용 제시 **suhag-ui hwal-yong jesi** "mathematical application providing" (providing application of mathematics)	Providing information and examples of how math can be used	"Why do we learn calculus? We go through a lot of changes in our lives. Running car speed, price fluctuations, seasonal changes, etc. There's a lot of order and rules involved, and it's a tool that finds the order and rules from increasing and decreasing states and deals with them mathematically is a function, and dealing with changes in that function is calculus.
수학 용어 정의하기 **suhag yong-eo jeong-uihagi** "mathematical terms defining" (defining mathematical terms)	The activity of making mathematical definitions of a term when it is first introduced into a textbook or class, or identifying the exact mathematical meaning of the term already in use;	"When point P on the line AB is AP:BP = m:n, point P is called line AB internalising to m:n. "The internal division is divided inside with a rate," explaining gives definition.
수학과 진로 지도 **suhaggwa jinlo jido** "mathematics related career guiding" (career guiding related to mathematics)	Informing and guiding students on how to connect to their future career related to mathematics learning.	The act of motivating students whose future goal is to become an architecture by guiding them to study mathematics because they need the ability to measure and calculate various shapes.
수학적 모델링 **suhagjeog modelling** (mathematical modeling)	An activity that students solve problems mathematically by representing real-life situations using mathematical concepts and language, and interpreting and applying them to real-life contexts.	"Now that we have to find the rate of change for the time of this object, what is the rate of change for time? The rate of change over time indicates how much the position of an object changes over time, right? So if you differentiate the position from time, you'll see a rate of change in time. So if we can differentiate between the time t that we're looking at here, we can measure the amount of time this object changes over time."

Term	Description	Examples
수학적 성질 탐구 **suhagjeog seongjil tamgu** "mathematical properties and features exploration" (investigating mathematical properties)	Exploration activities that reveal implicit mathematical properties and characteristics in a given task or problem situation	After learning about parallelograms, trapezoids, rectangles and squares, students explore the properties of these shapes and explore their inclusive relationships.
수학적 의사소통 **suhagjeog uisasotong** (mathematical communication)	Representing mathematical knowledge or ideas, the results of mathematical activities, problem-solving processes, beliefs and attitudes in mathematical words, texts, pictures, and symbols, and understanding of others' mathematical ideas.	An activity in which students share opinions among small group members or through a whole classroom discussion. For example, "Why don't we draw a perpendicular line?" "I think it's better to draw it in BC, not there."
순회지도 **suhoejido** (circuit instruction or circuit guiding)	When students do group and individual work in class, the teacher is walking through the classroom to check the student's learning status, paying attention to the students' activities and helping the students/groups in need.	Walking around in groups, teacher's response from each group or guided according to the level of activity. In the first group, teacher said, "Where is it? So, on this graph, you can draw the line like this." In the next group, "Comparing this with this point, the increase in x is a, the increase in y is b?"
탐구활동 **tamgu hwaldong** "inquiry and exploration activity" (exploration activity)	In contrast to knowledge transfer-oriented activities through teacher explanations, exploration activities are which students explore and use knowledge to construct, understand, and apply concepts on their own.	The activity in which students explore formulas of a sum of the power of natural numbers by using blocks.

(Continued)

TABLE 19.2 (*Continued*)

Term	Description	Examples
탐구한 성질 정리하기 **tamguhan seongjil jeonglihagi** "wrapping up and organising inquiry and exploration activity" (organising inquiry features)	The activity of organizing and summarizing in one's own language the mathematical properties, features and mathematical concepts that student(s) has discovered or understood from inquiry and exploration activity.	"Writing down in your own words on your notebook what you discovered from our inquiry activity using a term "slope"."
태도 점수 **taedo jeomsu** (attitude scores)	Scoring students' attitude or degree of participation of the class and reflect it in the evaluation.	"Group 2 had a great presentation, so I'm going to give them an attitude score."
토론하기 **tolonhagi** "debating and discussing" (making arguments)	Activities of speaking that justifies one's position on mathematically different ideas.	Activities in which students assert their ideas on topics such as whether the imaginary number is greater than or smaller than zero, or whether the comparison is possible.
또래교수/동료 멘토링 **ttolaegyosu/ donglyo mentoling** (peer-instruction/ peer-mentoring)	In class, activities in which students explain(teach) math to each other.	An activity in which a student explains mathematical knowledge to a friend, such as leaving three in group and going one after another, explaining to a mate, etc.
연구수업 **yeongusueob** "researched lesson" (research lesson)	Research lesson is conducted for the purpose of both improving teachers' instruction and students' learning by investigating lesson materials, planning lesson, observed the lesson by colleagues, and reflection on the lesson. Research lesson observation is usually attended by colleagues, vice principals, and principals.	
융복합 수업 **yungboghab sueob** "convergence and fusion lesson" (interdisciplinary integration lesson)	A class that links and blends various curricular or other subjects.	To know the principles of the music system through investigating "the rate of numbers" and to draw and play music, which is convergent between math and music.

TABLE 19.3 Korean Lexicon – Terms by category.

Kyosu-haksup Hwaldong (Instruction-learning activity)	answering, calling students, checking comprehension, circuit instruction, collecting student thinking, discussing, driving student presentation, encouraging, explaining, explaining contexts, exploring activity, guiding activity sheet, informing about next lesson, introducing learning goals, listening, making argument, motivating, note-taking, organising inquiry features, peer-mentoring, problem-solving, questioning, reading a core theorem together, reading a problem, recalling attention, reconfirming learning objectives, rewarding (individual rewarding, group rewarding), scoring, selecting a presenter, summarising core concepts, thought-provoking questioning, solving individual worksheet, wrapping up activity/work sheet, writing down, writing on blackboard
Suhak Kyosu-haksup Hwaldong (Explicit strategy related to mathematics instruction-learning activity)	career guiding related to math, concepts and principles focused learning, connecting math concepts with previous lessons', connecting to real-life contexts, defining mathematical terms, discovering learning, EBS math, inductive investigating, informing difficulty of problems, investigating mathematical properties, justifying, making connections among ideas, mathematical communication, providing application of math, providing types of problems, storytelling math, mathematical modeling, students' writing solutions on blackboard
Kyo-sil Soo-up Gae-sun (Improving instruction-learning practice)	action research, classroom observation, instructional supervision, peer-supervision, research lesson, researching instruction, self-supervision, teachers' council in schools
Soo-up Hyungtae (Lesson format)	activity-centered lesson, collaborative learning, convergence lesson, flipped learning, learner-centered lesson, lecture-style lesson, multimedia lesson, presentation lesson, technology-enhanced lesson
Pyung-ga (Assessment)	assignment assessment, attitude scores, class participation, evaluation of communication, evaluation of performance, formative assessment, group scores, mathematical essay problems, mid-term and final exam, peer-assessment, problems, process-focused assessment, providing a range of an exam, self-assessment
Soo-up Gu-jo (Lesson structure)	introductory phase of lesson, lesson finale, main phase of lesson, reminding prior lessons, small group activity, whole group learning
Soo-up Junbi (Lesson preparation)	lesson plan, reconstituting textbook, researching lesson materials
Ki-ta (Others)	accelerated learning, after school, alignment of curriculum, alignment of lesson and assessment, humanising education, lesson, lesson and assessment, math club, mathematical competency, individual learning

Acknowledgement

This work was funded by the 2018 Hongik University Research Fund.

Bibliography

Presentations and Papers

The work of the Korean national team has been presented locally and internationally at the following meetings and conferences:

- International Group for the Psychology of Mathematics Education (PME) Annual Conference 2018
- The International Conference of the Korean Society of Mathematics Education 2018
- The Korean Society of Educational Studies in Mathematics 2018

A selection of peer-reviewed publications from the Korean research team include:

Cho, H. & Kim, H-j. (2018). Identifying and documenting pedagogical terms in Korean mathematics classroom. *Journal of Korea Society Educational Studies in Mathematics: School Mathematics*, 20(3), 463–481.
Kim, H-j. & Cho, H. (2018). The international classroom lexicon project: Identifying and documenting Korean middle school mathematics classroom practices. In E. Bergqvist, M. Österholm, C. Granberg, & L. Sumpter (Eds). *Proceedings of the 42nd Conference of the International Group for the Psychology of Mathematics Education* Vol. 5 (p. 86). Umeå, Sweden: PME.

20

A LEXICAL SNAPSHOT

An investigation into the evolving terminology of middle-school mathematics teachers in the United States

Tracy E. Dobie, Miriam Gamoran Sherin and Sarah L. White

Introduction

Language plays an important role in regulating individuals' experiences and interactions while navigating the world (Cole & Engeström, 1993). The language that different communities develop provides information about what is meaningful and important for those groups to describe. In the United States, researchers have highlighted the lack of a "common technical vocabulary" (Lortie, 1975, p. 73) for describing teaching and called for the development of a framework that includes "well-defined common terms for describing and analysing teaching" (Grossman & McDonald, 2008, p. 187). A first step in establishing such shared vocabulary is documenting the current terminology available to teachers in the United States for discussing their practice. This chapter describes initial work to do just that.

The research presented here takes place as part of The International Classroom Lexicon Project, composed of teams of researchers and teachers across ten countries. Since vocabularies for describing teaching and learning vary across cultures (Clarke, 2010, 2013), a primary motivation of the project is to identify similarities and differences in the language available to teachers in a range of countries. Doing so allows us to explore what aspects of a teacher's practice are named across countries, as well as notable features of that language. In creating such documents for comparison, we, of course, also have an opportunity to gain a deeper understanding of the language for describing middle-school mathematics classrooms in our home countries.

In this chapter, we share the work of the United States team in documenting local teachers' terminology for describing what happens in their classrooms. We began this work knowing that it would not be feasible to develop a single lexicon that represents the entirety of the United States. Rather than attempting to create a comprehensive lexicon, our goal was to document some of the language that is widely familiar to teachers and to understand more about the importance of language for teachers. This version of the U.S. Lexicon highlights a first set of terms that emerged as widely familiar among a group of mathematics teachers in grades 6–9 (students aged 11–15 years old). Certainly, not all teachers use – or even know – all of these terms, and there are of course terms in widespread use that are absent from our lexicon, as well as terms that are not widely used but are particularly important among certain groups of teachers. However, we believe that this first step in our work provides a snapshot of teachers' vocabulary that allows us to raise questions about what teachers do and do not name, and to reflect on the development and trajectory of language for teaching and learning in the United States.

The need for a lexicon

As we consider future directions for teaching in the United States, it is important to begin with an understanding of where we are now and how teachers conceive of their practice. Since there is a strong connection between thinking and language use (Vygotsky, 1934/1987), exploring teachers' professional language is one way to learn about their practice. By unpacking the language that teachers use to describe what happens in their classrooms, we can gain insight into what events and activities rise to the level of being "nameworthy" (Carroll, 1980, p. 321), including what terms have been sustained over time and what new concepts and vocabulary have gained recent widespread acceptance among teachers.

Research on teacher learning communities has emphasized the importance of discourse in helping teachers to grow and improve their practice (e.g., McLaughlin & Talbert, 2006). For example, van Es (2012) worked with a group of elementary teachers who reflected on video clips of their own mathematics classrooms. She found that as the teachers developed into a high-functioning learning community, they engaged in conversations that involved multiple participants, questioned each other and pressed each other's thinking, and were guided by discourse norms that encouraged inquiry and justifications grounded in evidence. While such research highlights the key role of discussion among teachers, there remains a lack of widespread professional language for teaching in the United States (e.g., Grossman & McDonald, 2008; Lortie, 1975).

Recently, some work has begun to identify shared language among teachers and consider the importance of having shared vocabulary. For example, Milewski and Strickland (2016) asked secondary mathematics teachers in the United States to reflect on the language they use to describe instructional practices that involve responding to student contributions. The teachers together developed a framework of this language and, using the framework, were subsequently able to more effectively shift their practice in the classroom. Such research highlights the potentially powerful role of specific shared language in a teaching community. Moving forward, more work is needed to uncover what that language looks like and how it emerges. It is our hope that the present research will contribute to the conversation around teachers' professional language and future directions for the language and practice of teaching in the coming years.

Features of the U.S. context

Before introducing our research methodology, we discuss three features of the educational context in the United States, highlighting challenges involved in developing a U.S. Lexicon and potential influences of these features on teachers' language development. First, schools come in many shapes and sizes across the United States. U.S. schools are situated in urban, suburban, and rural settings and, as such, serve a wide array of communities. These communities vary in terms of features such as racial composition and socioeconomic status, which have been shown to have a considerable impact on student achievement in the U.S. due to systemic inequalities in the education system (Rumberger & Palardy, 2005). Additionally, schools have access to very different kinds of resources – financial, material, and personnel (Gamoran, 2003) – leading to varied opportunities for teaching and learning across the country. Schools also vary in size and grade level distribution, the latter of which is particularly important when considering eighth grade, the focus of The International Classroom Lexicon Project. Eighth grade is sometimes included in an elementary school building with students from kindergarten through grade eight. In other cases, eighth grade is part of "middle school," which can include grades six to eight, seven and eight only, or seven to nine. These myriad school contexts provide different work environments for teachers (De Lima, 2008; Simon & Johnson, 2015), which may affect the language that teachers use in conversations with students, teachers, and others in the community.

Second, schools across the United States are not required to follow a national mathematics curriculum, though a majority have chosen to adopt the Common Core State Standards for Mathematics (National Governors Association Center for Best Practices, 2010). These standards specify the concepts and procedures to be covered at each grade level and introduce eight standards for mathematical practice that should be developed in students at all grade levels. Even with this widespread adoption of the Common Core State Standards, there is still a great deal of local control over how mathematics is taught. Individual school districts select which textbooks and supplemental materials to use, as well as how to sequence the mathematics content at each grade. In some cases, teachers do not use a specific published curriculum and instead have autonomy to choose from or adapt a wide selection of existing materials, or to develop new lessons themselves. In particular, the content covered in eighth-grade mathematics classes varies within schools as well as across schools. In many schools, the content that a student is taught at the eighth grade depends on the student's assessed ability (Gamoran, 2004), which dictates whether the student is placed in an advanced, regular, or remedial mathematics class. These class types typically differ in the composition of students, subject matter, and relative speed at which that subject matter is taught (Cogan, Schmidt & Wiley, 2001). Because different curricular materials may use different terminology, the language used in classrooms may differ, as well.

Third, differences in teacher education and professional development programs in the U.S. might contribute to the variety of language used by mathematics teachers. Requirements for teacher certification differ state by state. Furthermore, universities can satisfy these requirements using different courses and with different emphases (Cochran-Smith & Zeichner, 2009), which have been shown to affect the practices and behaviors that teachers attend to in their everyday practice (Grossman, 1990). Beyond teacher preparation programs, teachers in the U.S. have different opportunities for collaboration and professional development that may affect their professional lexicons. For example, a school-based professional development program may equip teachers within that school with shared language and practices (Borko, Jacobs & Koellner, 2010). Outside of these within-school communities, U.S. teachers may also be members of mathematics teacher associations, such as the National Council of Teachers of Mathematics (NCTM), and thus use terminology put forward by those organizations. Increasingly popular among U.S. teachers is communication through online platforms such as Twitter, which have their own conversation norms (Lemon, 2016; Carpenter & Krutka, 2014). Access to and participation in these different communities may affect teachers' familiarity with language through the emergence of communities of practice that develop their own terminology (Tusting, 2005).

Given the variety of school contexts, curricula, and professional communities in the U.S., it is unsurprising that the teaching profession currently lacks a common lexicon (Grossman & McDonald, 2008). Each of these factors likely diversifies the language that teachers use in their everyday lives and presents challenges for developing a shared lexicon. At the same time, we suspect that our efforts to identify elements of a U.S. lexicon will help us to better understand some implications of these factors.

Research methodology

Here we discuss the four phases of U.S. Lexicon development: initial generation of terms, local review and refinement, national review and refinement, and lexicon organization. In phase one, the U.S. team worked to develop an initial list of potential terms for the U.S. lexicon. First, two middle-school mathematics teachers and two education researchers (the first two authors) watched videos of eighth-grade mathematics lessons from eight countries[1] (the United States and seven other countries), identifying events and interactions that took place for which they had a name. Subsequently, approximately 20 former and current mathematics teachers brainstormed short lists of terms that they would use to describe what happens in middle-school mathematics classrooms. In alignment with all of the countries participating in The International Classroom Lexicon Project, we excluded terms that describe mathematical content areas (e.g., subtraction, algebra) and technical mathematical vocabulary (e.g., dividend, ratio, graph). To further narrow our focus, we also excluded terms that describe physical mathematical tools (e.g., ruler, pencil, calculator). Taking these restrictions into account, 157 terms were identified. To conclude this phase, the initial team of two mathematics teachers and two researchers composed definitions for the 157 terms.

The second phase of development included a process of local review and refinement to narrow the list of proposed terms and agree on definitions for the terms. The research team conducted three focus groups with middle-school mathematics teachers in the midwestern United States, who had a range of teaching experience. The first focus group included three teachers from different schools (a suburban public school and two urban private schools). The second focus group was composed of four mathematics teachers from a single suburban public school, and the third group was composed of four mathematics teachers who taught at an urban religious-affiliated private school.[2]

Each focus group meeting began with the teachers completing a Q-sort task (Block, 2008) in which they sorted potential lexicon terms according to whether each term was very familiar, somewhat familiar, or unfamiliar. Participating teachers then provided feedback on the definitions of those terms with which they were very familiar. In a second activity, teachers viewed definitions of terms that have multiple names and identified the term that each definition described, allowing us to determine the most familiar name. Finally, teachers were invited to propose additional terms for inclusion in the lexicon.

Focus group discussions were recorded and transcribed, and a spreadsheet was created to indicate the number of teachers who rated each term as unfamiliar, somewhat familiar, or very familiar. We then eliminated any terms that were very familiar to three or fewer of the 11 participating teachers in order to identify widely known terms. After removing the less familiar terms, which varied in type, we were left with 103 terms in the lexicon. Definitions of those terms were then edited by the researchers based on teachers' suggestions.

In the third phase of development, we distributed a national survey via email and social media, asking teachers to review a subset of terms in the proposed lexicon and rate their familiarity with and usage of each term. Teachers were also asked to recommend changes to the term definitions and invited to propose new terms. The survey was completed by 241 teachers. Of those who reported their gender identification, 73% ($n = 131$) identified as female and 27% ($n = 49$) identified as male. Seventeen percent ($n = 31$) reported 1–3 years of teaching experience, 46% ($n = 84$) reported 4–14 years of teaching experience, and 37% ($n = 68$) reported 15 or more years of teaching experience. Survey respondents were from 28 of the 50 states in the U.S., with 40% ($n = 67$) teaching in suburban schools, 36% ($n = 59$) in urban schools, and 24% ($n = 40$) in rural schools. To determine the final lexicon, we eliminated any terms that were not rated as extremely familiar or very familiar (1 or 2 on a 5-point scale) by at least 66% of responding teachers, the cut-off agreed upon by members of The International Classroom Lexicon Project. Additionally, we reviewed all feedback provided on the definitions of the terms and accepted those suggestions that we believed would help to increase the clarity of the definitions.

In the final phase of lexicon development, the two original researchers worked with a new member of the research team and a local teacher from one of our focus groups to organize the remaining 99 terms according to theme. This cyclical process was repeated with each member of the team weighing in until a stable set of categories was identified (Miles & Huberman, 1994). During this process of categorization, the team members decided that some terms have nearly equivalent claims to two categories. In such cases, the term is cross-listed in both categories, making the total number of terms listed by category greater than 99. To be clear, many terms in the lexicon serve multiple purposes; however, only terms that are strongly connected to two categories are cross-listed. Furthermore, we acknowledge that the categorization we present in this chapter is, of course, just one possible way to sort the terms. Nevertheless, we believe it reveals interesting features of the overall lexicon.

In presenting this U.S. Lexicon we wish to mention one caveat. It was quite challenging at times to determine the specific form of the word to use in the lexicon. For example, we heard teachers use both "scaffold" and "scaffolding" to describe one way that teachers support students. To address this issue, we chose to include in the lexicon the form of the word that teachers used most often – or reported using most often – in their interactions with us. Finally, we note that terms in the English language are gender-neutral. In the lexicon examples included in the companion chapter (Chapter 21), we also use the gender-neutral pronoun "they" to refer to the actors who enact different terms.

Exploring the lexicon

The lexicon we present here is an artifact of a particular time and process and thus represents only some of the terms that are widely familiar to middle-school mathematics teachers in the United States. In this section, we explore the structure, key features, and size of this U.S. Lexicon, connecting back with notable aspects of the U.S. context when relevant.

Structure

This U.S. Lexicon includes 99 terms organized into eight categories: Administration, Assessment, Classroom Climate, Discussion, Engaging in Learning, Engaging with Mathematics, Participation Structures, and Pedagogical Tools and Approaches. The category that houses the most terms (*n* = 23) is Pedagogical Tools and Approaches. This category includes terms that describe instructional techniques, strategies, tasks, and activities that mathematics teachers use in the classroom (e.g., "differentiation," "open-ended question"). Engaging in Learning has the second largest number of terms (*n* = 17), followed by Assessment (*n* = 16). Engaging in Learning includes terms that refer to students' actions and experiences when they are learning and are typically not specific to mathematics (e.g., "brainstorm," "critical thinking"). Meanwhile, the Assessment category includes terms that are related to formal or informal activities and approaches used to check students' knowledge or understanding (e.g., "study guide," "correct mistakes"). Next, 13 terms fall into the Engaging with Mathematics category, which includes terms that refer specifically to students' mathematics classroom practices (e.g., "justify," "compare multiple strategies"), while 12 terms categorized as Discussion describe students' and teachers' actions when engaging in conversation in the mathematics classroom (e.g., "clarify," "ask questions"). Finally, the category Classroom Climate includes 11 terms that refer to the atmosphere and culture of the classroom (e.g., "positive feedback," "respect"), while Participation Structures and Administration each include eight terms. In the Participation Structures category, terms describe how students and teachers are organized for learning (e.g., "partner work," "think-pair-share"), while terms in Administration refer to routines for managing the mathematics classroom (e.g., "take attendance," "distribute materials").

Key features

When we consider the range of terms included in this U.S. Lexicon, several features stand out to us: an emphasis on verbal communication; a focus on students, including student participation and engagement in activities; somewhat generic, non-mathematics-specific vocabulary; and a focus on assessment. We describe each feature below.

First, we note that a significant number of terms in this lexicon focus on verbal communication. There are 12 terms in the Discussion category that describe verbal contributions in mathematics classrooms, with particular attention to students' communication (e.g., "justify," "share," "prove"). Additionally, there is a strong emphasis on doing and discussing mathematics with others in the Participation Structures category, which includes terms such as "think-pair-share," "collaborate," and "group work." This emphasis on discussion and collaboration reflects both recent literature focused on the importance of orchestrating discussions among students in the mathematics classroom (Chapin, O'Connor, & Anderson, 2003; Smith & Stein, 2011) and the Common Core State Standards' mathematical practices, one of which focuses on students' ability to critique arguments and justify their thinking to others (National Governors Association Center for Best Practices, 2010).

Related, we notice many terms describing the work that students are doing in the classroom, which we suspect represents a shift from several decades ago. The Engaging in Learning and Engaging with Mathematics categories, in particular, focus on the many ways in which students participate in the mathematics classroom (e.g., "creative thinking," "modeling," "use manipulatives"), and the Participation Structures category describes different formats in which that participation occurs (e.g., "student presentation," "partner work"). Additionally, some terms in the Pedagogical Tools and Approaches category suggest a focus on teachers engaging students in activities and actively involving students in their learning (e.g., "hands-on activity," "problem-based learning," "investigation," "student-centered"). The presence of these terms might reflect the research and standards described above, as well as recent professional development initiatives focusing on student mathematical thinking (e.g., Stein, Engle, Smith & Hughes, 2015). Additionally, documents released by the National Council of Teachers of Mathematics – such as its *Principles to Actions* book, describe effective mathematics teaching practices that focus on eliciting student thinking, engaging students in meaningful discourse, promoting reasoning and problem-solving among students, and giving students a chance to grapple with mathematics (NCTM, 2014).

Third, we notice a rather generic, non-mathematics-specific focus of terms in this U.S. Lexicon. We only identified 13 of the 99 terms as being specific to mathematics, and even among those there was some debate about just how math-specific they were. For example, while we felt that actions such as "justify" and "make connections" happen in very particular ways in mathematics and so included them in the Engaging with Mathematics category, we recognize that one might engage in those actions in other subject areas, as well. Furthermore, the remaining 86 terms are all generic enough that they could be applied to other domains. Certainly, there is mathematical terminology describing specific content and tools, which is not included in this lexicon, but of the terms that describe the practices of learning and teaching mathematics, very few are specific to mathematics.

Finally, this U.S. Lexicon contains a range of terms related to assessment practices. Some terms refer to more formal methods of assessment, such as "test" and "quiz," while others refer to newer and more informal methods, such as "exit slip" and "formative assessment." Given the increased emphasis on assessment and testing in the U.S. over the last few decades (Shepard, 2000, 2015), the presence of such terms is unsurprising. However, as with all patterns noticed, questions remain about the degree to which these terms are used and practices are enacted as teachers carry out their daily mathematics practice.

Size

Lastly, we highlight the number of terms not only included in this lexicon but proposed as candidates. As previously mentioned, we started out with 157 potential terms (not including those that did not meet inclusion criteria). It was challenging to narrow down these terms, as nearly every term was very familiar to at least one of our focus group teachers. Additionally, during the national review and refinement phase, many more terms were proposed by teachers, and the research team continually identified terms that have recently taken hold in some teacher communities but are left out of this lexicon. Thus, a very large number of terms were identified by teachers and researchers throughout this process. Given the previously described wide range in school types and structures, adherence to the Common Core State Standards, classroom practices and expectations, and mathematics curricula used, it is perhaps to be expected that such a breadth of terms would be proposed. While this range of terms could be used to explore interesting variation between contexts, our task was to create a single lexicon illustrating common language across contexts. Below as we reflect on this process, we highlight some challenges associated with that endeavor and propose directions for future research.

Reflection

This chapter reports on the development of a lexicon that illustrates ways in which middle-school mathematics teachers in the United States discuss mathematics teaching and learning. The artifact that we have created and present in this chapter is not without limitations. We have learned a great deal throughout this process, and the challenges we have faced along the way have inspired new ideas about directions to pursue for future research. For example, one shortcoming of the chosen methodology is that we only included in our lexicon terms that are familiar to at least 66% of teachers, thus excluding terms that are especially important to particular communities of teachers even if not widely known. As a result, this lexicon documents the majority and, in some sense, perpetuates the status quo, rather than emphasizing specialized or newly emerging language, or the language of teachers whose voices are not typically heard. Given this gap, we are beginning to explore language that is important to specific groups of teachers. One starting place for this work is interviewing teachers who practice culturally relevant mathematics instruction (e.g., Ladson-Billings, 1995), a teaching approach that was developed to help teachers acknowledge the strengths and resources of African American students. By conducting these interviews, we hope to create a space where we can explore important terminology for teaching equitably that is currently left out of this lexicon and from which others might learn.

Another limitation of this lexicon relates to yet a different aspect of the methodology – the use of self-report data. Rather than asking teachers to report on the language they use, we might have actually observed teachers in their classrooms, meetings with colleagues, and school professional development sessions to capture the terminology used in those settings. This approach would tell us more about the language that teachers use in practice, while our methodology tells us more about the language that teachers view as familiar. In fact, we question whether some of the terms included in this lexicon are actually used by teachers in practice, or if teachers simply rated them as familiar because they could envision using the term to describe some actions they take. For example, teachers might frequently ask for student volunteers in their classrooms, but one might question whether "ask for a student volunteer" is truly a technical term that should be included in a U.S. lexicon of mathematics teaching and learning. Thus, while our existing work provides insights into teachers' perspectives on their own language use, future analyses might take a different approach that utilizes observations and audio or video recordings to capture teacher language use in the moment.

Despite the limitations of this lexicon, we believe it represents a first step in uncovering the shared professional language of middle-school mathematics teachers in the United States. At the same time, it raises new questions to consider: How might a lexicon of elementary or secondary mathematics teaching differ from this one? How can this lexicon be used, if at all, to help teachers improve their practice? What might we learn from asking teachers about the terms that are important to their practice, rather than the terms they are familiar with? How might this lexicon change in 5 years? 10? 20?

These questions inspire many interesting directions for future research. As mentioned previously, we view this lexicon as a brief snapshot of an evolving document. Thus, we expect new language to be uncovered and added, as well as to develop, moving forward. In fact, we have compiled a list of 20 additional terms that were suggested by multiple teachers during our national review and that appear from pilot testing to be widely familiar to teachers. As such, we can already see possibilities for expanding this lexicon. Additionally, moving forward we hope to use this lexicon to further reflect on our language use in the United States and the significance of professional language for teachers. As one example, we are beginning to explore the importance of language in teachers' noticing abilities (Sherin, Jacobs & Philipp, 2011). In this work, we inquire as to whether having a name for an action, idea, or other phenomenon helps teachers to see it happening in their classrooms. This question takes up one way in which we might use the lexicon in its current form, even if not comprehensive, to explore affordances of language for middle-school mathematics teachers.

Furthermore, as part of our work exploring teachers' usage of lexicon terms, we are investigating how language might differ depending on the medium through which one communicates. Given the increasing presence of online teacher communities, it is likely that written language is growing in importance and influence. While the present study solely examined teachers' verbal communication, we conducted another set of analyses that explores teachers' language use in written online contexts (Dobie & Sherin, 2021). To do so, we utilized the tools of machine learning and computational text processing (Baker & Yacef, 2009; Siemens & Long, 2011) to assess language use in journal articles written by teachers and teacher educators. Moving forward, we also hope to explore teachers' language use in digital venues such as Twitter, online professional learning courses, and blogs.

While these are a few directions we are currently exploring, there are certainly many more questions to ask of this lexicon, and of the role of professional language in mathematics teachers' experiences in the United States. We hope that our efforts will help to catalyze future research that investigates mathematics teachers' professional language and considers how a focus on language can help us to improve teaching in the United States.

Acknowledgements

This project was supported by the Alumnae of Northwestern University. We wish to thank all of the teachers who participated and made this work possible.

Notes

1 The eight countries included Australia, Chile, China, Czech Republic, Finland, France, Germany, and the United States. When the language spoken was not English, subtitles on the video were provided.
2 To be clear, while we chose to select participants from different school contexts, this was not due to any hypotheses on our part concerning how language differs across these spaces. In fact, part of what we intend to emphasize in this chapter is the difficulty in making claims about differences by school type or geographic region in the U.S., as many initiatives are decided at the state, city, district, or even school level, and there are differing levels of financial support, teacher training, and curriculum use within regions and school types. (For more information see, for example, Cohen and Spillane, 1992).

References

Baker, R., & Yacef, K. (2009). The state of educational data mining in 2009: A review and future visions. *Journal of Educational Data Mining*, 1(1), 3–17.

Block, J. (2008). *The Q-sort in character appraisal: Encoding subjective impressions of persons quantitatively*. American Psychological Association.

Borko, H., Jacobs, J., Koellner, K. (2010). Contemporary approaches to teacher professional development. In P. Peterson, E. Baker, & B. McGaw (Eds.), *International encyclopedia of education* (Vol. 7, pp. 548–556). Oxford, England: Elsevier.

Carroll, J. M. (1980). Naming and describing in social communication. *Language and Speech*, 23(4), 309–322.

Carpenter, J. P., & Krutka, D. G. (2014). How and why educators use Twitter: A survey of the field. *Journal of Research on Technology in Education*, 46(4), 414–434.

Chapin, S. H., O'Connor, C., & Anderson, N. C. (2003). Classroom discussions using math talk in elementary classrooms. *Math Solutions*, 11, 1–3.

Clarke, D. J. (2010). *The cultural specificity of accomplished practice: Contingent conceptions of excellence*. In Y. Shimizu, Y. Sekiguchi, & K. Hino (Eds.), *In search of excellence in mathematics Education (Proceedings of the 5th East Asia regional conference on mathematics education*: pp. 14–38). Tokyo, Japan.

Clarke, D. J. (2013). Contingent conceptions of accomplished practice: The cultural specificity of discourse in and about the mathematics classroom. *ZDM*, 45(1), 21–33.

Cochran-Smith, M., & Zeichner, K. M. (Eds.). (2009). *Studying teacher education: The report of the AERA panel on research and teacher education*. Mahwah, NJ: Lawrence Erlbaum.

Cogan, L. S., Schmidt, W. H., & Wiley, D. E. (2001). Who takes what math and in which track? Using TIMSS to characterize US students' eighth-grade mathematics learning opportunities. *Educational Evaluation and Policy Analysis*, 23(4), 323–341.

Cohen, D. K., & Spillane, J. P. (1992). Chapter 1: Policy and practice: The relations between governance and instruction. *Review of Research in Education*, 18(1), 3–49.

Cole, M. & Engeström, Y. (1993) A cultural-historical approach to distributed cognition. In G. Salomon (Ed.), *Distributed cognitions: Psychological and educational considerations* (pp. 1–46). New York: Cambridge University Press.

De Lima, J. A. (2008). Department networks and distributed leadership in schools. *School Leadership and Management*, 28(2), 159–187.

Dobie, T. E. & Sherin, B. L. (2021). The language of mathematics teaching: a text mining approach to explore the zeitgeist of US mathematics education. *Educational Studies in Mathematics*. https://doi.org/10.1007/s10649-020-10019-8.

Gamoran, A. (Ed.). (2003). *Transforming teaching in math and science: How schools and districts can support change*. New York, NY: Teachers College Press.

Gamoran, A. (2004). Classroom organization and instructional quality. In M.C. Wang & H.J. Walberg (Eds.), *Can unlike students learn together? Grade retention, tracking, and grouping* (pp. 141–155). Greenwich, CT: Information Age Publishing.

Grossman, P. L. (1990). *The making of a teacher: Teacher knowledge and teacher education*. New York, NY: Teachers College Press.

Grossman, P., & McDonald, M. (2008). Back to the future: Directions for research in teaching and teacher education. *American Educational Research Journal*, 45(1), 184–205.

Ladson-Billings, G. (1995). Toward a theory of culturally relevant pedagogy. *American Educational Research Journal*, 32(3), 465–491.

Lemon, N. (2016). Pre-Service teachers engaging with Twitter as a professional online learning environment. In I. Management Association (Ed.), *Professional development and workplace learning: concepts, methodologies, tools, and applications* (pp. 1434–1464). Hershey, PA: IGI Global.

Lortie, D. C. (1975). *Schoolteacher: A sociological study*. Chicago, IL: University of Chicago Press.

McLaughlin, M. W., & Talbert, J. E. (2006). *Building school-based teacher learning communities: Professional strategies to improve student achievement*. New York, NY: Teachers College Press.

Miles, M. B. & Huberman, A. M. (1994). *Qualitative data analysis*. Thousand Oaks, CA: Sage Publications.

Milewski, A., & Strickland, S. (2016). (Toward) developing a common language for describing instructional practices of responding: A teacher-generated framework. *Mathematics Teacher Educator*, 4(2), 126–144.

National Council of Teachers of Mathematics (2014). *Principles to actions: Ensuring mathematics success for all*. Reston, VA: National Council of Teachers of Mathematics.

National Governors Association Center for Best Practices, Council of Chief State School Officers. (2010). *Common Core State Standards in Mathematics*. Washington, DC: National Governors Association Center for Best Practices, Council of Chief State School Officers.

Rumberger, R. W., & Palardy, G. J. (2005). Does segregation still matter? The impact of student composition on academic achievement in high school. *Teachers College Record*, 107(9), 1999–2045.

Shepard, L. A. (2000). The role of assessment in a learning culture. *Educational Researcher*, 29(7), 4–14.

Shepard, L. A. (2015). If we know so much from research on learning, why are educational reforms not successful? In M. J. Feuer, A. I. Berman, and R. C. Atkinson (Eds.), *Past as prologue: The national academy of education at 50* (pp. 41–51). Washington, DC: National Academy of Education.

Sherin, M., Jacobs, V., & Philipp, R. (Eds.). (2011). *Mathematics teacher noticing: Seeing through teachers' eyes*. New York, NY: Routledge.

Siemens, G., & Long, P. (2011). Penetrating the fog: Analytics in learning and education. *Educause Review*, 46(5), 30–32.

Simon, N. S., & Johnson, S. M. (2015). Teacher turnover in high-poverty schools: What we know and can do. *Teachers College Record*, 117(3), 1–36.

Smith, M. S., & Stein, M. K. (2011). *Five practices for orchestrating productive mathematics discussions*. Reston, VA: National Council of Teachers of Mathematics.

Stein, M. K., Engle, R. A., Smith, M. S., & Hughes, E. K. (2015). Orchestrating productive mathematical discussion: Helping teachers learn to better incorporate student thinking. In L.B. Resnick, C.S.C. Asterhan, and S.N. Clarke (Eds.), *Socializing intelligence through academic talk and dialogue* (pp. 357–388). Washington, DC: American Educational Research Association.

Tusting, K. (2005). Language and power in communities of practice. In D. Barton and K. Tusting (Eds.), *Beyond communities of practice: Language, power and social context* (pp. 36–54). New York, NY: Cambridge University Press.

van Es, E. A. (2012). Examining the development of a teacher learning community: The case of a video club. *Teaching and Teacher Education*, 28(2), 182–192.

Vygotsky, L. (1987). *Thought and language*. Cambridge, MA: MIT Press. (Original work published 1934).

21

UNITED STATES LEXICON

Tracy E. Dobie, Miriam Gamoran Sherin, Sarah L. White and Katie M. Mayle

A United States Lexicon

The lexicon presented in this chapter consists of 99 terms that were widely familiar to middle-school mathematics teachers in the United States. The authors, along with several teachers, developed operational definitions for each term, as well as brief examples. Note that when an individual student or teacher is referenced in an example, the gender-neutral pronoun *they/their* is used to refer to that individual. Certainly, these examples depict only one possible enactment of the term; many additional examples are possible. Similarly, a categorization scheme for the terms is provided, though we acknowledge that the terms could be organized in other ways.

The first lexicon table (see Table 21.1) lists only the terms in this lexicon. The second lexicon table (see Table 21.2) includes additional information arranged into three columns. The first column lists the **Term** itself, the second column provides a **Description,** and the third column presents **Examples** of what it might look like to see the term in action in the classroom.

To help make sense of the types of terms included in this lexicon, the authors engaged in an iterative process to organize the terms (see Chapter 20 in this volume for additional details), which resulted in the following eight categories: Administration (8 terms), Assessment (16 terms), Classroom Climate (11 terms), Discussion (12 terms), Engaging in Learning (17 terms), Engaging with Mathematics (13 terms), Participation Structures (8 terms), and Pedagogical Tools and Approaches (23 terms). Note that some terms were determined to have nearly equivalent claim to two categories and so are cross-listed and italicized (see Table 21.3).

TABLE 21.1 United States Lexicon – Terms.

access prior knowledge	agenda	"aha" moment	answer questions
ask for student volunteer	ask questions	assess	assign homework
assign seats	brainstorm	build rapport	calculate
challenge	check answers	clarify	classroom environment
classroom management	collaborate	compare multiple strategies	confusion
correct mistakes	creative thinking	critical thinking	differentiation
distribute materials	effort	establish routines	example
exit slip	explain	explore	extra credit
formative assessment	games	give directions	go over answers
go over homework	group work	growth mindset	guess and check
guided notes	guided practice	hands-on activity	high expectations
hint	homework check	investigation	justify
listening	make connections	mastery	math practices
memorize	modeling	note-taking	observe
offer feedback	open-ended question	partner work	pattern recognition
positive feedback	practice	pre-assessment	problem-based learning
problem-solving	prove	quiz	real-world connections
reasoning	redirect	reflection	remediation
respect	review	scaffolding	self-advocate
share	show	skills practice	struggling
student accountability	student as teacher	student presentation	student strategies
student-centered	study guide	take attendance	test
testing	think-pair-share	thinking	try
use manipulatives	wait time	warm-up	whole-class discussion
word problem	worked example	writing	

TABLE 21.2 United States Lexicon – Terms with operational definitions.

Term	Description	Examples
access prior knowledge	Draw out from students what they already know.	The teacher asks students to share what they have previously learned about adding fractions.
agenda	Plan for the day's lesson, sometimes presented to students at the start of the lesson.	The teacher writes on the board an ordered list of activities to be completed during that day's lesson.
"aha" moment	Sudden discovery made by a student that indicates new understanding.	After grappling with a concept, a student has an epiphany about the work they are doing and better understands the material.
answer questions	Respond to questions posed by either a teacher or a student.	A student raises their hand to respond to an inquiry the teacher posed to the class.
ask for student volunteer	Make a request for a student who is willing to participate, usually to answer a question or come up to the board.	When it is time to discuss solutions to a problem, the teacher asks if anyone would like to share their solution with the class.
ask questions	Attempt to elicit information from others by posing inquiries, rather than making statements.	As the teacher solves a problem at the board, a student raises their hand and inquires, "Why did you multiply both sides of the equation by twelve?"
assess	Evaluate what a student knows or can do.	While students are working independently, the teacher observes their work to determine their knowledge of the content.
assign homework	Tell students which problems or work they need to complete prior to the next class.	The teacher tells students to finish a worksheet at home before tomorrow's class.
assign seats	Direct students to sit at specific desks or tables.	The teacher tells students where to sit in class based on who the teacher thinks would work well together.
brainstorm	Propose a range of new ideas or approaches for solving a problem.	To start a unit on surface area and volume, the teacher asks students to think of as many three-dimensional shapes as they can.
build rapport	Develop relationships with and among students to help everyone feel comfortable with and respected by each other.	The teacher talks to students about their plans for the upcoming school break to show that the teacher cares about the students outside of the classroom.
calculate	Compute using mathematical operations.	Students multiply to find the total amount of money earned from ticket sales for the school play.

Term	Description	Examples
challenge	Demanding or stimulating task that often encourages students to think differently.	A student who has quickly mastered how to calculate perimeter and area is presented the more difficult task of maximizing the area of a rectangle with a given perimeter.
check answers	Student determines whether answers are correct, often by comparing with another student, the teacher, or an answer key.	After students complete an in-class activity, the teacher provides a list of solutions so that the students can verify whether they solved the problems correctly.
clarify	Provide additional detail or alternate wording to make a statement or question more easily understandable.	A student helps their classmates understand their earlier comment better by saying, "What I meant was..."
classroom environment	Social and emotional climate of a classroom. Often refers to a positive feeling in the classroom that teachers try to create.	A teacher builds a positive connection with students by asking them about their interests outside the classroom. The students recognize the caring attention the teacher gives everyone and attempt to treat their classmates in the same respectful manner.
classroom management	Efforts by teachers to ensure students act in accordance with expectations.	If students become distracted during a whole-class discussion, the teacher redirects their behavior to get them back on track.
collaborate	Students work jointly on an activity.	Students work together to solve a task by sharing ideas and problem-solving strategies.
compare multiple strategies	Share various solution methods for solving a problem. Contrasting elements may or may not be explicitly discussed.	The teacher asks students to share their solution strategies to a task and identify similarities and differences in their approaches.
confusion	Being unclear or uncertain about, or misunderstanding, an idea.	A student is unsure how to decide if a slope of a line is positive or negative.
correct mistakes	Identify and fix errors, such as in reasoning or incorrect solutions.	After finding miscalculations in their solutions, students rework the problems to ensure their answers are error-free.
creative thinking	The act of students applying what they have learned to unfamiliar situations in innovative ways, or considering problems from a fresh or imaginative perspective.	When asked to find the area of an irregular figure, students decide to separate the figure into triangles and quadrilaterals so that they can easily calculate the total area.

(*Continued*)

TABLE 21.2 (*Continued*)

Term	Description	Examples
critical thinking	The process of students analyzing, applying, or evaluating information they have gathered and using it to make reasoned judgments.	When solving a system of linear equations, a student considers features of two equations in order to decide whether substitution or elimination will be the most efficient approach.
differentiation	The act of providing materials, tasks, or instruction to students according to their learning needs.	A teacher makes modifications to their lesson plan to include an activity for students who need additional support.
distribute materials	Pass out materials that students will use during class.	At the start of an activity, the teacher gives each student a worksheet, protractor, and ruler that they will need to complete the task.
effort	Mental exertion that students apply when working on an activity.	When a student's approach to solving a perimeter problem by generating an equation does not work, the student works to figure out a new approach.
establish routines	Set up a system of procedures to be followed regularly in the classroom.	A teacher begins each day's lesson with a warm-up so that students know what to expect when they arrive in class.
example	Problem or task that has already been solved and is meant to serve as a sample case.	The textbook includes a solved problem that illustrates how to find the slope of a line given two points.
exit slip	Problem or question given to students at the end of class, either on a piece of paper or digitally, to gauge student understanding of the day's lesson. Students are expected to complete the problem before leaving class.	At the end of a lesson, the teacher asks students to answer three problems pertaining to the lesson's objectives. The teacher reviews student responses to determine what needs to be reviewed in class the following day.
explain	Make an idea, task, solution, or strategy clear by describing it in detail and providing relevant information.	While working with a group of peers, one student outlines the steps they took to solve an inequality and details the process of graphing their solution on a number line.
explore	Inquire into an idea, method, or task, often by trying out different strategies or solution pathways.	Students are asked to maximize the volume of a fish tank given a total surface area allowed. Some students create models, while others try making a table to determine the optimal dimensions.

Term	Description	Examples
extra credit	Supplementary questions or tasks provided by teachers for students to earn additional points, often in the context of an assessment.	At the end of an assessment, the teacher provides students with optional, more difficult problems that, when answered correctly, will earn them additional points towards their exam grade.
formative assessment	Ongoing, often informal, evaluation of student learning during instruction.	The teacher walks around the room, observing conversations as students work and monitoring students' understanding of the learning objectives.
games	Activities that are used to review or practice math problems in a fun–and sometimes competitive–way.	The teacher creates a trivia competition that is used as practice before an assessment.
give directions	Provide students with instructions about a task to be completed.	The teacher describes a set of steps each student is expected to take in order to complete their work.
go over answers	Review answers to a problem or set of problems as a whole class. Occasionally takes place without teacher involvement.	Students and their teacher check answers to previously solved problems and discuss solution strategies when there is a discrepancy.
go over homework	Examine correctness of homework, assess which problems were difficult, and/or provide feedback on homework problems, typically at the beginning of class.	Class begins with students presenting their solutions to the previous night's homework. Students have the opportunity to ask for guidance on problems that were difficult, and the teacher reviews problems with which many students struggled.
group work	Work done with other students, typically in groups of three or more.	Students work in groups of four to try to identify the slopes of different lines.
growth mindset	Belief that students' intelligence is not fixed and can improve through hard work.	After receiving a poor grade on an exam, a student works to learn from their mistakes, believing they can do well on future exams. The student does not assume that they are not good at math.
guess and check	Problem-solving strategy in which a student makes a conjecture about the answer and then examines whether the conjecture is correct.	A pair of students tries to figure out how many pieces of candy they can purchase with $10. They make a conjecture and test it. When they discover the number they chose is too high, they try again with a lower number.

(*Continued*)

TABLE 21.2 (*Continued*)

Term	Description	Examples
guided notes	Teacher-prepared materials for students that outline information but leave blanks for students to fill in the answers.	A teacher gives students a handout with formulas for the surface area of different shapes and blank lines for students to label the variables in each equation. The teacher reviews the sheet with the class.
guided practice	Interactive instruction in which teachers lead students to a solution as they work through a problem together.	A teacher asks their class to help solve a linear equation. Students provide suggestions for next steps, and the teacher uses the students' answers to reach a solution.
hands-on activity	Task in which students use manipulatives or are engaged through tactile stimulation.	Students measure the dimensions of a cereal box to explore surface area and volume.
high expectations	Ambitious standards to which teachers hold students; the assumption that students will achieve success.	A teacher presents students with complex tasks and believes they are capable of reaching the challenging goals the teacher set for them.
hint	Tip provided to students intended to help them solve a problem without giving the answer.	When students struggle to graph linear equations, their teacher does not tell them how to do so. Instead, the teacher encourages them to consider the slope of the line, hoping this will move them in the right direction.
homework check	Process used by teachers to determine who has completed homework, often at the start of class. Might also refer to the process of going over homework (see "going over homework").	The teacher circulates around the room and records which students completed the previous night's homework.
investigation	Open-ended activity that involves student exploration rather than rote problem-solving.	While learning about transformations of absolute value functions, students use trial and error on a graphing calculator and look for similarities and differences in the equations and graphs produced.
justify	Provide evidence to support explanations, ideas, or solutions in order to illustrate that one's solutions or explanations are reasonable.	A student claims that a function is non-linear and describes characteristics of non-linear functions in order to support their claim.
listening	Attending quietly to what others say.	While a student is speaking, their classmates are silently paying attention to what the student is saying.

Term	Description	Examples
make connections	Identify similarities between strategies, approaches, or tasks–especially between a new concept and prior knowledge or experience.	When discussing a new concept, the teacher asks, "Isn't this similar to something you learned last year?"
mastery	Comprehensive knowledge or skill in a particular area that students work toward achieving.	A student demonstrates their comprehension of a set of learning objectives by correctly answering 95% of the questions on a unit exam.
math practices	Processes students engage in when doing mathematics, such as critiquing others' reasoning and justifying one's solutions.	During class discussion, students share their reasoning and listen to each other's ideas, identifying points of agreement and disagreement.
memorize	Learn by rote; often contrasted with learning through deep understanding.	Students recite their multiplication and division facts repeatedly in order to promote quick recall.
modeling	1. Demonstrating an approach while thinking out loud to provide an example of how students might solve similar problems. 2. Making sense of the world through a mathematical perspective (i.e., modeling with mathematics).	1. While students watch, the teacher works through a problem at the board, describing their thought process and helping students understand the reason behind each step. 2. Students explore a word problem relating the speeds of two children in a race. One group depicts each child's progress using a number line to better understand the situation.
note-taking	Recording information in written form. Might include copying or summarizing information from a source or highlighting or annotating text.	A teacher presents information, while students write down key ideas in their notebooks.
observe	Watch and see what others do.	As one student explains their solution to a problem, their classmates watch and follow along with the student's ideas.
offer feedback	Evaluate or provide comments on students' ideas or solutions.	As students work in small groups, one student remarks that another student's approach includes unnecessary steps and proposes a more efficient strategy.
open-ended question	Type of inquiry, typically posed by a teacher, that does not have a single correct answer.	When comparing bank loans offered by different companies, students are asked to state what factors would influence their decision to choose one company over another.

(Continued)

TABLE 21.2 (*Continued*)

Term	Description	Examples
partner work	Work done together in groups of two.	Two students collaborate on an assigned task.
pattern recognition	Noticing repetition or commonalities among a group of numbers or other objects.	When studying arithmetic sequences, the teacher asks students to relate the numbers in the sequence "3, 6, 9, 12, 15, 18, 21" and identify the next three numbers in the sequence.
positive feedback	Approval or admiration provided by teachers to students to highlight students' effort, approach, or accuracy.	As a student completes a problem at the board, the teacher comments, "I really like how you are showing each step."
practice	Repeated exercise by students in a particular area.	Students complete a set of problems in which they use different approaches to find the equation of a line.
pre-assessment	Task given to students by teachers to evaluate their knowledge of a topic before it is taught.	The teacher gives students five questions to complete prior to a new unit on dividing fractions in order to gauge students' retention of the topic from the previous year.
problem-based learning	Student-centered style of instruction in which students work for an extended period of time to investigate a problem related to everyday life.	Student groups design, manufacture, market, and sell friendship bracelets, exploring how to maximize the profit of their company.
problem-solving	Finding solutions using non-rote strategies.	Students experiment with new approaches to find the vertex of an absolute value function.
prove	Demonstrate to be true by providing evidence.	A student uses recently learned theorems to explain how they concluded that two triangles are congruent.
quiz	Assessment of student knowledge on a specific set of information. Typically shorter than and counts for less than a test.	After students learn about subtracting fractions, they complete a ten-question independent assessment.
real-world connections	Illustrations of how specific math activities and concepts are used outside of math class.	A teacher asks students to decide which bank to use based on interest and savings options.
reasoning	Thinking about a problem or idea in a logical way in order to reach a solution.	To determine whether a triangle with given side lengths represents a right triangle, students discuss the right triangle theorem and the relationship between side lengths of right triangles.

Term	Description	Examples
redirect	Shift students' attention in attempt to get them back on task.	While comparing baseball statistics, students begin to discuss an upcoming game. The teacher reintroduces a previous question to shift student attention back to the original task.
reflection	Giving additional thought or consideration to a method, strategy, or solution.	After solving a percent change problem, students consider whether their solutions are reasonable and contemplate the real-life applications of the task.
remediation	Additional support provided to students who are struggling or whose achievement is behind others.	A teacher sits with struggling students during independent work time to help strengthen their skills in factoring polynomials.
respect	Behave with regard to the feelings of others.	During small group discussion, students listen intently and use thoughtful and appropriate language to respond to each other's ideas.
review	Go over material that was previously learned or studied, sometimes prior to an upcoming assessment.	A teacher works with students to complete practice problems in preparation for an upcoming assessment.
scaffolding	A series of supports provided to move students forward on a problem, assignment, or concept.	Students are given suggestions for how to proceed or simpler problems to build the skills needed to solve a complex task.
self-advocate	Take responsibility for one's own learning by communicating interests and needs to others, speaking up for oneself, and seeking out help as needed in order to succeed.	A student who has been sick asks their teacher for extra time to work on assignments.
share	Describe or demonstrate one's methods or thinking with others.	A student presents their solution strategy to the whole class.
show	Demonstrate or illustrate one's methods or thinking to others, often by writing or using manipulatives.	When solving a linear equation, a student writes additional notes along the side of the problem to indicate the steps taken.
skills practice	Repetition with a specific procedure.	Students work through a set of thirty problems, all focused on finding the sum of two fractions.

(Continued)

TABLE 21.2 (*Continued*)

Term	Description	Examples
struggling	1. Having difficulty with a problem or task. Might be used to describe students who have fallen behind in class. 2. Persisting while working on a challenging task. Sometimes discussed in terms of the teacher's goal of getting students to engage in positive "productive struggle."	1. When presented with a system of equations, a pair of students cannot figure out how to solve it, as they have had trouble solving equations all year. 2. When encountering a challenging task, a student tries various strategies to get a correct solution, not giving up despite the difficulty of the task.
student accountability	Taking responsibility for one's own actions and learning.	After a student fails an exam, they acknowledge that they did not adequately prepare for it and begin attending lunchtime tutoring sessions to get extra help.
student as teacher	Student acts as instructor and provides help or teaches concepts to peers.	A student guides their peers through whole-class instruction on how to graph a two-variable linear equation.
student presentation	Demonstration or display of a student's ideas or methods, typically at the front of the room.	After determining how much pizza is needed to feed the whole school, students take turns going to the front of the room and explaining their solutions using verbal and visual models.
student strategies	Methods used by students to approach and/or solve a problem.	To decide which company offers a more economical gym membership, student makes a table to compare the costs at different points in time.
student-centered	Instructional approaches that focus and build on students' own learning needs, interests, and backgrounds.	A teacher creates stations for students to move through that focus on different skills and problems that students have struggled with throughout the unit.
study guide	Document listing topics and/or sample questions that students should review, often for an upcoming exam.	The teacher provides students with an outline of concepts that will be included in an upcoming assessment.
take attendance	Record which students are present in class and which students are absent or tardy.	At the start of class, the teacher identifies if any students are missing or arrive late, recording the information for future reference.
test	Assessment of students' knowledge using a series of questions/problems. Often occurs at the end of a chapter or unit.	The teacher gives students a range of problems designed to gauge students' mastery of a set of learning objectives. Students individually complete this graded assignment during an allotted timeframe in class.

Term	Description	Examples
testing	Assessing students' knowledge using a standardized exam that is administered at a district, state, national, or international level. At times, can also refer to assessing at the classroom level (see "Test").	All students in a district take a standardized exam that is used to measure mastery of district and state learning objectives.
think-pair-share	Three-phase instructional strategy in which students first work individually on a problem, then discuss ideas with a partner, and finally share their thoughts with the whole class.	The teacher asks students to think of three things they know about the slopes of lines. Then students compare ideas with a partner. Finally, the whole group comes together to discuss characteristics of slope.
thinking	Directing one's mind toward something; having an idea.	Students are asked to decide whether or not a triangle with given side lengths represents a right triangle. Each student recalls information they have learned and formulates an explanation of why the triangle is or is not a right triangle.
try	Make a sincere attempt or effort to do something.	After the teacher provides students with a problem to solve, students begin thinking about an approach to take and writing out their ideas.
use manipulatives	Work with physical or virtual objects, such as cubes or counters, to support math learning.	After being introduced to the concept of compound probability, students roll dice to investigate chances of winning a board game.
wait time	A pause or period of silence after a question is asked that allows students time to think about the question.	After posing a question to the class, a teacher counts to seven in their head to give more students the chance to process the question before calling on a student to respond.
warm-up	Brief mathematical activity for students at the beginning of class.	When students enter the classroom, they begin working on two problems written on the board, which will prepare them for the day's lesson.
whole-class discussion	Participation structure in which teachers and students engage in conversation together, with teachers generally facilitating student conversation.	A teacher engages students in a conversation about the characteristics of quadrilaterals, calling on students to share their ideas and asking questions to prompt thinking.

(Continued)

TABLE 21.2 (*Continued*)

Term	Description	Examples
word problem	Math task that places content into a concrete or applied context.	Students read a scenario about a baseball player's performance across three games and are asked to determine the player's batting average.
worked example	Step-by-step demonstration of how to solve a problem, often provided by teachers to students as a model.	The teacher presents an inequality to their class and models the steps needed to solve it as students follow along.
writing	Composing text, often in order to describe one's thinking or methods.	Students reflect on their knowledge of a concept by recording their thoughts in a journal.

TABLE 21.3 United States Lexicon – Terms by category.

Administration	agenda, assign homework, assign seats, distribute materials, give directions, homework check, take attendance, *warm-up*
Assessment	assess, check answers, correct mistakes, exit slip, extra credit, formative assessment, go over answers, go over homework, mastery, *offer feedback,* pre-assessment, quiz, review, study guide, test, testing
Classroom Climate	build rapport, classroom environment, classroom management, establish routines, growth mindset, high expectations, positive feedback, redirect, respect, self-advocate, student accountability
Discussion	answer questions, ask questions, clarify, explain, *justify,* listening, *offer feedback*, *open-ended question, prove, reasoning,* share, *wait time*
Engaging in Learning	"aha" moment, brainstorm, confusion, creative thinking, critical thinking, effort, explore, memorize, note-taking, observe, practice, reflection, show, struggling, thinking, try, writing
Engaging with Mathematics	calculate, compare multiple strategies, guess and check, *justify,* make connections, math practices, *modeling,* pattern recognition, problem-solving, *prove, reasoning,* student strategies, *use manipulatives*
Participation Structures	ask for student volunteer, collaborate, group work, partner work, student as teacher, student presentation, think-pair-share, whole-class discussion
Pedagogical Approaches and Tools	access prior knowledge, challenge, differentiation, example, games, guided notes, guided practice, hands-on activity, hint, investigation, *modeling, open-ended question,* problem-based learning, real-world connections, remediation, scaffolding, skills practice, student-centered, *use manipulatives, wait time, warm-up,* word problem, worked example

Note: Terms in italics are in more than one category.

The lexicon documented in this chapter is the result of extensive discussion and reflection among both mathematics education researchers and middle-school teachers of mathematics. Certainly, this lexicon is not without limitations. Many additional terms exist, and alternate categorizations of the terms are possible. Still, we view this lexicon as a productive first step in uncovering the shared professional language of middle-school mathematics teachers in the United States and an opportunity to begin reflecting more systematically on our language use, including practices that have been left unnamed. We hope that this lexicon contributes to conversations around mathematics teachers' professional language and catalyzes future research investigating how a focus on language can enhance mathematics teaching in the United States.

Acknowledgements

This work was supported by the Alumnae of Northwestern University. We wish to thank all of the teachers who participated and made this work possible.

Bibliography

Presentations and papers

The work of the United States national team has been presented locally and internationally at the following meetings and conferences:

- American Educational Research Association (AERA) Annual Meeting 2018
- International Group for the Psychology of Mathematics Education (PME) Annual Conference 2017
- National Council of Teachers of Mathematics Regional Conference 2019
- North American Chapter of the International Group for the Psychology of Mathematics Education 2017

Peer-reviewed conference publications from the United States research team include the following:

Dobie, T.E., White, S., & Sherin, M. (2017) *Towards a shared language of instruction: Exploring teachers' lexicon for mathematics teaching and learning. Proceedings of the 39th Annual Meeting of the North American Chapter of the International Group for the Psychology of Mathematics Education* (pp. 1254–1257). Indianapolis, IN: PME-NA.
Mesiti, C., Clarke, D.J., Dobie, T., White, S., & Sherin, M. (2017). "What do you see that you can name?" Documenting the language teachers use to describe phenomena in middle school mathematics classrooms in Australia and the USA. In B. Kaur, W.K. Ho, T.L. Toh, & B.H. Choy (Eds.), *Proceedings of the 41st Annual Meeting of the International Group for the Psychology of Mathematics Education* (Vol. 2, pp. 241–248). Singapore: PME.

INDEX